ASHRAE GreenGuide

Design, Construction, and Operation of Sustainable Buildings

This publication was developed under the auspices of ASHRAE Technical Committee (TC) 2.8, Building Environmental Impacts and Sustainability. TC 2.8 is concerned with the impacts of buildings on the local, regional, and global environment; means for identifying and reducing these impacts; and enhancing ASHRAE member awareness of the impacts.

Tom Lawrence, PhD, PE, LEED AP, is the chair of the editorial revision committee and coeditor of *ASHRAE GreenGuide*, Fifth Edition. He is a senior public service associate with the University of Georgia and has more than 35 years of experience in engineering and related technical fields. He is a past chair of ASHRAE Technical Committee (TC) 2.8 and is a member of the committee that wrote and maintains ANSI/ASHRAE/USGBC/IES Standard 189.1, *Standard for the Design of High-Performance Green Buildings Except Low-Rise Residential Buildings*. Lawrence is an ASHRAE Fellow and Director at Large for 2016-2019. As an ASHRAE Distinguished Lecturer, he gives seminars and workshops on high-performance building design at venues around the world.

Abdel K. Darwich, PE, CEng, LEED AP, HFDP, is a coeditor of *ASHRAE Green Guide*, Fifth Edition. He is an associate principal with Guttmann and Blaevoet Consulting Engineers in their Sacramento, CA, office. Darwich has more than 20 years of experience in the design of mechanical systems for health care, commercial, industrial, higher education, and mixed-use projects, with emphasis on low/zero-energy design and indoor air quality. He is a member of ASHRAE TC 2.8 and Standing Standard Project Committee 62.1, an ASHRAE-certified Healthcare Facility Design Professional, and a recipient of two ASHRAE Technology Awards (2013 and 2016) and an ASHRAE Distinguished Service Award (2017).

Janice K. Means, PE, LEED AP, FESD is the third coeditor of *ASHRAE GreenGuide*, Fifth Edition. She is an associate professor in the College of Architecture and Design at Lawrence Technological University in Southfield, MI. Means has more than 30 years of experience in the industry. She is a trained presenter for ANSI/ASHRAE/IESNA 90.1-2001 and -2004; a voting member of the ASHRAE Student Activities Committee, Guideline Project Committee (GPC) 34, and TC 6.7; a former voting member of TC 2.8; Region V Student Activities Regional Vice Chair and a contributing author to Chapter 36 of the 2008 *ASHRAE Handbook—HVAC Systems and Equipment* and the new ASHRAE Guideline 34P, *Energy Efficiency for Historic Buildings*.

ASHRAE STAFF

SPECIAL PUBLICATIONS

Mark S. Owen, Editor/Group Manager of Handbook and Special Publications
Cindy Sheffield Michaels, Managing Editor
James Madison Walker, Managing Editor of Standards
Lauren Ramsdell, Assistant Editor
Mary Bolton, Editorial Assistant
Michshell Phillips, Editorial Coordinator

PUBLISHING SERVICES

David Soltis, Group Manager of Publishing Services and Electronic Communications
Jayne Jackson, Publication Traffic Administrator

PUBLISHER

W. Stephen Comstock

Updates and errata for this publication will be posted on the ASHRAE website at www.ashrae.org/publicationupdates.

ASHRAE GreenGuide

Design, Construction, and Operation of Sustainable Buildings

Fifth Edition

Atlanta

ISBN 978-1-939200-80-8 (hardback)
ISBN 978-1-939200-81-5 (PDF)

1791 Tullie Circle, NE
Atlanta, GA 30329
www.ashrae.org

First edition published 2003. Second edition published 2006.
Third edition published 2010. Fourth edition published 2013.
Fifth edition published 2018.
Printed in the United States of America

Cover design by Megan Joyce. Cover image by Costas A. Balaras.

Library of Congress Cataloging-in-Publication Data

Names: ASHRAE (Firm)
Title: ASHRAE greenguide : design, construction, and operation of sustainable buildings.
Other titles: ASHRAE green guide
Description: Fifth edition. | Atlanta, GA : ASHRAE, [2018] | Includes bibliographical references and index.
Identifiers: LCCN 2017039504| ISBN 9781939200808 (hardback) | ISBN 9781939200815 (PDF)
Subjects: LCSH: Sustainable buildings--Design and construction. | Sustainable buildings--United States--Design and construction. | Sustainable architecture. | Buildings--Environmental engineering. | Sustainable construction.
Classification: LCC TH880 .A83 2018 | DDC 720/.47--dc23 LC record available at https://lccn.loc.gov/2017039504

Tomorrow's Child

Without a name, an unseen face,
And knowing not the time or place,
Tomorrow's Child, though yet unborn,
I saw you first last Tuesday morn.
A wise friend introduced us two,
And through his shining point of view
I saw a day, which you would see,
A day for you, and not for me.
Knowing you has changed my thinking,
Never having had an inkling
That perhaps the things I do
Might someday threaten you.
Tomorrow's Child, my daughter-son,
I'm afraid I've just begun
To think of you and of your good,
Though always having known I should.
Begin I will to weigh the cost
Of what I squander, what is lost,
If ever I forget that you
Will someday come to live here too.

by Glenn Thomas, ©1996

Reprinted from
Mid-Course Correction:
Toward a
Sustainable Enterprise:
The Interface Model
by Ray Anderson.
Chelsea Green
Publishing Company, 1999

Contents

Section 1: Basics

Section 2: The Design Process

GREENTIPS

DIGGING DEEPER SIDEBARS

FOREWORD

originally written by the late William Coad
edited by Tom Lawrence for the fifth edition

ASHRAE is the preeminent technical society representing engineers practicing in the fields of heating, refrigeration, and air conditioning—the technology that uses approximately one-third of the global nonrenewable energy consumed annually.

ASHRAE membership has actively pursued more effective means of utilizing precious nonrenewable energy resources for many decades, from the standpoints of source availability, efficiency of utilization, and technology of substituting with renewable sources. One significant publication in *ASHRAE Transactions* is a paper authored in 1951 by G.W. Gleason, Dean of Engineering at Oregon State University, titled "Energy—Choose it Wisely Today for Safety Tomorrow" (Gleason 1951). The flip side of the energy coin is the environment and, again, ASHRAE has historically dealt with the impact that the practice of the HVAC&R sciences have had upon both the indoor and the global environment.

However, the engineering community, to a great extent, serves the needs and desires of accepted economic norms and the consuming public, a large majority of whom have not embraced an energy/environmental ethic. As a result, much of the technology in energy effectiveness and environmental sensitivity that ASHRAE members have developed over this past century has had limited impact on society.

In 1975, when ASHRAE published ASHRAE Standard 90-75, *Energy Conservation in New Building Design* (ASHRAE 1975), that standard served as our initial outreach effort to develop an awareness of the energy ethic and to extend our capabilities throughout society as a whole. Since that time, updated revisions of Standard 90 have moved the science ahead. In 1993, the chapter on energy resources was added to the 1993 *ASHRAE Handbook—Fundamentals*. In 2002, ASHRAE entered into a partnering agreement with U.S. Green Building Council (USGBC), and it is intended that this and future editions of this design guide will continue to assist ASHRAE in its efforts at promoting sustainable design, as well as the many other organizations that have advocated for high-performance building design.

The consuming public and other representative groups of building professionals continue to become more and more aware of the societal need to provide buildings that are more energy resource effective and environmentally compatible. The topics involved with "green building" or "high-performance green buildings" are much more than just energy as well. These include water efficiency, indoor environmental quality, and the materials used in building construction, for example. ASHRAE has been working toward being recognized as one of the preeminent authorities on green buildings in the industry, for example by taking the lead in the creation of ANSI/ASHRAE/USGBC/IES Standard 189.1, *Standard for the Design of High-Performance Green Buildings Except Low-Rise Residential Buildings*. In addition, newer emerging technologies, such as the smart grid or smart buildings, are coming forth. This publication, authored and edited by ASHRAE volunteers, is intended to complement the efforts in all these topical areas and to help the membership keep up to date.

PREFACE TO THE FIFTH EDITION

by Tom Lawrence

This new edition of *ASHRAE GreenGuide* represents another significant update and revision to what has become one of ASHRAE's primary contributions toward sustainable design of the built environment.

The new fifth edition is being released as we close in on the fifteenth anniversary of the release of the first edition in January 2004. A lot has changed in the industry in general, and in green design practices in particular, during those years. Many of the concepts that seemed "out there" in the time period that the first edition was published are now considered mainstream and, in some cases, are now actually required by code. We have tried to incorporate these trends in making revisions to this fifth edition as well as take a look into the crystal ball and anticipate future trends. It has been ASHRAE's plan to update and maintain this book on a regular basis because of this rapidly evolving field and to support ASHRAE's committed goal of global leadership in sustainability and technical education. This also follows the example set by ASHRAE's rebranding in 2012.

Since the release of the fourth edition of the *GreenGuide*, a number of developments have occurred in the green building arena. ASHRAE has continued to refine and modify ANSI/ASHRAE/USGBC/IES Standard 189.1, with a new release of that standard published in late 2014 and another in process for late 2017. A recent agreement between ASHRAE and the International Code Council (ICC) will result in the merging of Standard 189.1 and the *International Green Construction Code* (IgCC) beginning with the 2018 code cycle. ASHRAE is responsible for the technical content and ICC is responsible for the administrative provisions. This will hopefully result in further adoption and use of this code, and other organizations and jurisdictions are using these as the basis for their own codes and design standards. Thus, the industry is witnessing the continued evolution of green building programs from strictly voluntary to being both more mainstream within the industry as well as mandatory in jurisdictions that have adopted these standards for their building codes.

This fifth edition features four entirely new chapters, as well as updated information and GreenTips. The specific Building-Type GreenTips have been located in a new standalone Chapter 18. The new Chapter 19 covers green building design as it relates to existing buildings, and the new Chapter 17 discusses differing aspects of green design for residential structures. Finally, Chapter 20, Emerging Trends and Epilogue, briefly covers new topics that are becoming key to the industry. Chapter 13, Smart Building Systems, including information on building automation systems, has also been extensively revised and expanded in scope.

No one person, or even a small committee of people, can be expected to have the breadth and depth of expertise to create a book that covers the wide range of issues governing the design, construction, and operation of green buildings. Thus, many people with various backgrounds were recruited to help review and edit the existing chapters and to write the new chapters. One of the goals for the editorial committee for this revision was to bring in a large number of additional outside reviewers. We truly appreciate those that contributed their time and talents in reviewing, editing, and writing. Finally, the current edition could not have been possible without all the hard work and dedication put into it by others who created the previous editions. This book truly represents the collaborative nature of the work done by dedicated volunteers within ASHRAE. All work performed—by the authors, editors, developing subcommittees, other reviewers, and technical committee participants—was strictly voluntary.

WHO SHOULD USE *ASHRAE GREENGUIDE*

The original stated purpose for the *ASHRAE GreenGuide* was created primarily for HVAC&R designers, but time has shown that it is also a useful reference for architects, owners, building managers, operators, contractors, students, and others in the building industry who want to understand the technical issues regarding high-performance design from an integrated building systems perspective. Considerable emphasis is placed on teamwork and close coordination between parties.

The *GreenGuide* was originally intended for use by younger engineers or architects, or more experienced professionals about to enter into their first green design projects. However, a survey of those who purchased one of the earlier editions of this publication revealed that it was primarily being used by more experienced individuals. The survey also indicted a higher percentage of the readership from countries outside of North America, perhaps reflecting the growing internationalization of ASHRAE. These trends have been taken into account with the revision process for later editions.

HOW TO USE *ASHRAE GREENGUIDE*

This book is intended to be used more as a reference than as something one would read in sequence from beginning to end. The table of contents is the best place for any reader to get an overall view of what is covered in this publication. Throughout the *GreenGuide*, numerous techniques, processes, measures, or special systems are described succinctly in a modified outline or bullet form. These are called

ASHRAE GreenTips. Each GreenTip concludes with a listing of other sources that may be referenced for greater detail. (Lists of GreenTips and Digging Deeper sidebars can be found in the table of contents.)

All readers should take the time to review Chapter 1, Introduction and Background, which provides some essential definitions and meanings of key terms. Chapter 2 describes green rating systems and the relevant standards and paths to compliance as they relate to the work of the mechanical engineers. The project initial phases are covered, with Chapter 3 providing an overview of project strategies and the early stages of the design process and Chapter 4 covering the commissioning process.

The bulk of this book covers the design process, starting with the architectural design and planning impacts in Chapter 5. These chapters are essential reading for all who are interested in how the green design process works. Other topics in this section include conceptual engineering design, sustainable sites, indoor environmental quality, energy conversion systems, energy and water resources, lighting, water efficiency, smart building systems, and the process for completing design and documentation.

The final chapters finish the book with discussions on construction and operations and maintenance. This final section also includes the chapters on residential buildings, GreenTips for specific building types and existing buildings, and ends with the emerging trends chapter. In prior editions of this book, these tips were contained at the end of the chapter on the architectural design process; however, it is felt that these are important enough to justify as a separate chapter and not get lost in the back pages of one particular chapter.

At the end of the guide is a comprehensive References and Resources section, which compiles all the sources mentioned throughout the guide, and an index for rapid location of a particular subject of interest.

HISTORICAL BACKGROUND ON *ASHRAE GREENGUIDE*

The idea for the publication of this guide was initiated by 1999–2000 ASHRAE President Jim Wolf and carried forward by then President-Elect (and subsequently President) William J. Coad. Members of that first subcommittee were David L. Grumman, Fellow ASHRAE, chair and editor; Jordan L. Heiman, Fellow ASHRAE; and Sheila Hayter, chair of TC 1.10 (a precursor to the TC 2.8 of today).

The *GreenGuide* subcommittee responsible for the second and third editions consisted of John Swift and Tom Lawrence, along with the people noted in the Acknowledgments section. Work on the fourth and fifth editions was overseen by a subcommittee of TC 2.8 led by Tom Lawrence.

ACKNOWLEDGMENTS

The following individuals served as coeditors on this edition of *ASHRAE GreenGuide,* provided written materials and editorial content, and formed the Senior Editorial Group of the ASHRAE TC 2.8 *GreenGuide* subcommittee for the fifth edition:

Thomas Lawrence
University of Georgia
Athens, GA

Abdel K Darwich
Guttmann and Blaevoet Consulting Engineers, Sacramento, CA

Janice K. Means
Lawrence Technological University
Southfield, MI

Dunstan Macauley
WSP
Arlington, VA

In addition, the following individuals contributed new written materials on various topics for the fifth edition of *ASHRAE GreenGuide*. All or portions of these contributions were incorporated, with minor editing.

Constantinos A. Balaras
Kevin Brown
Kevin Cross
Philip Fairey
Dave Grumman
Paul Haydock
Mark MacCracken
Farhan Mehboob
Jason Morosko
Lisa Ng
Ashish Rakheja
Max Sherman
Annie Smith
Stanton Stafford
Mitchell Swann
Ted Tiffany
Paul Torcellini
Fatih Turan
Tania Ullah

ASHRAE extends its appreciation to the International Code Council (ICC) staff, Misty Guard and Mike Pfeiffer, and Solar Rating & Certification Corporation (ICC-SRCC) staff, Eileen Prado, and Shawn Martin, for their high-level review and comments that served to further enhance this publication. ICC, ICC-SRCC and their membership shall not be responsible for any errors, omissions, or damages arising out of information contained in this publication.

The committee is deeply thankful for all the individuals who helped with or contributed to the first four editions of the *GreenGuide*. The fifth edition would not be where it is without their help. However, because of space constraints, their names cannot be listed in the fifth edition.

Section 1: Basics

CHAPTER ONE

INTRODUCTION AND BACKGROUND

INTRODUCTION

There continues to be a growing awareness about the impact of the built environment on the natural environment. The use of sustainable engineering concepts has evolved quite rapidly in recent years and is now well recognized in HVAC&R and related engineering professions. This in turn is being encouraged by increased client demand for more sustainable buildings, commonly called *green buildings*.

Interest in sustainable or green buildings (the distinction between the two is discussed below) has been particularly evident in the concern about energy and water resource consumption, but also includes broader concerns such as indoor environmental air quality, material use, and "smart" development and planning. Many countries in the world now have green-building rating systems (voluntary) and/or codes (mandatory) in some form or other. Organizations devoted to green buildings now exist in most countries. Even as the concept of green design is reaching mainstream acceptance, these organizations continue to promote these concepts, exhort the industry and society to action, strive to motivate industry practitioners and building owners, warn of consequences from ignoring these concepts, and instruct how to achieve green design.

ASHRAE identified a need for guidance on green building concepts specifically directed toward practicing professionals involved on a day-to-day basis in the mechanical/electrical/plumbing (MEP) building system design process. However, readers may find that this guide may also serve other needs—for example, as the basis for a university course in sustainable building design. From a survey conducted in 2011, a wider range of people now use the *ASHRAE GreenGuide*, including students and other professionals in related disciplines. The topics covered in this guide are global and thus there has been an effort to keep this guide applicable internationally.

Green is one of those words that can have many meanings, depending on the circumstances. One of these is the greenery of nature in the flora around us. This symbolic reference to nature is the meaning this term relates to in this publication. The difference between a green and sustainable design is the degree to which the design helps to minimize the building impact on the environment while simultaneously providing a healthy, comfortable indoor environment. When the term *green* is used, is commonly is thought of as focusing on the energy and resources involved, while *sustainable* is broader in scope and considers the three Ps: people, profit and planet. However, some may not recognize a difference between the two terms and use them interchangeably; this is also the general approach taken in this book. This guide is not intended to cover the full breadth of sustainability, as this would require and extensive series of volumes, but it is a good overview of the main topics and issues involved. For additional key characteristics and detailed discussion of sustainability in buildings and the built environment, refer to the "Sustainability" chapter in the *ASHRAE Handbook—Fundamentals* (ASHRAE 2017a).

It is important to note that the definition of *green buildings* places an emphasis on integrated design of mechanical, electrical, architectural, and other systems. Specifically, a green/sustainable building design is one that achieves high performance, over the full life cycle, in the following areas:

- Minimizing natural resource consumption through more efficient utilization of nonrenewable energy and other natural resources, land, water, and construction materials, including utilization of renewable energy resources to strive to achieve net zero energy consumption.
- Minimizing emissions that negatively impact our global atmosphere and ultimately the indoor environment, especially those related to indoor air quality (IAQ), greenhouse gases, global warming, particulates, or acid rain.
- Minimizing discharge of solid waste and liquid effluents, including demolition and occupant waste, sewer, and stormwater, and the associated infrastructure required to accommodate removal.
- Minimizing negative impacts on the building site.
- Optimizing the quality of the indoor environment, including air quality, thermal regime, illumination, acoustics/noise, and visual aspects to provide comfortable human physiological and psychological perceptions.
- Optimizing the integration of the new building project within the overall built and urban environment. A truly green/sustainable building should not be thought of or considered in a vacuum, but rather in how it integrates within the overall societal context.

Ultimately, even if a project does not have overtly stated green/sustainable goals, the overall approaches, processes, and concepts presented in this guide provide a design philosophy useful for any project. Using the principles of this guide, an owner or a team member can document the objectives and criteria to include in a project, forming the foundation for a collaborative integrated project delivery approach. This can lower design, construction, and operation costs, resulting in a lower total cost for the life of the project.

RELATIONSHIP TO SUSTAINABILITY

The related term *sustainable design* is very commonly used, almost to the point of losing any consistent meaning. While there have been some rather varied and complex definitions put forth (see the Digging Deeper sidebar titled “Some Definitions and Views of Sustainability from Other Sources”), a simple one is adapted in this guide: sustainability is providing for the needs of the present without detracting from the ability to fulfill the needs of the future.

The preceding discussion suggests that the concepts of green design and sustainable design have no absolutes—that is, they cannot be defined in black-and-white terms. These terms are more useful when thought of as a mindset: a goal to be sought and a process to follow. This guide is a means of (1) encouraging designers of the built environment to employ strategies for developing a green/sustainable design, and (2) setting forth some practical techniques to help practitioners achieve the goal of green design, thus making a significant contribution to sustainability.

Another method for assessing sustainability is through the concept of the triple bottom line (Savitz and Weber 2006). This concept advances the idea that monetary cost is not the only way to value project design options. The triple bottom line concept advocates for the criteria to include economic, social, and environmental impacts of building design and operations decisions.

COMMITMENT TO GREEN/SUSTAINABLE HIGH-PERFORMANCE PROJECTS

Green projects require more than a project team with good intentions; they require commitment from the owner and the rest of the project team, early documentation of sustainable/green goals recorded by the Owner’s Project Requirement (OPR), and the designer’s documented basis of design. The most successful projects incorporating green design are ones with dedicated, proactive owners who are willing to examine (or give the design team the freedom to examine) the entire spectrum of ownership—from design to construction to long-term operation of their facilities. These owners understand that green buildings require more planning, better execution, and better operational procedures, requiring a firm commitment to changing how building projects are designed, constructed, operated, and maintained to achieve a lower total cost of ownership and lower long-term environmental impacts.

Implementing green/sustainable practices could indeed raise the initial design soft costs associated with a project, particularly compared to a code minimum building design. First cost is an important issue and often is a stumbling block in moving building design from the code minimum ("good or adequate design") to one that is more truly sustainable. Implementing the commissioning process early in the predesign phase of a project adds an initial budget line item but can often actually reduce overall total design/construction costs and the ultimate cost of ownership.

In addition, significant savings and improved productivity of the building occupants can be realized for the life of the building, lowering the total cost of ownership and/or providing better value for tenants. To achieve lifelong benefits also requires operating procedures for monitoring performance, making adjustments (continuing commissioning) when needed, and appropriate maintenance.

WHAT DRIVES GREEN PROJECTS

Green-building advocates can cite plenty of reasons why buildings should be designed utilizing integrated green concepts. The fact that these reasons exist does not make it happen in routine building projects, nor does the existence of designers—or design firms—with green design experience. The main driver of green-building design is the motivation of the owner—the one who initiates the creation of a project, the one who pays for it (or who carries the burden of its financing), and the one who has (or has identified) the need to be met by the project in question. If the owner does not believe that green design is needed, thinks it is unimportant, or thinks it is of secondary importance to other needs, then it will not happen. In addition, recent trends in the industry are moving toward green-building practices being made mandatory, either through local adoption of new codes and standards or through an organizational policy. These trends are discussed in more detail in Chapter 2.

THE IMPACT OF CARBON CONSIDERATIONS

The attention paid to concerns about greenhouse gas emissions has certainly increased in much of the world. During the first decade of the twenty-first century, two organizations issued challenges to the industry to design and implement buildings that had a significantly lower energy consumption compared to current typical designs. The Architecture 2030 Challenge (see the "References and Resources" section at the end of the chapter for more information) is one of these. Architecture 2030 was initiated by Edward Mazria in 2002, setting a goal of net zero energy and net zero carbon buildings by the year 2030. This goal is to be realized by achieving substantially better building energy performance on a sliding scale from 2010 through 2030. The near-term focus of the challenge was adopted by the American Institute of Architects (AIA). The Architecture 2010 Imperative achieved a goal of constructing new buildings that show a 50% improvement in energy efficiency compared to those built using the 1999 version of ANSI/ASHRAE/IES Standard 90.1, *Energy Standard for Buildings Except Low-Rise Residential Buildings*

(ASHRAE 1999). In Europe, there is an ongoing parallel effort to meet an ambitious goal of nearly zero energy buildings after 2020, according to the Energy Performance of Buildings Directive (EPBD) (European Commission 2010).

There is bad news and good news when we look at how buildings are involved with greenhouse gas (GHG) emissions. First, the bad news: buildings (commercial and residential) are responsible for approximately 30% of the GHG emissions in the United States and most developed countries, and the trend is also holding up in key developing nations. The good news is that buildings have also been identified as the economic sector with the best potential for cost-effective mitigation of GHG emissions, as highlighted in Figure 1-1. In this figure, the carbon dioxide (CO_2) emissions by sector and total non-CO_2 GHG emissions across sectors are shown in the baseline scenario at top while the bottom portion of Figure 1-1 shows the net result from mitigation scenarios that reach an average of about 450 ppm (in a range of 430 to 480 ppm) CO_2 equivalent (CO_2-eq) emissions (likely to limit warming to 3.6°F [2°C] above preindustrial levels) with CO_2 capture and storage (CCS, left) and without CCS (right). The difference between the baseline and mitigation scenarios in this figure represent the net emissions decrease possible for each sector, and the buildings sector represents one of the highest potential options. Therefore, the buildings industry can and should take responsibility for reducing GHG emissions, primarily through a reduction in energy consumption for new construction, in refurbishing existing buildings, and planning for the operation and maintenance to maintain the high level of efficiency.

The Conference of the Parties meeting in Paris in late 2015 (COP21) was recognized as a breakthrough event where the first significant changes to a global approach to address climate concerns were made in nearly 25 years. More importantly for the buildings industry, the important role of buildings in addressing this problem was recognized at this conference and a new organization, the Global Alliance for Buildings and Construction, was created. This alliance has a four-step strategic approach to: (1) reduce the energy demand of buildings (in particular, existing buildings), (2) decarbonize the energy and power supply for buildings, and (3) reduce the embodied greenhouse gases in materials and equipment through life cycle analysis, and increase resiliency by adaptations against climate change and associated other risks.

ASHRAE took the lead in meeting these challenges in several ways. To address the Architecture 2010 Imperative, significant effort was put into modifying Standard 90.1 (ASHRAE 1999) to drastically improve energy efficiency. The 2010 version of Standard 90.1, in essence, met the AIA challenge for 2010 by introducing requirement changes that were developed and introduced during that decade. Subsequent versions of Standard 90.1 continue to increase the minimum energy efficiency requirements. Although the specific requirements may differ in some cases, ANSI/ASHRAE/USGBC/IES Standard 189.1, *Standard for the Design of High-Performance Green Buildings Except Low-Rise Residential Buildings* (ASHRAE

Sectoral CO_2 and non-CO_2 GHG emissions in baseline and mitigation scenarios with and without CCS

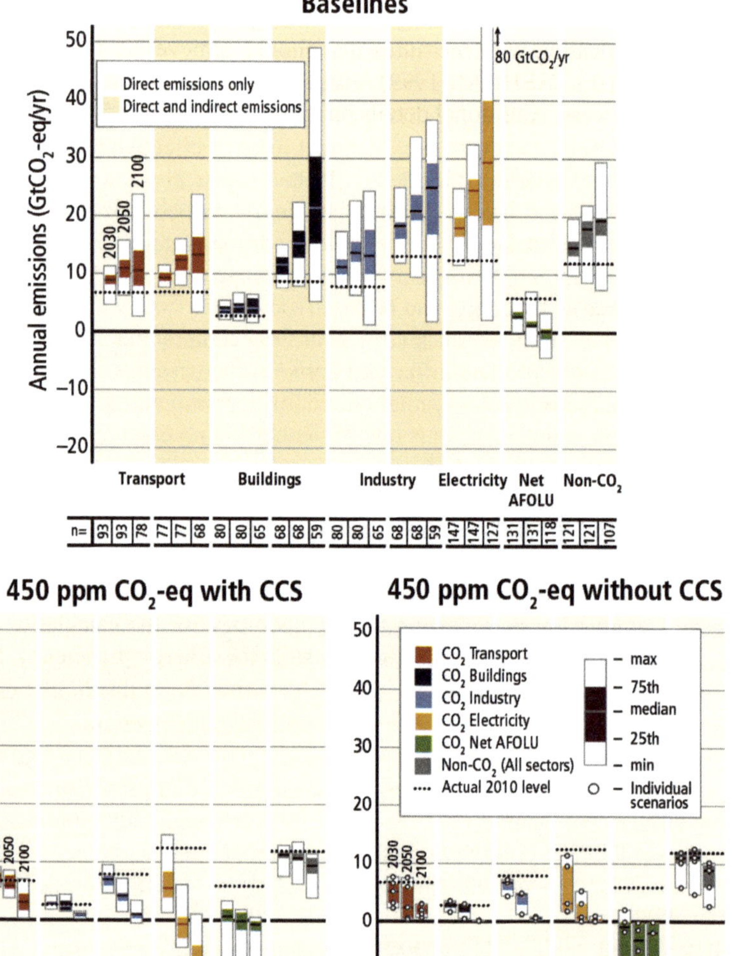

Source: IPCC (2014)

Figure 1-1 Cross-sectoral mitigation strategies.

2017b) and the *International Green Construction Code* (IgCC) have energy efficiency levels that exceed the minimum code values in Standard 90.1.

Another way this is being accomplished is through the production of the ASHRAE Advanced Energy Design Guide (AEDG) series. The series covers prescriptive measures that result in significant energy efficiency improvements, with the first series dealing with measures that should achieve 30% and 50% savings over Standard 90.1 (ASHRAE 1999) with even more strenuous improvements in the planning process. Additional details on these guides can be found in Chapter 2.

The HVAC&R engineer can provide a significant benefit to society (as well as to the building project's owners) via CO_2 emissions reduction associated with lower energy consumption. All new building projects that wish to comply with green principles should at least estimate the CO_2 equivalent emissions footprint of the building (of which a large part is produced through energy consumption). Using publicized emissions factors, these calculations are not complicated and can provide insight. For existing buildings, the goal is to compute the reduction in emissions associated with proposed energy conservation measures. In both cases, the GHG emissions factor used should be based on source energy and not on energy consumed on site alone. A good initial reference source for emissions factors is a 2007 National Renewable Energy Laboratory (NREL) report (Deru and Torcellini 2007). The emissions factors for site electricity consumption in the United States have been updated by the U.S. Environmental Protection Agency (EPA) to reflect recent trends away from coal-based electricity production to more natural gas and other renewable energy systems such as solar and wind. Along the same lines, the European Commission is reviewing the methodology for the calculation of a primary energy factor in the context of revising the Energy Efficiency Directive (EED) and preparing the upcoming legislative proposals on the 2030 Climate and Energy Framework. Currently, a default coefficient of 2.5 may be used for converting kilowatt-hour electricity (EED 2006/32/EC), although EU Member States may apply a different coefficient provided they can justify it. The ongoing efforts underline the need to regularly revise the conversion factor for electricity and that the methodology adequately reflects the strong efforts of the European power sector to decrease the carbon footprint and increase the share of renewables in the power generation mix. European informative default values for various energy carriers are available in ISO 52000-1, *Energy Performance of Buildings—Overarching EPB Assessment*

SUSTAINABILITY IN ARCHITECTURE

The emergence of green-building engineering is best understood in the context of the movement in architecture toward sustainable buildings and communities. Detailed reviews of this movement appear elsewhere and fall outside the scope of this document. A brief review of the history and background of the green design movement is provided, followed by a discussion of its applicability. Several leading

Digging Deeper

INTERNATIONAL PERSPECTIVE: REGULATIONS AND COMMENTARY

Society has recognized that previous industrial and developmental actions caused long-term damage to our environment, resulting in loss of food sources and plant and animal species, and changes to the Earth's climate. As a result of learning from past mistakes and studying the environment, the international community identified certain actions that threaten the ecosystem's biodiversity, and, consequently, it developed several governmental regulations designed to protect our environment. Thus, in this sense, the green design initiative began with the implementation of building regulations. An example is the regulated phaseout of fully halogenated chlorofluorocarbons (CFCs) and partially halogenated refrigerant hydrochlorofluorocarbons (HCFCs).

In Europe, the main regulatory instrument for tackling the energy consumption of buildings is the Energy Performance of Buildings Directive (EPBD) recast (European Commission 2010), which took effect in 2012 and replaced the original EPBD Directive (European Commission 2002). All EU member states introduced national laws, regulations, and administrative provisions for setting minimum requirements on the energy performance of new and existing buildings that are subject to major renovations and for energy performance certification of buildings. Additional requirements include regular inspection of boilers and air conditioning systems in buildings, an assessment of the existing facilities, and provision of advice on possible improvements and alternative solutions. Moreover, the EPBD recast strengthens the energy performance requirements and clarifies and streamlines some of the original EPBD provisions to reduce the large differences between EU member states' practices. In particular, it requires that EU member states lay down the requirements so that new buildings are nearly zero energy by 2020 (2018 for public buildings) and the application of cost-optimal levels for setting minimum energy performance requirements for both the building's thermal envelope and technical systems.

Energy performance certificates (EPC) are issued when buildings are constructed, sold, or rented out. The EPC documents the energy performance of the building and is expressed as a numeric indicator or a letter grade that allows benchmarking of primary energy consumption. The certificate also includes recommendations for cost-effective improvement of the energy performance, and is valid for up to ten years. National efforts and examples of EPCs are detailed in the works by Arcipowska et al. (2014) and Maldonado (2016).

The Concerted Action EPBD launched by the European Commission provides updated information on the implementation status in the various European countries (www.epbd-ca.org).

Digging Deeper

Buildings are at center stage of the ambitious European efforts and energy strategies for secure, competitive, and sustainable energy toward 2020, 2030, and 2050 (https://ec.europa.eu/energy/en/topics/energy-strategy).

It is not just in developed countries that green-building design and energy efficiency concerns are taking hold. The later part of the past decade has seen an explosion of adopting building energy efficiency standards and green-building design programs. For example, India was the first expansion of the Leadership in Energy and Environmental Design® (LEED®) Green Building Rating System programs outside of the United States, with the establishment of the India Green Building Council in 2003. India also created a nationwide energy efficiency standard, the Energy Conservation Building Code, in 2008. This code was based on ANSI/ASHRAE/IES Standard 90.1 but modified for the local climates and situations. Similarly, ASHRAE's Standard 90.2 was used as the basis for energy efficiency in residential construction in Kuwait (knowns as Kuwait 90.2), as modified for Kuwait's local climate and construction practices.

Energy efficiency standards throughout the world generally adopt two approaches. One is to have a set of mandatory requirements and then offer a prescriptive or a performance-based path for compliance (examples include the approaches taken for energy codes in the U.S., Canada, India, and Australia). Another approach is to have a set of mandatory items, then build on this with a point system for other features, with a minimum number of points required. This approach is the one taken by Japan and South Korea, for example.

As documented by the existing and emerging, green-building programs around the world (Mills et al. 2012), it is evident that the green-building movement is not just a fad, but truly is transforming the marketplace worldwide.

methodologies for performing and evaluating green-building design efforts are reviewed.

Prior to the industrial revolution, building efforts were often directed throughout design and construction by a single architect—the so-called *master builder model.* The master builder alone bore full responsibility for the design and construction of the building, including any engineering required. This model lent itself to a building designed as one system, with the means of providing heat, light, water, and other building services often closely integrated into the architectural elements. Sustainability, semantically if not conceptually, predates these eras, and some modern unsustainable practices had yet to arise. Sustainability in itself was not the goal of yesteryear's master builders. Yet some of the resulting structures appear to have achieved an admirable combination of great longevity and sustainability in construction, operation, and maintenance. It is interesting to compare the ecological

Digging Deeper

SOME DEFINITIONS AND VIEWS OF SUSTAINABILITY FROM OTHER SOURCES

- "The best chance we have of addressing the combined challenges of energy supply and demand, climate change and energy security is to accelerate the introduction of new technologies for energy supply and use and deploy them on a very large scale." —Thomas Friedman, *Hot, Flat and Crowded* (Friedman 2008)
- "Humanity must rediscover its ancient ability to recognize and live within the cycles of the natural world." —*The Natural Step for Business* (Nattrass and Altomare 1999)
- Development is sustainable "if it meets the needs of the present without compromising the ability of future generations to meet their own needs." —*Our Common Future* (Brundtland Report) (WCED 1987)
- To be sustainable, "a society needs to meet three conditions: Its rates of use of renewable resources should not exceed their rates of regeneration; its rates of use of nonrenewable resources should not exceed the rate at which sustainable renewable substitutes are developed; and its rates of pollution emissions should not exceed the assimilative capacity of the environment." —Herman Daly (Nattrass and Altomare 1999)
- "Sustainability is a state or process that can be maintained indefinitely. The principles of sustainability integrate three closely intertwined elements—the environment, the economy, and the social system – into a system that can be maintained in a healthy state indefinitely." —*Energy Management Handbook,* eighth edition (Doty and Turner 2012)
- "In this disorganized, fast-paced world, we have reached a critical point. Now is the time to rethink the way we work, to balance our most important assets." —Paola Antonelli, Curator, Department of Architecture and Design, New York City Museum of Modern Art (Antonelli 2008)

footprint (a concept discussed later in this book) of Roman structures from two millennia ago heated by radiant floors to a twentieth century structure of comparable size, site, and use.

In the nineteenth century, as ever more complicated technologies and the scientific method developed, the discipline of engineering building systems and design emerged separate from architecture. This change was not arbitrary or willful, but rather was due to the increasing complexity of design tools and construction technologies and a burgeoning range of available materials and techniques. This complexity continued to grow throughout the twentieth century and continues today. With the architect transformed from master builder to lead design consultant, most HVAC&R engineering practices performed work predominantly as a subcontract to the architect, whose firm, in turn, was retained by the client. Hand-in-hand with these trends emerged the twentieth century doctrine of *buildings over nature*, an approach still widely demanded by clients and supplied by architectural and engineering firms.

Under this approach (buildings are designed under the architect, who is prime consultant, following the buildings over nature paradigm) the architect conceives the shell and interior design concepts first. Only then does the architect turn to structural engineers, then HVAC&R engineers, then electrical engineers, etc. (Not coincidentally, this hierarchy and sequence of engineering involvement mirrors the relative expense of the subsystems being designed.)

With notable exceptions, this sequence has reinforced the trend toward buildings over nature: relying on the brute force of sizable HVAC systems that are resource-intensive—and energy-intensive to operate—to build and maintain conditions acceptable for human occupancy. In this approach to the design process, many opportunities to integrate architectural elements with engineered systems are missed—often because it's too late. Even with an integrated design team to bridge back over the gaps in the traditional design process, a sustainable building with optimally engineered subsystems will not result if not done by professionals with appropriate knowledge and insight.

When Green Design is Applicable

Perhaps the obvious answer is "When is green design *not* applicable?" However, practicalities do exist in the design process, funding, and expectations of stakeholders in the process that may, in some people's opinion, preclude consideration of green design. This book is intended to help overcome these impediments.

One leading trend in architecture, especially in the design of smaller buildings, is to invite nature in as an alternative to walling it off with a shell and then providing sufficiently powerful mechanical/electrical systems to perpetuate this isolation. This situation presents a significant opportunity for engineers today. Architects and clients who take this approach require fresh and complementary engineering approaches, not tradition-bound engineering that incorporates extra capacity to overcome the natural forces a design team may have invited into a building. Natural ventilation and hybrid mechanical/natural ventilation, radiant heating, radiant cooling, and solar-assisted air conditioning are just a few examples of the tools with which today's engineers are increasingly required to acquire fluency. Some of these "new" techniques have been well known for centuries and used around the world. In the green-building era, they can be enhanced with new capabilities allowed by technology advancements, better understanding of the physical processes involved, and for modern buildings.

Fortunately, there is a great deal of information available about green-building design, including this *GreenGuide*. Further, tools for understanding and defending engineering decisions in such projects are available, for example, ANSI/ASHRAE Standard

55, *Thermal Environmental Conditions for Human Occupancy* (ASHRAE 2017a), includes an adaptive design method that is more applicable to buildings that interact more freely with the outdoor environment. ASHRAE Standard 55 also accommodates an increasing variety of design solutions intended both to provide comfort and to respect the imperative for sustainable buildings.

Another more widely demanded approach to green HVAC engineering presents a significant opportunity for engineers. This approach applies to projects ranging from flagship green-building projects to more conventional ones where the client has only a limited appetite for green design. The demand for environmentally conscious engineering is evidenced by the expansion of engineering groups, either within or outside architectural practices, that have built a reputation for a green approach to building design. In addition, many younger engineers and architects just entering the profession are more committed to the concept of sustainable design than their more established predecessors.

Green HVAC engineering can be provided, for its own sake, independent of any client or architect demand. Ideally, the end result is an energy-efficient system that is more robust and provides for better thermal control and indoor environment than the cookie-cutter conventional design. The appetite for environmentally conscious engineering must be carefully gaged, and opportunities to educate the design team carefully seized. In this way, engineers can bring greater value to their projects and distinguish themselves from competing individuals and firms.

Embodied Energy and Life-Cycle Assessment

Building materials used in the construction and operation of buildings have energy embodied in them due to the manufacturing, transportation, and installation processes of converting raw materials to final products. The material selection process should consider the environmental impact of demolition and disposal after the service life of the products. Another new type of building life-cycle assessment (beyond life cycle costing) focuses on the environmental impact of products and processes. This is termed *life-cycle environmental assessment* or simply *life-cycle assessment* (LCA). This is a cradle-to-grave approach that evaluates all stages of a product's life to determine its cumulative environmental impact. In the case of a building, the structure is a product itself, but it also is comprised of a large number of other individual products (e.g., materials, equipment). Thus, the combined impact of the entire building as a system should be quantified. Different issues and priorities should be considered. For example, the selection of building materials with a lower embodied energy may be desirable for the construction phase, but it may be more environmentally conscious to consider a more energy-intensive one if it results in higher operational energy savings during the building's life cycle or the product's lifetime, whichever comes first. Similarly, during building refurbishment and the evaluation of different energy conservation measures, one should also consider the embodied energy in the new building materials (e.g., adding thermal insulation) or the replacement of building elements or systems (e.g., replacing windows or boilers), by comparing them against the resulting operational energy savings.

Several international efforts are underway to develop a standard approach, along with professional and easy-to-use tools to overcome cumbersome calculations and facilitate performing an LCA in routine, day-to-day projects. This is critical to ensure that these issues are addressed during the early stages of the decision-making process, when most critical decisions are made, for new building design or refurbishment. For large projects, use of building information modeling (BIM) can support LCA by reducing the time to reenter data (e.g. material quantities) and facilitate calculations, once ontology and semantic issues for the exchange of data and relevant information are properly handled.

A detailed description of the LCA approach has been developed by the EPA, and more detail can be found on their website on LCA, included in the Online Resources section at the end of this chapter. LCA databases and tools are used to calculate and compare the embodied energy of common building materials and products. Designers should give preference to resource-efficient materials and reduce waste by recycling and reusing whenever possible.

The building design team has a variety of options to consider in conducting an LCA analysis. The International Organization for Standardization (ISO) 14000 series of standards on environmental management serves as a method to govern the development of these tools. LCA tools are available from private commercial as well as governmental or public domain sources. The Building for Environmental and Economic Sustainability (BEES) tool was developed by the National Institute for Standards and Technology in the United States, with support from the U.S. Environmental Protection Agency. The Tools for the Reduction and Assessment of Chemical and Other Environmental Impacts (TRACI) from the EPA focuses on chemical releases and raw materials usage in products. In Canada, the Athena life-cycle inventory database contains detailed, high-quality and regional construction data and complies with ISO 14040/14044 standards (Athena 2014). Some commercial firms also offer LCA tools.

The European Commission has long recognized that LCA provides the best framework for assessing the potential environmental impacts of products and underlined the need for more consistent data and consensus LCA methodologies. A European platform from the European Commission Joint Research Centre is available to facilitate the availability of quality-assured life-cycle data. The European reference Life Cycle Database (ELCD) is composed of life-cycle inventory data from front-running European business associations and other sources for key materials, energy carriers, transport, and waste management. Data sets can be used free of charge and also distributed to third parties.

In summary, tremendous growth has occurred in the development of green building programs and practices over the past couple of decades. This *GreenGuide* provides insight into the making the design and operation of buildings more sustainable.

REFERENCES AND RESOURCES

Published

AIA. 1996. *Environmental Resource Guide*. Edited by Joseph Demkin. New York: John Wiley & Sons.

Antonelli, P. 2008. *Design and the Elastic Mind*. New York: The Museum of Modern Art.

ASHRAE. 1999. ANSI/ASHRAE/IES Standard 90.1-1999, *Energy Standard for Buildings Except Low-Rise Residential Buildings*. Atlanta: ASHRAE.

ASHRAE. 2016. ANSI/ASHRAE/IES Standard 90.1-2016, *Energy Standard for Buildings Except Low-Rise Residential Buildings*. Atlanta: ASHRAE.

ASHRAE. 2017a. ANSI/ASHRAE Standard 55-2017, *Thermal Environmental Conditions for Human Occupancy*. Atlanta: ASHRAE.

ASHRAE. 2017b. *ASHRAE Handbook—Fundamentals*. Chapter 35. Atlanta: ASHRAE.

ASHRAE. 2017c. ANSI/ASHRAE/USGBC/IES Standard 189.1-2017, *Standard for the Design of High-Performance Green Buildings Except Low-Rise Residential Buildings*. Atlanta: ASHRAE.

Athena. 2014. *User manual and transparency document: Impact estimator for buildings* v. 5. Ottawa: Athena Sustainable Materials Institute. https://calculatelca.com/wp-content/uploads/2014/10/IE4B_v5_User_Guide_September_2014.pdf

Deru, M., and P. Torcellini. 2007. *Source Energy and Emissions Factors for Energy Use in Buildings*. Technical Report NREL/TP_550-38617, National Renewable Energy Laboratory, Golden, CO.

Doty, S., and W.C. Turner. 2012. *Energy management handbook*, 8th edition. Lilburn, GA: Fairmont Press.

European Commission. 2002. Energy Performance of Buildings, Directive 2002/91/EC of the European Parliament and of the Council. *Official Journal of the European Communities*, Brussels.

European Commission. 2010. Energy Performance of Buildings, Directive 2010/31/EU of the European Parliament and of the Council. *Official Journal of the European Communities*, Brussels.

Friedman, T. 2008. *Hot, flat, and crowded*. New York: Farrar, Straus and Giroux.

Grondzik, W.T. 2001. The (mechanical) engineer's role in sustainable design: Indoor environmental quality issues in sustainable design. HTML presentation available at www.nettally.com/gzik/ieq/ieq-sust_files/frame.htm.

IPCC. 2014. Climate Change 2014: Synthesis Report. Contribution of Working Groups I, II and III to the Fifth Assessment Report of the Intergovernmental Panel on Climate Change R.K. Pachauri and L.A. Meyer (eds). pp. 151. Geneva: IPCC.

ISO. 2017 (under development). ISO 52000-1. *Energy Performance of Buildings—Overarching EPB Assessment*. Part 1: General Framework and Procedures. Geneva

Nattrass, B., and M. Altomare. 1999. *The natural step for business*. Vancouver: New Society Publishers.
Savitz, A.W., and K. Weber. 2006. *The Triple Bottom Line*. San Francisco: John Wiley & Sons, Inc.
WCED. 1987. *Our common future*. Oxford, England: Oxford University Press.
William McDonough Architects. 1992. *The Hannover Principles: Design for Sustainability*. Charlottesville, VA: William McDonough & Partners. www.mcdonough.com/wp-content/uploads/2013/03/Hannover-Principles-1992.pdf

Online

The American Institute of Architects
www.aia.org.
Architecture 2030 Challenge
www.architecture2030.org.
Advanced Energy Design Guides
www.ashrae.org/aedg.
Building for Environmental and Economic Sustainability (BEES)
www.nist.gov/el/economics/BEESSoftware.cfm.
BuildingGreen
www.greenbuildingadvisor.com.
Building Research Establishment Environmental Assessment Method (BREEAM®) rating program
www.breeam.org.
European Commission, Concerted Action Energy Performance of Buildings Directive
www.epbd-ca.org.
European Commission Joint Research Centre. European Life Cycle Database.
http://eplca.jrc.ec.europa.eu/ELCD3/
European Commission Joint Research Centre. European Platform on Life Cycle Assessment.
http://eplca.jrc.ec.europa.eu/.
Green Globes
www.greenglobes.com.
GreenSpec® Product Guide
https://www.buildinggreen.com/product-guidance.
International Initiative for a Sustainable Built Environment.
www.iisbe.org/.
International Living Building Institute, The Living Building Challenge
https://living-future.org/lbc/.

International Organization for Standardization family of 14000 standards,
www.iso.org/iso/home/standards/management-standards/iso14000.htm.
Intergovernmental Panel on Climate Change
www.ipcc.ch.
Lawrence Berkeley National Laboratories, Environmental Energy Technologies Division
http://eetd.lbl.gov/.
Minnesota Sustainable Design Guide
www.sustainabledesignguide.umn.edu.
National Renewable Energy Laboratory, Buildings Research
www.nrel.gov/buildings/.
Natural Resources Canada, Evaluation of the Built Environment Portfolio.
/www.nrcan.gc.ca/evaluation/reports/2014/16317
The Natural Step
www.thenaturalstep.org.
New Buildings Institute
www.newbuildings.org/.
Rocky Mountain Institute
www.rmi.org.
Tools for the Reduction and Assessment of Chemical and Other Environmental Impacts (TRACI)
https://www.epa.gov/chemical-research/tool-reduction-and-assessment-chemicals-and-other-environmental-impacts-traci.
U.S. Green Building Council, Leadership in Energy and Environmental Design, Green Building Certification System
www.usgbc.org/leed.
The Whole Building Design Guide
www.wbdg.org.

CHAPTER TWO

GREEN RATING SYSTEMS, STANDARDS, AND OTHER GUIDANCE

Rapid growth in interest in green buildings over the past two decades has occurred with corresponding growth in the number, depth, and breadth of green-building resources available.

There are three general types of programs or resources that exist to encourage green-building design. The first type is composed of green-building rating systems (sometimes referred to as *green building label* programs), such as the LEED program. Second are general guidelines or resources, such as this guide, that have been created and published to encourage and assist designers in achieving green-building design. Third is the more recent trend of green-building practices incorporated as part of design standards and the code enforcement process. This chapter provides a brief summary of each type and cites several specific examples.

GREEN-BUILDING RATING SYSTEMS—INTRODUCTION

Various rating systems, developed by organizations around the world, strive to indicate how well a building meets prescribed requirements and to determine whether a building design is green and to what level. They all provide useful tools to identify and prioritize key environmental issues. These tools incorporate a coordinated method for accomplishing, validating, and benchmarking sustainably designed projects. As with any generalized method, each has its own limitations and may not apply directly to every project's regional, political, and owner design-intent-specific requirements.

There are a wide variety of labeling and certification programs available to measure the environmental impact and performance of existing buildings. These programs can be divided into two general categories: those narrowly focused on energy (and in some cases, water) use, and those more broadly focused on those and other sustainability categories, such as indoor environmental quality (IEQ). Key programs in the first category include the U.S. Environmental Protection Agency's (EPA) ENERGY STAR® Portfolio Manager, ASHRAE's Building EQ, and country-specific energy performance certification established in Europe. Key programs in the second category include U.S.

Green Building Council's Leadership in Energy and Environmental Design® (LEED®) rating system and the Building Research Establishment (BRE) Global Environmental Assessment Method (BREEAM®).

While this guide does not endorse or recommend use of any one particular green-building rating system or program, it does encourage their use when the application will produce an exceptional green design and encourages the building operators to maintain and operate the building in a manner that provides its occupants a continuing healthy and energy-efficient living/work space.

It could easily be said that the green-building movement really started in earnest with the initial establishment of the BREEAM rating system in 1990. BREEAM is a creation of the Building Research Establishment (BRE) in the UK. This is a voluntary, consensus-based, market-oriented assessment program. With one mandatory and two optional assessment areas, BREEAM encourages and benchmarks sustainably designed office buildings. The mandatory assessment area is the potential environmental impact of the building; the two optional areas are design process and operation/maintenance. Several other countries and regions have developed or are developing related spinoffs inspired by BREEAM, and BREEAM has been adopted in other countries. Although initially focused on specific building types, it is been adapted to include a wider range of different types of buildings. Similar to what has happened with other green rating systems, BREEAM has been adapted to various type of programs (called *schemes*), such as BREEAM for new construction, domestic refurbishment, communities, in-use (existing buildings), and for homes (EcoHomes).

The rating method primarily used in the United States is the LEED program, created by U.S. Green Building Council (USGBC). USGBC started offering this system in the 1990s and it is intended to be a voluntary, consensus-based, market-driven green-building certification system. It evaluates environmental performance from a “whole-building” perspective over a building’s life cycle, providing a numerical standard for what constitutes a green building. USGBC’s goal has been to raise awareness of the benefits of building green, and it has transformed the marketplace. Additional discussion on LEED can be found in the following section.

Another rating method that was originally developed in Canada and is being used in the United States is the Green Globes program. Green Globes is an online auditing tool that includes many of the same concepts as LEED. While both aim to help a building owner or designer develop a sustainable design, Green Globes is primarily a self-assessment tool (although third-party assessment is an option) and also provides recommendations for the project team to follow for improving the sustainability of the design. In the United Kingdom, Green Globes is known as the Global Environmental Method program.

Other green-building rating programs exist in countries throughout the world, for example, Australia’s Green Star and National Australian Built Environment Rating System (NABERS), Japan’s Comprehensive Assessment System for Built Environment Efficiency (CASBEE), Hong Kong's Building Environmental Assessment Method (BEAM), and the Estidama program in the United Arab Emirates.

The procedures used by those organizations and governments providing building rating systems vary. Many building rating programs use static building labels—that is, the building's energy (or other green attributes) is evaluated once, the label is applied, and the providing organization does no reassessment to determine if the building continues to actually meet the original specifications. The application of dynamic labeling is preferred. A few of the building rating programs actually do include a reassessment (e.g., every two years), and buildings not continuing to perform lose their building label (Means and Walters 2010).

THE LEED RATING SYSTEM

Since its development and introduction in the late 1990s, the LEED program has become a major factor in the advancement of green buildings in the United States and elsewhere, as well as an influence on how all buildings are thought of in the design and construction process. LEED has been applied to numerous projects over a range of project certification levels, and its use has grown rapidly over the past several years. The LEED rating system started out with a basic program for new construction, but because a large majority of buildings already exist, a LEED for existing buildings was released in 2004 and has been revised several times since. LEED rating systems have also been developed for a variety of specific building types. These include building core and shell and commercial interiors for project developers and tenants (respectively), as well as schools, retail, hospitality, data centers, warehouses and health care, and homes. Growing in importance is the role of LEED for Existing Buildings Operations and Maintenance (EBOM). The LEED program and registered building projects have already been, or are being, established in other countries including India, Australia, Canada, and China, and new project registrations in countries outside the United States make up about 40% of the total, according to recent trends. Many other countries have developed their own green building rating systems following the basic criteria in LEED or other rating systems.

LEED is a voluntary program that uses a point-based rating system for a given building project. USGBC has been continuously working to update and modify the LEED program since its initial release. Over the years the revisions have, among other things, reworked the point ratings to provide a more effective focus to drive positive environmental and health benefits. Part of the revisions were based on using the Tool for the Reduction and Assessment of Chemical and Other Environmental Impacts (TRACI) developed by the U.S. Environmental Protection Agency (EPA).

Another change to the LEED program since its initial release was the inclusion of regional priority credits. These are used to put additional emphasis on design

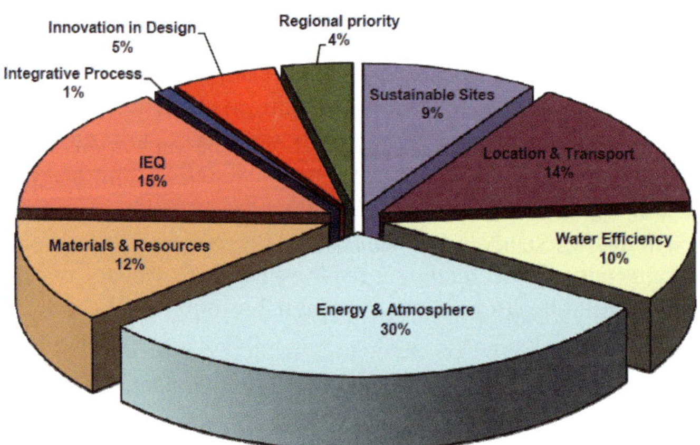

Image courtesy of Brendan Owens, U.S. Green Building Council.

Figure 2-1 Credit point distribution for LEED for new construction, v4.

features that are particularly important in relation to the local climate and region where the project is to be located. For example, in areas with known stormwater problems, the Sustainable Sites Credit 6.1 for Stormwater Design—Quantity Control may be included in this priority list. A full list of the Regional Priority credits can be found on the USGBC website (see the References and Resources section at the end of the chapter).

The fourth version of the LEED program is now known as LEED v4, and was finalized in 2013. Achieving 40 points will earn LEED Certified status, and the higher levels of silver, gold, and platinum can be achieved with 50, 60, and 80 points, respectively. The percentage distribution of these credits is illustrated in Figure 2-1. "Energy and Atmosphere" is the category with the highest number of points available, and this is in the direct purview of the HVAC&R engineer.

Further discussion on how LEED may evolve to more closely align with other green-building codes is given in the section on ANSI/ASHRAE/USGBC/IES Standard 189.1 and the *International Green Construction Code*.

OTHER GUIDELINES, RESOURCES, AND CERTIFICATION PROGRAMS

Beside the rating systems discussed above, there are other types of programs that are available to measure the environmental impact and performance of buildings (existing or new construction). Some of these are more broadly focused on the broad range of sustainability categories and IEQ. Other key programs that narrowly focus on energy (and in some cases, water) use include the EPA Portfolio Manager, ASHRAE's Building EQ, and mandatory country-specific energy performance certification in the European Union. Key programs that are more broadly focused on energy and water use, along with other sustainability categories and IEQ, include the LEED and the BREEAM rating systems, discussed earlier in this chapter.

This section outlines just some of those additional resources that the green building designer can access. ASHRAE has publications that can help the inexperienced and experienced industry professional alike. These include the following:

- ANSI/ASHRAE/IES Standard 202, *Commissioning Process for Buildings and Systems* (ASHRAE 2013).
- Advanced Energy Design Guide series. (The 30% Guides were developed initially and were intended to offer a 30% savings compared to ANSI/ASHRAE/IES Standard 90.1-1999, while the more recent guides are designed for 50% and more savings compared to the 2004 version of Standard 90.1. These design guides are available for free download courtesy of a collaboration between ASHRAE and the U.S. EPA.
- *Indoor Air Quality Guide: Best Practices for Design, Construction and Commissioning* (ASHRAE 2009a).
- *The ASHRAE Guide for Buildings in Hot and Humid Climates*, 2nd edition (ASHRAE 2009b).
- *Cold-Climate Buildings Design Guide* (ASHRAE 2015).
- ASHRAE also developed user's manuals for many of its most widely used standards, such as Standards 62.1, 62.2, 90.1 and 189.1.
- *High Performing Buildings* magazine. This publication was created by ASHRAE to provide real-world, case study examples for reference.

Many of these are also discussed elsewhere in this guide but are mentioned here as a reminder.

The Advanced Energy Design Guide series is a series of books that provide a set of prescriptive technical approaches to achieve significant energy savings. The documents are focused on specific building types, typically smaller building projects that may not have resources available for much engineering study and analysis of energy saving technologies (e.g., small retail stores). Recommendations are provided based on the climate zone the project is located in. The initial series of guides was targeted toward achieving 30% energy savings compared to Standard 90.1-1999, and the intent is to repeat the series with increasing efficiency levels leading to net zero designs. The more recent 50% Guides are designed for 50% savings compared to the 2004 version of Standard 90.1. Additional guides with a higher efficiency target are in the planning process, and funding considerations will determine how far they go (e.g., a net zero target by 2020 for the AEDG series is one consideration).

These energy guides were produced in collaboration with the American Institute of Architects (AIA), the Illuminating Engineering Society of North America (IES), and USGBC, with assistance provided by the U.S. Department of Energy (DOE). These guides are available for free from the ASHRAE website and have been widely distributed since their release.

The *Indoor Air Quality Guide* was released in January of 2010 and was developed in conjunction with AIA, the Building Owners and Managers Association (BOMA), the Sheet Metal and Air Conditioning Contractors National Association (SMACNA), the U.S. EPA and USGBC. A summary of the guidance offered is available for free download from the ASHRAE website, while the detailed document is available for purchase.

BUILDING ENERGY QUOTIENT (Building EQ)

ASHRAE unveiled a building energy labeling program known as the Building Energy Quotient (Building EQ) program in 2009. The program provides a method to rate a building's energy performance both "As Designed" (Asset Rating) and "As Operated" (Operational Rating).

ASHRAE's Building EQ assessment is designed to be performed by either an ASHRAE-certified Building Energy Assessment Professional (BEAP) or by a licensed Professional Engineer. This assessment includes both utility bill analysis and an ASHRAE Level 1 Energy Audit, which is described in the next section of this chapter. ASHRAE produces an energy rating ranging from "A+" to "F", with a "C" representing median energy performance in comparison to the building's peers (Figure 2-2). The building owner also receives a report with recommendations for how to reduce energy and water use while maintaining acceptable IEQ. Note that ASHRAE offers a related "as designed" Building EQ rating, so that owners can compare the actual performance of their building to its potential performance (Figure 2-3).

ASHRAE's Building EQ program provides the general public, building owners and tenants, potential owners and tenants, and building operations and maintenance staff with information on the potential and actual energy use of buildings. This information is useful for the following reasons:

- Building owners and operators can see how their building compares to peer buildings to establish a measure of their potential for energy performance improvement.
- Building owners can use the information provided to differentiate their building from others to secure potential buyers or tenants.
- Potential buyers or tenants can gain insight into the value and potential long-term cost of a building.
- Operation and maintenance staff can use the results to inform their decisions regarding maintenance activities, influence building owners and managers to

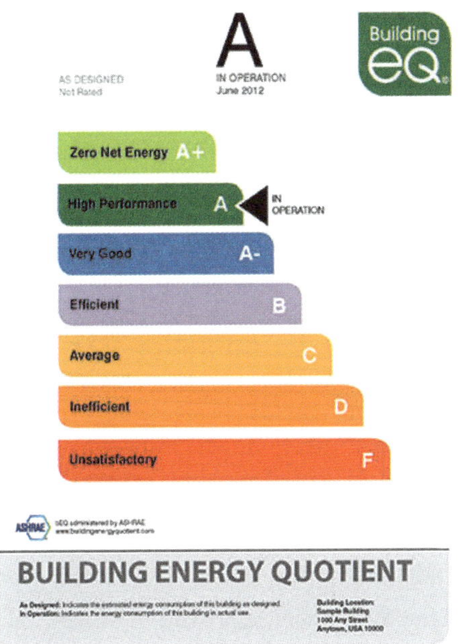

Figure 2-2 Sample Building EQ plaque.

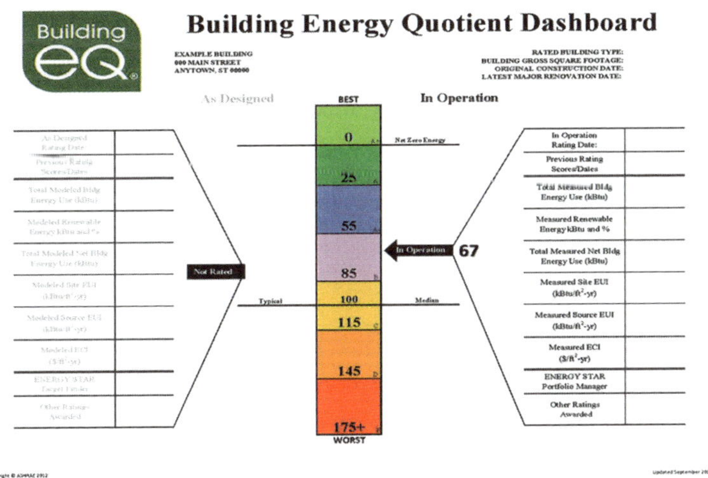

Figure 2-3 Sample Building EQ dashboard.

pursue equipment upgrades, and demonstrate the return on investment for energy efficiency projects.

Beyond the benefit received by individual building owners and managers, the increased availability of building data (specifically the relationship between the design and operation of buildings) will be a valuable research tool for the building community.

The EPA Portfolio Manager is an online tool that uses utility bills and basic building information to develop building energy performance reports, water performance reports, and greenhouse gas emissions reports. It is intended for use by building owners and managers. The tool assigns a normalized ENERGY STAR score of 1 to 100 to each building, with a score of 50 indicating median energy performance compared to its peers. Buildings receiving a score of 75 or above are eligible for ENERGY STAR certification, which must be obtained by a Professional Engineer or a Registered Architect. This tool also makes high-level recommendations on how to reduce water and energy use.

EUROPEAN PROGRAMS

In 2002, the European Parliament approved the Energy Performance of Buildings Directive (EPBD), which required member nations to develop methodologies for the calculation of the energy performance of buildings, to establish minimum energy performance requirements for both new buildings and existing "large" buildings subject to "major" renovation, and to develop energy performance certification programs. An energy performance certificate (EPC) must be issued when any building (commercial or residential) is constructed, sold, or rented to a new tenant. The EPC documents the building's energy performance, expressed as letter score from A to G, based on primary energy consumption, carbon dioxide emissions, or energy cost per unit floor area to facilitate comparisons between similar buildings. The EPC also includes cost-effective recommendations for improving the building's energy performance, which specify the initial cost, estimated annual energy and carbon dioxide emissions savings, and the simple payback period.

New buildings should meet minimum energy performance requirements (i.e., energy class B), as should existing buildings or building units that undergo major renovation where technically and economically feasible. Most national schemes have implemented an asset rating system based on calculated energy use for new and small existing nonpublic buildings; some have chosen an operational rating system based on billed energy for large and complex nonresidential buildings. The majority of the national schemes require a certified energy inspector to perform a building energy audit to collect all relevant data, which are then used to perform the calculations for issuing a certificate. Implementation and oversight of the certification process is typically the responsibility of the national government. Additional information concerning European EPCs is provided in Arcipowska et al. (2014).

Finally, BRE Global developed and maintains the BREEAM "in-use" standard. This program encompasses nine sustainability and IEQ categories: energy, water, materials, pollution, land use and ecology, health and well-being, waste, transport, and management. Points may be obtained in each category and are then converted to a weighted score ranging between 0% and 100%. Assessments range upward from "Acceptable" (one star) to "Outstanding" (six stars). Organizations either self-assess or obtain formal certification. Obtaining formal certification requires engaging the services of an assessor licensed by BRE Global.

IMPLEMENTATION THROUGH GREEN BUILDING STANDARDS AND CODES

Since the middle part of the past decade, there has been a movement to make green-building practices a more mandatory part of the normal building code process. Several cities in the United States now require LEED certification for building projects above a certain size or classification (such as a government building). In addition, ASHRAE has initiated a process to create a series of new standards for high-performance green buildings, releasing in early 2010 the initial version of Standard 189.1, *Standard for the Design of High-Performance Green Buildings Except Low-Rise Residential Buildings*. This Standard is under continuous maintenance and has had several subsequent new releases, most recently (as of this writing) in 2017 (ASHRAE 2017).

ANSI/ASHRAE/USGBC/IES STANDARD 189.1 AND THE *INTERNATIONAL GREEN CONSTRUCTION CODE* (IgCC)

In 2006, ASHRAE (in conjunction with USGBC and IES) began a process to create a standard that would address a growing need within the industry for a code-language document for green buildings suitable for adoption as part of building codes. ANSI/ASHRAE/USGBC/IES Standard 189.1-2009, *Standard for the Design of High-Performance Green Buildings Except Low-Rise Residential Buildings* (ASHRAE 2009) was developed during a more than three-year process with extensive public review and was initially published in early 2010. This standard is in a continuous maintenance process, and an updated version was released in 2017.

The standard differs from LEED or other products in that it is not a rating system, nor is it a design guideline per se. The purpose of Standard 189.1 is to provide minimum requirements for the siting, design, construction, and plans for operation of high-performance green buildings, while attempting to balance environmental responsibility, resource efficiency, occupant comfort and well-being, and community sensitivity. One key point of this is that Standard 189.1 is not targeted for any building project, but rather specifically for high-performance building projects. This document is intended to help fill a perceived gap in the evolving building codes in this area, as localities begin to adopt green-building designs as a requirement.

Image courtesy of Tom Lawrence and ASHRAE

Figure 2-4 Relation of Standard 189.1 to other ASHRAE Standards.

While many of the topics and criteria may overlap or seem similar to LEED for new construction, Standard 189.1 differs in that it establishes mandatory, minimal requirements across all topical areas. Besides the obvious intent of providing a vehicle for adoption into building codes, this standard may also be used by developers, corporations, universities, or governmental agencies to set requirements for their own building projects.

Standard 189.1 is not intended to do away with other ASHRAE standards. Rather, it builds upon key ASHRAE standards and adopts these with modifications when considered necessary to develop a document that deals with high-performance green buildings, as illustrated in Figure 2-4.

This standard includes mandatory criteria in all topical areas (for example, water or energy) and provides for two compliance paths. The prescriptive path includes simple compliance criteria; simple in the sense that they are more like a checklist of technologies or system requirements. The performance path is more complicated in that it requires more analysis to verify that compliance is indeed achieved.

A brief overview of some key criteria in Standard 189.1 that the typical ASHRAE member or design professional should be aware of is given below. This is just a brief overview and is not intended to be an all-inclusive summary.

Sustainable Sites

In addition to a cool-roof requirement for cooling-dominated climates (Climate Zones 0–3), the standard also has provisions for building walls to be shaded or to have a minimum solar reflective index value for opaque wall materials in all but the coldest climate zones.

Water Use Efficiency

Standard 189.1 puts limits on the number of cycles of water through a cooling tower. It also requires condensate collection on air-handling units above 5.5 tons (19 kW) of cooling capacity in more humid regions (areas with a design wet-bulb temperature greater than 72°F [22°C]).

Standard 189.1 is also unique in that it requires the installation of meters with data storage and retrieval capability on systems and areas above a given threshold in water usage. Similar provisions are included in the energy section for energy use and IEQ section for monitoring of outdoor airflow.

Energy Efficiency

While the LEED program, through Energy and Atmosphere Credit 1, provides for a sliding scale of points awarded for energy efficiency improvements above Standard 90.1, Standard 189.1 began with the intended goal of providing mandatory measures that would result in buildings using 30% less energy than what is designed according to the existing Standard 90.1 at that time, including process loads.

During the development of Standard 189.1, many concepts and requirements were considered for improvements. At the same time, addenda to Standard 90.1 were developed that increased the overall efficiency levels of Standard 90.1. The net result is that the difference in overall average energy utilization index (EUI) for buildings designed according to Standard 189.1 and Standard 90.1-2016 is not that great. It has been ASHRAE's intent to have the energy efficiency levels for Standard 189.1 improve at a faster rate than those for Standard 90.1, with an ultimate goal of having Standard 189.1 reach nearly net zero or cost-effective net zero status by the year 2020 (although the definition of an optimum nearly net zero value has yet to be established).

One of the key considerations when developing the energy requirements for Standard 189.1 was whether to include on-site renewable energy and, if so, to what extent. Many renewable energy systems are not yet fully cost competitive with conventional energy sources, depending on the local energy costs and particularly when excluding incentive programs that may go away at any time. Therefore, the standard only includes the provisions for being "renewable ready" with provisions for allowing ease of future installation of renewable energy systems as a mandatory requirement. Exceptions are included for areas with low solar incidence or local shading. In the prescriptive path, on-site renewable energy is included, but a project can still comply with this standard using other methods that would have equivalent benefits or the engineers and designers may elect to go with the performance compliance path.

Energy metering is required for key systems (e.g., HVAC) above certain thresholds, because even when a building is initially designed to be energy efficient, it can quickly slip into having less than stellar energy efficiency if not continuously monitored and well maintained.

Standard 189.1 makes numerous modifications to the requirements in Standard 90.1 regarding HVAC systems. The following is a summary of key points:

- The threshold for occupancy levels requiring demand-controlled ventilation is lowered.
- The minimum size requirement for economizers is reduced. Other specific exception and requirement changes are included, but a description is beyond the scope of this guide.
- Fan power limits (per volume of air moved) are lowered.
- The requirements for energy recovery from exhaust air are expanded and the minimum effectiveness of the energy recovery device is set at 60%.
- Levels of duct insulation are increased.
- Unoccupied hotel/motel guest room controls are included.

Depending on the project approach, additional requirements for equipment efficiency beyond Standard 90.1 may also be incorporated. In addition, requirements are set for automated peak demand reduction of the building.

The performance path for showing compliance with the energy section of ANSI/ASHRAE/USGBC/IES Standard 189.1 includes demonstrating equivalent performance in terms of both energy cost and CO_2 equivalent emissions, compared to if the building project had been designed strictly to the criteria in the prescriptive path.

IEQ

Several aspects of indoor environmental quality are relevant to HVAC design: tobacco smoke control, outdoor air monitoring, filtration /air cleaning, and determination of outdoor airflow rate.

The minimum ventilation design for outdoor airflow is to be according to ANSI/ASHRAE Standard 62.1 (ASHRAE 2016b) using the Ventilation Rate Procedure. Outdoor air monitoring is to be done using permanently mounted, direct outdoor airflow measurement devices. In contrast to LEED, CO_2 monitoring in densely occupied zones is not included as part of Standard 189.1.

Tobacco smoke control is achieved by simply banning smoking within the building and near entrances, outdoor air intakes, or operable windows.

Materials and Resources

A number of requirements included in this section parallel those included with the LEED program for new construction. Items of particular note include a construction waste management provision to divert a minimum of 50% of nonhazardous waste and demolition debris from being sent to a landfill, a ban on CFC-containing equipment, and for fire suppression systems to contain no ozone-depleting substances.

Construction and Plans for Operation

Standard 189.1 includes provisions for not only how a building should be constructed, but also for planning for how it should be operated once occupied. Since the standard is written and intended for adoption into building codes, only items that would be expected to be developed and in place at the time a certificate of occupancy is issued could reasonably be considered for inclusion in this standard. The approach taken within Standard 189.1 is to set requirements for the development of plans for operation in critical areas.

This standard includes requirements for building acceptance testing and/or commissioning, erosion control, IAQ, moisture control, and idling of construction vehicles to be implemented during construction. Commissioning is to be done according to requirements that in essence parallel those in ASHRAE guidelines. IAQ requirements during construction and before occupancy are similar to those in the LEED program but not identical and when different are generally more stringent.

Plans for operation are required in key areas that would be needed to help ensure the building performs as would be expected for a high-performance green building. These include criteria in setting up long-term monitoring and verification of water and energy use, as well as IAQ through provisions such as outdoor air monitoring.

Maintenance and service life plans are required as well, and these involve equipment and systems relevant to the HVAC&R or MEP engineer.

International Green Construction Code (IgCC)

Soon after the initial release of Standard 189.1, ASHRAE and the International Code Council (ICC) reached an agreement whereby the standard would be included as an appendix to the *International Green Construction Code* (IgCC). The IgCC was first released in March 2012 and initially specified Standard 189.1 as a compliance option. By that it was meant that the project team had a choice for compliance: they can comply with the IgCC or with Standard 189.1.

After several years of a dual set of standards, ASHRAE and ICC agreed that the two standards should be merged. Starting with the 2018 code cycle, ASHRAE will be the subject matter expert for technical content while ICC will ensure that the resulting standard/code will be responsible for the administrative provisions in the 2018 IgCC. The 2018 version of the IgCC will reflect this merger. This "IgCC powered by 189.1" will hopefully lead to more widespread adoption of green building codes.

USGBC is also expected to review the measures in the IgCC in 2018 by comparing these to the LEED requirements, hopefully leading to closer alignment of LEED to the IgCC.

RESIDENTIAL BUILDINGS

The situations surrounding the residential building market vary widely around the world based on local situations. The joint ICC/ASHRAE 700 National Green Building Standard was released in 2015 (ICC 2015), and since then ASHRAE has expanded its emphasis with respect to the residential market. A local jurisdiction may elect to include compliance with ICC 700 as part of their adoption of the IgCC. The 2018 version of the IgCC will be "deemed to comply" for residential construction.

OTHER BUILDING CODES

In 2010, California became the first state in the Untied States to adopt a green building code (known as CALGreen). See the accompanying Digging Deeper sidebar for more information on CALGreen.

A compilation showing where in the United States states and jurisdictions have surpassed minimum energy code requirements and an outline of several green rating systems is available online by the Building Codes Assistance Project (http://bcapcodes.org). Major cities across the Untied States that have taken exceptional steps towards increasing the energy efficiency of their buildings include Chicago, New York, San Francisco, and Washington, DC. An overview of international building energy efficiency policies (e.g., codes, incentives, and labels for all types of buildings including mandatory, model code, and voluntary programs) is maintained by the IEA in a publicly available database (www.iea.org/beep/).

Other Resources

Further information regarding these resources can be found in the References and Resources section at the end of the chapter. Other guides and methods include the following:

CALGREEN CODE: AMERICA'S FIRST STATEWIDE GREEN-BUILDING CODE

In 2010, the California Green Building Standards Code (CALGreen) was developed to promote the design of efficient and environmentally responsible residential and nonresidential buildings in California. The CALGreen code is part of the overall California Building Standards Code and is the first statewide green code established in the United States. It was developed, in part, in an effort to meet the provisions of Assembly Bill (AB) 32, which requires a cap on greenhouse gas emissions by 2020, with mandatory reporting. The 2010 CALGreen Code became effective January 1, 2011 and a modified 2016 version took effect in January 2017.

To reduce the overall environmental impact of new buildings constructed in California, and to meet their maximum environmental efficiency targets, the CALGreen Code adopts many green-building practices as mandatory building code requirements. The CALGreen Code includes requirements (divisions) for planning and design, energy efficiency, water efficiency and conservation, material conservation and resource efficiency, and environmental quality. The code also requires building commissioning to verify and ensure that all building systems operate as designed to meet their maximum energy efficiency targets.

Some similarities to LEED programs include standards for stormwater pollution prevention, light pollution reduction, indoor and site water savings, construction waste management, energy performance, outdoor air delivery, carbon monoxide monitoring, and materials selection. In some cases, CALGreen has stricter targets than LEED, in others, LEED is stricter, and in many others, the requirements are identical. There are several CALGreen requirements not found in LEED, such as installing water meters on buildings with area greater than 50,000 ft^2 (4600 m^2), providing weather-resistant exterior walls and foundation envelopes, defining the type of fireplace that can be installed, and employing acoustical control (interior and exterior).

In addition to the mandatory statewide CALGreen requirements, a city or county may adopt local ordinances to require more restrictive standards that go above and beyond the mandatory measures. These packages of voluntary measures, called Tier 1 and Tier 2, include a set of provisions from each code division. These provisions are additional measures that are stricter than the mandatory codes. For instance, building energy performance must exceed the California Energy Code (Title 24) by 15% and 30% for Tier 1 and Tier 2, respectively. Additionally, Tier 1 includes one additional elective from the water efficiency division, whereas Tier 2 includes 3. Some of the cities that have adopted Tier 1 include Burlingame, Napa, and Santa Rosa. As of the writing of this edition, only Palo Alto has adopted Tier 2.

- The Whole Building Design Guide
- *The Living Building Challenge* V2.0 (An updated version 2.1 of this program was released in May 2012)
- Green Building Advisor
- California Collaborative for High Performance Schools
- Minnesota Sustainable Design Guide
- New York High Performance Building Guidelines (Released in 1999 by the non-profit Design Trust for Public Space, but still has relevant information for today)

Work referred to by architects includes:

- *The Hannover Principles* (William McDonough Architects 1992)
- GreenSpec® Product Guide
- Information from The Natural Step (a nonprofit organization)
- International Organization for Standardization (ISO 14000 family of standards)
- Building for Environmental and Economic Sustainability (BEES). (This software tool to analyze the impact of a building uses a life-cycle assessment approach as specified in ISO 14040. An online version was released in 2010.)
- Tool for the Reduction and Assessment of Chemical and Other Environmental Impacts (TRACI)

REFERENCES AND RESOURCES

Published

AIA. 1996. *Environmental Resource Guide*. Edited by Joseph Demkin. New York: John Wiley & Sons.

Arcipowska A., Anagnostopoulos F., Mariottini F., Kunkel S. 2014. Energy performance certificates across the EU. Brussels: Buildings Performance Institute Europe. http://bpie.eu/publication/energy-performance-certificates-across-the-eu/

ASHRAE. 2009a. *Indoor Air Quality Guide: Best Practices for Design, Construction and Commissioning*. Atlanta: ASHRAE.

ASHRAE. 2009b. *The ASHRAE Guide for Buildings in Hot and Humid Climates*. Atlanta: ASHRAE.

ASHRAE. 2012. ANSI/ASHRAE Standard 52.2-2012, *Method of Testing General Ventilation Air-Cleaning Devices for Removal Efficiency by Particle Size*. Atlanta: ASHRAE.

ASHRAE. 2013. ASHRAE/IES 202-2013. *Commissioning Process for Buildings and Systems*. Atlanta: ASHRAE.

ASHRAE. 2015. *Cold-Climate Buildings Design Guide*. Atlanta: ASHRAE.

ASHRAE. 2016a. ANSI/ASHRAE Standard 55-2016, *Thermal Environmental Conditions for Human Occupancy*. Atlanta: ASHRAE.

ASHRAE. 2016c. ANSI/ASHRAE Standard 62.1-2016, *Ventilation for Acceptable Indoor Air Quality*. Atlanta: ASHRAE.

ASHRAE. 2016b. ANSI/ASHRAE/IES Standard 90.1-2016, *Energy Standard for Buildings Except Low-Rise Residential Buildings.* Atlanta: ASHRAE.

ASHRAE. 2017. ANSI/ASHRAE/USGBC/IES Standard 189.1-2017, *Standard for the Design of High-Performance Green Buildings Except Low-Rise Residential Buildings.* Atlanta: ASHRAE.

Deru, M., and P. Torcellini. 2007. *Source Energy and Emissions Factors for Energy Use in Buildings.* Technical Report NREL/TP_550-38617, National Renewable Energy Laboratory, Golden, CO.

ICC. 2015. ICC/ASHRAE National Green Building Standard. eISBN-13: 978-0-86718-751-9.

Jarnagin, R.E., R.M. Colker, D. Nail and H. Davies. 2009. ASHRAE Building EQ. *ASHRAE Journal* 51(12):18–21.

McDonough, W. 1992. *The Hannover principles: Design for sustainability*. Presentation, Earth Summit, Brazil.

Means, J.K. and F.H. Walters. 2010. Do High Performance-Labeled Buildings Really Perform at the Promised Levels? CLIMA 2010 REHVA World Congress 10, Antalya, Turkey. ISBN 978-0975-6907-14-6: Proceedings CD-ROM.

Taylor, S.T. 2005. LEED and Standard 62.1. *ASHRAE Journal Sustainability Supplement*, September.

USGBC. 2009. *LEED 2009 Green Building Design and Construction Reference Guide*. Washington, D.C.: U.S. Green Building Council.

William McDonough Architects. 1992. *The Hannover Principles: Design for Sustainability*. Charlottesville, VA: William McDonough & Partners. www.mcdonough.com/wp-content/uploads/2013/03/Hannover-Principles-1992.pdf

Online

The American Institute of Architects
www.aia.org.

ASHRAE Building Energy Quotient
www.buildingenergyquotient.org/.

ASHRAE Advanced Energy Design Guides
www.ashrae.org/standards-research--technology/advanced-energy-design-guides.

BuildingGreen
www.greenbuildingadvisor.com.

Building Codes Assistance Project
http://bcapcodes.org

Building Research Establishment Environmental Assessment Method (BREEAM®) rating program
www.breeam.org.

Center of Excellence for Sustainable Development, Smart Communities Network
www.smartcommunities.ncat.org.

Collaborative for High Performance Schools
www.chps.net.

European Commission—Nearly Zero Energy Buildings
https://ec.europa.eu/energy/en/topics/energy-efficiency/buildings/nearly-zero-energy-buildings.

Estidama (United Arab Emirates)
https://www.upc.gov.ae/en/estidama/estidama-program.

Green Building Advisor
www.greenbuildingadvisor.com.

Green Globes
www.greenglobes.com.

GreenSpec® Product Guide
https://www.buildinggreen.com/product-guidance.

Green Star (Australia)
www.gbca.org.au/green-star/.

International Code Council (ICC).
https://www.iccsafe.org/

International Energy Agency, Building Energy Efficiency Policies Database
http://www.iea.org/beep/.

International Initiative for a Sustainable Built Environment.
www.iisbe.org/.

International Living Building Institute, The Living Building Challenge
http://living-future.org/lbc.

International Organization for Standardization family of 14000 standards,
https://www.iso.org/iso-14001-environmental-management.html.

Intergovernmental Panel on Climate Change
www.ipcc.ch.

International Performance Measurement and Verification Protocol
www.evo-world.org.

Lawrence Berkeley National Laboratories, Environmental Energy Technologies Division
http://eetd.lbl.gov.

Minnesota Sustainable Design Guide
www.sustainabledesignguide.umn.edu.

National Australian Built Environment Rating System (NABERS)
https://nabers.gov.au/public/webpages/home.aspx.

National Institute for Standard and Technology, Building for Environmental and Economic Sustainability (BEES)
www.nist.gov/el/economics/BEESSoftware.cfm.

National Renewable Energy Laboratory, Buildings Research
www.nrel.gov/buildings.

Natural Resources Canada, Evaluation of the Built Environment Portfolio.
/www.nrcan.gc.ca/evaluation/reports/2014/16317

The Natural Step
www.naturalstep.org.

New Buildings Institute
www.newbuildings.org.

Rocky Mountain Institute
www.rmi.org.

Tools for the Reduction and Assessment of Chemical and Other Environmental Impacts (TRACI)
https://www.epa.gov/chemical-research/tool-reduction-and-assessment-chemicals-and-other-environmental-impacts-traci.

U.S. Green Building Council, Leadership in Energy and Environmental Design, Green Building Certification System
www.usgbc.org/leed.

U.S. Green Building Council, Regional Priority Credit Listing
www.usgbc.org/rpc.

The Whole Building Design Guide
www.wbdg.org.

Section 2:
The Design Process

CHAPTER THREE

Project Strategies and Early Design

INGREDIENTS OF A SUCCESSFUL GREEN PROJECT ENDEAVOR

The following ingredients are essential in delivering a successful green design:

- Commitment from the entire project team, starting with the owner.
- Establishing Owner's Project Requirements (OPR), including green design goals, early in the design process.
- Integration of team ideas.
- Effective execution throughout the project's phases—from predesign through the end of its useful service life.

Establishing Green Design Goals Early

Establishing goals early in the project planning stages is a key to developing a successful green design and minimizing costs. It is easy to say that goals need to be established, but many designers and owners struggle with what green design is and what green/sustainable goals should be established. The following are typical questions to ask:

- What does it cost to design and construct a green project?
- Where do you get the best return for the investment?
- How far should the team go to accomplish a green design?

Today there are many guides a team can use with ideas on which green/sustainable principles should be considered. Chapter 2 of this guide presents several rating systems and references on environmental performance improvement. The essence of these documents is to provide guidance on how to reduce the impact the building will have on the environment. While the approaches and goals contained in each differ, all suggest common principles that designers may find helpful to apply to their projects.

Integration of Team Ideas

No green project will be successful if the various project stakeholders are not included in the process. These stakeholders include the owner, the owner's operations staff, the commissioning authority, design disciplines, contractors, and users. These stakeholders, if known, should work in close coordination, beginning in the earliest stages. Use the commissioning process, as discussed in Chapter 4, to obtain input from the various stakeholders and define the owner's objectives and criteria based on which the integrated team must work together with a clear direction and focus to deliver the project. It is becoming less frequent that the mechanical, electrical, and plumbing engineers and landscape architects become involved only after the building's form and space arrangements are set (i.e., end of schematic design). That is far too late for the necessary interplay of ideas between engineering and architectural disciplines that must occur for green design to be effective.

THE OWNER'S ROLE

Of all the participants, it is the owner who is the most crucial when it comes to making a green building happen. With the owner's commitment, the design, construction, and operating teams will receive the motivation and empowerment needed to create a green design.

Key design team members can—and should—attempt to educate the owner on the long-term benefits of a sustainable/green design, particularly if the owner is unfamiliar with the concept. After all, experienced design team members are in the best position to sell the merits of green design. However, to be effective, such a commitment on the part of the owner must be made early in the design process.

Specific roles that an owner can fill in making a sustainable/green design effort successful include the following:

- Expressing commitment and enthusiasm for the green endeavor
- Establishing a basic value system (i.e., what is important, what is not)
- Selecting a commissioning authority (CxA)
- Participating in selection of design team members
- Setting schedules and budgets
- Participating in the design process, especially the early stages
- Maintaining interest, commitment, and enthusiasm throughout the project

The owner on a project could be a corporation or small business, hospital, university or college, office building developer, nonprofit organization, or an individual. In any case, that owner will have a designated representative on the building project team, presumably one who is very familiar with the owner's views and philosophy and can speak for that owner with authority.

INCENTIVES FOR GREEN DESIGN

For both individuals and firms in the design and construction profession, many incentives exist to develop green/sustainable projects. As with any aspect of business practice that adds value to a project, fees and client expectations must be carefully managed. While some clients may balk at added fees charged for the commissioning, additional coordination, and studies necessary to meet green/sustainable goals for systems that are part of a traditional project scope, others welcome these services. First, the appetite of the project client and full project team for a green/sustainable project must be gaged. Then, the commensurate level of commissioning, design, construction, and operational services can be provided.

Individuals and firms are finding that green project capabilities can positively impact building professionals' careers and the firms that employ them. Firms can enhance these capabilities by providing leadership on green issues, building individual competencies, providing ongoing support for professional development in relevant areas, rewarding accomplishments, marketing or promoting green success stories, and building their clientele's interest in green/sustainable design.

When properly delivered, green design/construction/operation capabilities can lead to enhanced service to clients, repeat business, increased public relations and marketing value, and increased demand for these green design services, especially among the architects or owners that represent a substantial proportion of many firms' billings. In addition, green capabilities can also improve employee retention and employee satisfaction. Finally, green/sustainable project competency can reduce risks in practice, meaning that knowledge of green issues is necessary to manage risk when participating in aggressively green projects.

Human Productivity. While difficult to measure, the benefits of improving the learning, living, and workplace environment (all aspects of indoor environmental quality [IEQ]) and feelings of well-being can yield big gains in human productivity. Figure 3-1 shows the typical relationship among the various categories of the costs of operating a business with human costs outstripping all other costs several times over.

Below are some examples of some data generated by private commercial organizations that support the concept that green design increases productivity:

Lockheed

- 15% rise in production
- 15% drop in absenteeism

West Bend Mutual Insurance

- 16% increase in claims processes

NG Bank

- 15% drop in absenteeism

Verifone

- 5% increase in productivity
- 40% drop in absenteeism

Similar results have been shown in learning and living environments. Some projects may lend themselves to the incorporation of green concepts without incurring additional costs beyond the business's own investment in the knowledge of, and experience with, green projects. Other sustainable design projects may require additional services for a greater number of design considerations not customarily included in ordinary design fees, such as extended energy analysis, daylighting penetration, and solar load analysis, or extended review (of materials and components, environmental impacts such as embodied energy, transportation, construction, composition, and indoor air quality [IAQ] performance).

Financial Incentives. The economic benefits of green design strategies as discussed earlier in this chapter are enhanced by identifying funding sources for the high-performance design concepts that are being considered for a building design project. Having the knowledge of where the opportunities can be found and assisting the team in obtaining the funding enhances the engineer's role and importance in the green design process and significantly improves the owner's return on investment (ROI). Funding sources include utility company rebates, state and federal grant programs for energy efficiency and renewable energy measures, and private foundation grants for design enhancements and green rating system implementation.

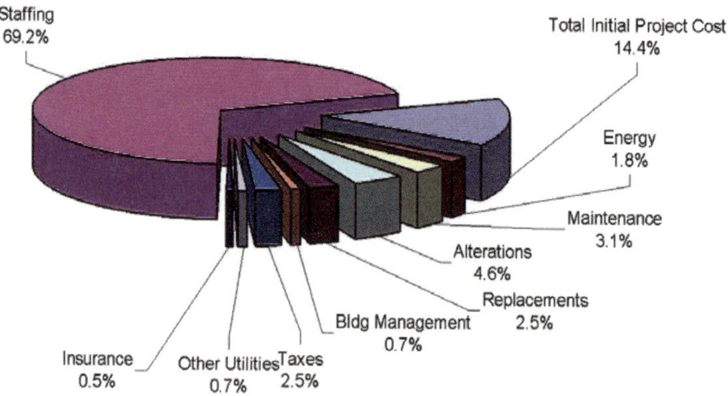

Image courtesy of Michael Dell'Isola, Faithful + Gould

Figure 3-1 People cost versus other costs in business operation.

Two broad categories of incentives, external and internal, can each encompass both direct and indirect financial benefits and nonfinancial advantages.

The familiar range of energy incentives offered by many utility-administered, state-mandated, demand-side management programs represents one obvious form of external financial incentive—the rebate check available to owners or utility ratepayers who pursue energy conservation measures in buildings and mechanical/electrical systems. Of course, the availability of these and other financial incentives varies from place to place and project to project. But many other external financial incentives can reduce project costs where applicable, such as the following:

- Sustainable design tax credits
- Marketable emissions credits
- Tax rebates
- Brownfield funds
- Historic preservation funds
- Community redevelopment funds
- Economic development funds
- Charitable foundation funds

In addition, future public policy initiatives associated with climate concerns may direct attention toward low-carbon or net zero carbon designs, as discussed later in this guide.

U.S. Examples. For example, a 26-story, 960-bed residence hall project completed in 2009 at Boston University (BU) used the NSTAR Comprehensive Design Rebate Program (NSTAR is a Massachusetts-based electric and gas utility). The program provides incentives for purchasing and installing high-efficiency equipment for use in commercial and industrial operations. NSTAR offers incentives of 90% of the incremental cost differential for comprehensive design or up to 75% of the incremental cost differential between standard base line and high-efficiency equipment. In addition, customers are eligible to receive cost sharing for engineering services and design and commissioning services.

On the BU housing project, NSTAR and the design team held an energy conservation measure charrette during the schematic design phase. Measures incorporated into the design included energy recovery for the 100% outdoor air ventilation units, chilled-water system energy enhancements, energy conservation measure motors and variable-speed control logic for the fan-coil units, upgraded glazing performance, and variable-speed drives on all pump motors (with integrated system control logic) on the project. The net result was that for an investment of just over $500,000 on a $130 million project, BU received a rebate from NSTAR for $240,000. In addition, the estimated energy savings from incorporation of these measures is estimated to be $210,000 annually. The result is an annual ROI of greater than 70%.

The Database of State Incentives for Renewables & Efficiency provides information on incentive programs for all 50 states: www.dsireusa.org/.

International. The European Union (EU) is aiming for a 20% cut in Europe's annual primary energy consumption by 2020 (http://ec.europa.eu/energy). The European Commission has proposed several measures to increase efficiency at all stages of the energy chain, targeting final consumption and the building sector, where the potential for savings is greatest. The European Directive 2002/91/EC on the energy performance of buildings (EPBD) and the EPBD recast (Directive 2010/31/EC) constitute the main legislative instruments for improving the energy efficiency of European building stock. EU member states are strengthening the energy performance requirements and setting more stringent goals for reducing the energy performance of buildings. Accordingly, by 2021 all new buildings must be "nearly zero energy buildings" while new buildings occupied/owned by public authorities should comply by 2019. National implementation efforts are underway to meet these ambitious goals while determining the optimal cost level for the life cycle of the buildings. All existing buildings that undergo major refurbishment (25% of building surface or value) should meet minimum energy performance standards, while national policies and specific measures should stimulate the transformation of refurbished buildings into "nearly zero energy buildings." Austria, Denmark, Germany, and Sweden have emerged as successful leaders in adopting and implementing the EPBD into national initiatives to promote green buildings. This is well aligned with the EU's goals set in Energy 2020, "a strategy for competitive, sustainable, and secure energy," known as the 20-20-20 targets: a reduction in EU greenhouse gas emissions of at least 20% below 1990 levels, an increase to 20% of RES contribution to EU's gross final energy consumption, and a 20% reduction in primary energy use by improving energy efficiency, by 2020.

Over the years, the Intelligent Energy Europe program has provided additional funding sources for projects that support the promotion of energy efficiency and the use of alternative energy sources in buildings. For example, the ecobuildings projects use a new approach when it comes to the design, construction, and operation of new and/or refurbished buildings. This approach is based on the best combination of the double approach: to reduce substantially and, if possible, to avoid the demand for heating, cooling, and lighting and to supply the necessary heating, cooling, and lighting in the most efficient way based as much as possible on renewable energy sources, polygeneration (combined heat and power), and trigeneration (combined heat, power, and cooling).

In 2009, the European Commission launched its BUILD UP initiative to increase the awareness of untapped savings potential for all parties in the building chain. BUILD UP promotes better and smarter buildings across Europe by connecting building professionals, local authorities, and citizens. Its interactive web portal supports

Europe's collective intelligence for an effective implementation of energy-saving measures in buildings. The BUILD UP web portal (www.buildup.eu) shares and promotes existing knowledge, guidelines, tools, and best practices for energy-saving measures, while it also informs and updates the market about the legislative framework in terms of goals, practical implications, and future revisions.

Mesures d'Utilisation Rationnelle de l'Energie (MURE, http://www.measures-odyssee-mure.eu) provides information on energy-efficiency policies and measures that have been carried out in EU member states, along with information on national financial instruments such as grants, subsidies, loans, and other related fiscal issues. The German Energy Agency's (Dena, www.dena.de/en/) Subsidy Overview EU-27 offers additional information on grant programs and regulatory frameworks in the EU member states. At this agency's website, information on technologies that generate heat from renewable energies is presented in concise table form.

Incentives that are internal to a project can be created through contractual arrangements. The simplest example is an added fee for added scope in doing the work needed to evaluate green design alternatives. At the other end of the spectrum, some proactive firms offer design review and commissioning services on a performance basis. Construction project insurance offerings have included rebates if no IAQ claims are made a predetermined number of years after construction is completed. If both client and firm are willing, there are many ways to create internal incentives for green engineering design on a project-specific basis.

THE DESIGN PROCESS—EARLY STAGES

The design process is the first crucial element in producing a green building. For design efficiency, it is necessary to define the owner's objectives and criteria, including sustainable/green goals, before beginning the design so as to minimize potential increased design costs. Once designed, the building must be constructed, its performance verified, and it must be operated in a way that supports the green concept. If it is not designed with the intent to make it green, the desired results will never be achieved.

Figure 3-2 conceptually shows the impact of providing design input at succeeding stages of a project, relative to the cost and effort required. The solid curve shows it is much easier to have a major impact on the performance (potential energy savings, water efficiency, maintenance costs, etc.) of a building if one starts at the very earliest stages of the design process. The available impacts diminish thereafter as one proceeds through the subsequent design and construction phases. A corollary to this is the cost of implementing changes to improve building performance rises at each successive stage of the project (cost is shown as the dotted curve of the graph).

Designers are often challenged and sometimes affronted by the idea of green design, as many feel they have been producing good designs for years. The experi-

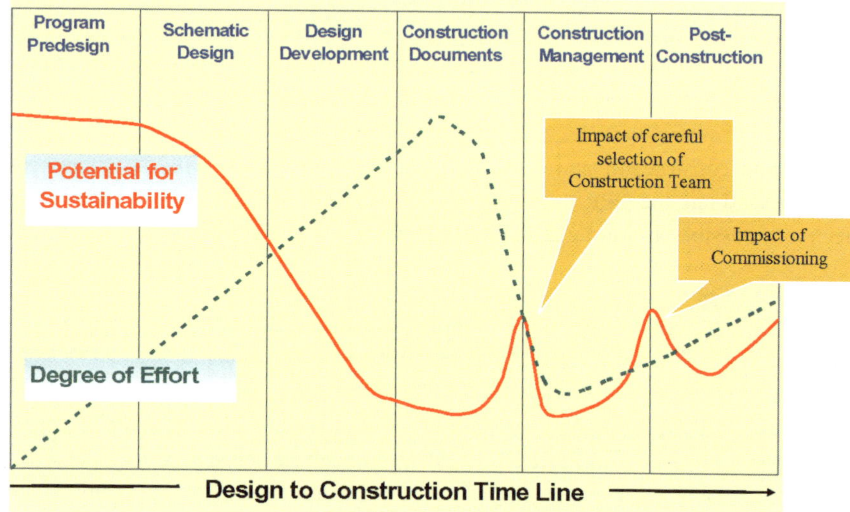

Image courtesy of Malcolm Lewis, CTG Energetics, Inc.

Figure 3-2 The changing potential for sustainability over a project timeline and the impact of commissioning.

ence of many is that they have been forced to design with low construction costs in mind, and when they offer opportunities to improve a building's design, they are often blocked by the owner because of budget constraints. The typical experience is that owners will not accept cost increases that do not show a return in potential savings in five years or less, and many demand 18 months or less. Many owners, especially owners who own the project for life, allow longer return on investments and use life-cycle cost parameters to lower the total cost of ownership.

Achieving green or sustainable design goals requires a different approach than has been customarily applied. Engineers and other designers are asked to become advocates, not just objective designers. Some have expressed the view that significant reductions in energy usage and greenhouse gas emissions will never occur by simply tweaking current practice. In other words, simply installing high-efficiency systems or equipment will not reduce energy usage sufficiently. Sustainable design requires designers to take a holistic approach and go beyond designing for just the owner and building occupants. They need to look at the long-term environmental impacts the development of a building will create. This may be uncomfortable because it seemingly asks designers to go beyond their area of primary expertise.

Both first cost and operating costs can be reduced by applying sustainable/green principles. Correct orientation and correct selection of glazing can reduce HVAC equipment size and cost, as can the use of recycled materials such as crushed concrete (in place of virgin stone) for soil stabilization and structural fill.

Starting in predesign and carrying through to postoccupancy is essential for the success of a sustainable/green design. It starts with examining every aspect of the process (e.g., the owner's site selection, building configuration, architectural elements, and efficient construction and operation) and can only occur with an integrated approach. Defining the OPR document containing the project goals, even before site selection if possible, is the suggested starting point.

At the very start of the project, green design goals have to be discussed, correlated to the owner's objectives and criteria, agreed to, and in fact embraced by the extended project team. This is often done in a charrette format or simply a session spent discussing the issues. As these goals are defined, they are included in the OPR document.

Goals for a project traditionally include the functional program, leaseable or usable area, capital cost, schedule, project image, and similar issues. The charrette simply puts environmental goals on a plane with the capital cost and other traditional goals.

One of the goals may be to achieve the environmental goals at the same or similar capital cost. (As with any goal, the environmental goals should be measurable and verifiable.) Another one of the goals may be a specific green-building rating system target and an energy target, or perhaps an energy target alone.

A typical set of goals for a green design project might be as follows:

1. Achieve a level of energy use at least 50% lower than the DOE-compiled average levels for the same building type and region, both projected and in actual operation. (Actual energy numbers may be adjusted for actual versus assumed climatic conditions and hours of usage.)
2. Achieve an actual peak aggregate electrical demand level not exceeding 4.5 kW/ft^2 (50 kW/m^2) of building gross area.
3. Provide at least 15% of the building's annual energy use (in operation) from renewable energy sources. (Such energy usage may be discounted from the aggregate energy use determined under the abovementioned Goal 1.)
4. Taking into account the determinations of Goals 1, 2, and 3, assess the impact of the lesser net energy use on raw energy resource use (including off-site energy use) compared to that of a comparable but conventional building, including the changed environmental impacts from that resource use, and verify that the aggregate energy and environmental impacts are no greater.
5. Achieve per capita (city) water usage that is 40% lower than the documented average for this building type and region.
6. Achieve an aggregate up-front capital cost for the project that does not exceed x dollars/ft^2 (m^2) of building gross area, which has been deemed by the project team to be no higher than 102% of what a conventional building would cost.

7. Recycle (or arrange for the recycling of) at least 60% of the aggregate waste materials generated by the building.
8. By means of postoccupancy surveys of building users conducted periodically over a five-year period, achieve an aggregate satisfaction level of 85% or better. Surveys solicit occupant satisfaction with the indoor environment as to the following dimensions: thermal comfort, air quality, acoustical quality, and visual/general comfort.
9. Obtain gold-level USGBC LEED certification for the building.

For an example of what one major firm has done, see the Digging Deeper sidebar, One Firm's Green-Building Design Process Checklist, featured later in this chapter.

THE DESIGN TEAM

Setting It Up

One of the first tasks in a sustainable/green design project is forming the design team and the commissioning team. This team should include the design team leader (often the architect), the owner, the CxA, the design engineers, and operations staff. Much of the design team's successful functioning depends not only on having ideas about what should go into the project but on the ability to analyze the impact of the ideas quickly and accurately. It is likely that a large part of this analysis will be completed by one of the engineers on the project.

A traditional project team includes the following members:

- Owner
- Project manager
- Architect
- HVAC&R engineer
- Plumbing/fire protection engineer
- Electrical engineer
- Lighting designer
- Structural engineer
- Landscaping/site specialist
- Civil engineer
- Code consultant

An expanded project team for a sustainable/green design with commissioning would also include the following members:

- Energy analyst
- Daylighting consultant
- Environmental design consultant

- CxA
- Construction manager/contractor
- Cost estimator
- Building operator
- Building users/occupants
- Acoustical consultant

The preceding lists the possible roles that might need to be filled on a reasonably large design project. Some roles may not be applicable or even needed on certain types of projects (e.g., civil engineer or landscaping/site specialist), and other roles may not be feasible in the early stages of project (e.g., building operator, building users, or code enforcement official). Further, the variety of roles does not mean that there needs to be an equal number of distinct individuals to fill them; one individual may fill several roles (e.g., the architect often serves as project manager, the HVAC&R engineer as plumbing engineer or energy analyst, the electrical engineer as lighting designer, and a contractor on the team as cost estimator). Likewise, depending on the type of project, there could be other specialists as well.

There are a few roles that are particularly important in green design, including energy analyst, environmental design consultant, and CxA, which are discussed as follows.

Energy Analyst. Although this role has existed for some time, it assumes a much more intense and timely function in sustainable/green design, as there is a need to quickly evaluate various ideas (and interactions between them) in terms of impact on energy. These can range from different building forms and architectural features to different mechanical and electrical systems. The person in this role must be intimately familiar with energy and daylight analysis modeling tools and able to provide feedback on ideas expressed reasonably quickly. In short, he or she is a much more integral part of a sustainable/green design team than in a traditional design effort. In this respect, for a sizable project, it might be difficult for a single person to fill this role if they are also serving in another role.

Environmental Design Consultant (EDC). As owners begin to request sustainable/green buildings from the design professions, a new discipline has emerged: the EDC. The role of this person is to help teams recognize design synergies and opportunities to implement sustainable and green features without increasing construction costs. When the CxA is also the EDC, the owner and designers benefit from improved integration of the design process across disciplines, with the intent of creating an outcome with much lower environmental impact and higher user satisfaction. Leading projects show that this can often be accomplished without adding cost. The EDC has input in areas such as site, water, waste, materials, IEQ, energy, durability, envelope design, renewable energy, and transportation Although this guide focuses primarily on those areas pertinent to the HVAC&R design professional, it is becoming evident that this profession must broaden its sphere of concern in order to contribute meaningfully to the creation of sustainable/green buildings.

The CxA already has a collaborative relationship with the designers and, as an EDC, works with the HVAC&R team and others throughout the process to meet the owner's objectives and criteria. For example, a design reviewer may raise the following questions for the team to consider:

- Is the building orientation optimized for minimum energy use?
- Is the combined system of building envelope, including glazing choices and the HVAC system, optimized for minimum energy use and lowest life-cycle cost?
- Have passive, active, and renewable strategies for building energy usage been optimized?
- Are the loads, occupancy, and design conditions properly described?
- Are the proposed analytical tools adequate for the task of computing life-cycle costs and guiding design decisions?
- Is the proposed integrated design approach going to deliver excellent air quality to occupants under all conditions?
- Is the proposed integrated design approach going to deliver thermal comfort to occupants under all conditions?
- Is the proposed integrated design approach going to be easy to maintain? Is there enough space for mechanical equipment and adequate access to service and perhaps to eventually replace it?
- Is the proposed mechanical approach going to give appropriate control of the system to users?
- Is the proposed integrated design approach going to consume a minimum amount of parasitic energy to run pumps and fans?
- Are there site or other conditions likely to impact the mechanical system in unusual ways?
- Has the impact of the building on the site and surroundings been identified and taken into account?
- Are the proposed systems properly sized for the loads?

While it may seem that the role of the EDC is very similar to that of the energy analyst, the roles differ in that the EDC is more of a question-asker or issue-raiser, similar to the CxA asking questions and suggesting strategies or solutions for the team to consider. The EDC's brief is broader and more comprehensive in scope; their role is to take a step back from the project and ask the broader questions regarding the environmental impact of the project.

CxA. (Please refer also to Chapter 4.)

The Team's Role

Green design requires owners to make decisions sooner, design documents to be more complete and comprehensive, the construction process to be better coordinated, and operators to be better trained in maintaining facilities. All of this will affect the viability and success of a green project endeavor. Contractors not familiar with this project model may sound the cost and schedule alarm because of their inexperience in the new procedure. First-time application of sustainable development principles can result in slightly higher first cost, but this phenomenon will reverse itself as teams gain experience and improve their learning curve. As the building industry becomes more familiar with applying these principles, lower costs of ownership will result.

In addition to the standard tasks associated with a design project, the design team is responsible for developing and implementing new concepts that will create a green project. For most, this will require learning on their own time and becoming familiar with new advances in software tools, green materials, and alternative systems. There is an abundance of information; advances are occurring daily in the development of green products and materials as well as processes. The speed of these changes requires designers to continuously add to their knowledge base.

The greatest challenge to accomplishing green design is creating a team organizational structure that provides the following:

- Criteria for assessing how green the project should be
- Strong leadership through the green design process to integrate team members
- Clearly defined objectives that result from careful examination of design alternatives, costs, and schedule impacts
- Documentation of success
- Strong leadership by experienced green-building practitioners leading the team through the decision process
- Definition of what tasks are required to accomplish green design
- Identification of who is responsible for each of the tasks
- Identification of when tasks must be completed, so as not to impede the design process or the project schedule
- Establishment of criteria for the selection of green design features considered for incorporation into a project
- Assistance with integrating selected green design goals into the construction documents
- Definition of the level of effort required for each of the green project goals
- Help to enable contractors overcome psychological and physical constraints
- Establishment of how to track, measure, and document the success of accomplished project goals

The designers must also help inform their clients that there are costs beyond the cost of extraction for the depletion of resources. The practice of looking only at simple payback when analyzing alternatives based on extraction cost has never been realistic because there is no way to replace many of these resources at any cost.

Currently, most design teams are eager to develop green designs but lack the experience of actually integrating green design into their projects. The addition of an experienced EDC will shorten the learning curve by helping them integrate their extensive knowledge, which will result in a cost-effective practical green design that meets the owner's requirements. In addition, most teams struggle with what makes a design green, how to incorporate green design principles, and the logistics of incorporating these principles into the design. Green design creates a need for a broader involvement of disciplines and an expanded range of experience to ensure that a wider range of input and participation is factored into the decision-making process.

The project team, from the initial concept to the construction documents, construction, and building operations, must work as an integrated unit to achieve the goals set by the owner's objectives and criteria to create better project performance, which is a basic principle of green design. This model will require the project team to investigate new approaches and process more information than ever before, as they strive to increase performance and lower the total cost of ownership. The decision-making process must change from the traditional hierarchical method, with an emphasis on lowest first cost, to an integrated method focused on life-cycle cost. To achieve this requires close collaboration of the project team combined with innovative thinking among all disciplines.

The design team's responsibility as part of the project team is to assist the owner with setting sustainability/green goals that often include the following:

- Life-cycle cost optimization of energy-consuming systems, materials, and maintenance
- Systems integration and maintainability
- Minimization of environmental impact
- Documenting basis of design
- Assisting with training of building operations and management staff during commissioning

THE ENGINEER'S ROLE

The HVAC&R engineer is a crucial player in the design of a green building. In fact, it is virtually impossible (and certainly not cost-effective) to design a green building without major involvement from that discipline. The HVAC&R engineer must escape the normal box in which they think and become more involved in the *why* of a design, as well as the *how*. This means moving beyond responding to questions asked by others and actively participating in the decision-making regarding how project goals will be achieved.

Engineers help analyze the various options to be considered, create mathematical computer models that are used to judge alternatives, provide creative input, and assist with development of new techniques and solutions. The HVAC&R engineer can be invaluable in helping the architect with building orientation considerations, floor plate form and dimension, and deciding which type of glazing will provide the maximum quantity of natural light, while at the same time analyzing the heat transfer characteristics of the glazing options. The HVAC&R engineer can also help the architect select structural systems and exterior walls to use thermal mass features to reduce equipment needs. Working with the electrical engineer and architect, the HVAC&R engineer can offer ideas and various options, such as incorporating daylighting and lighting controls to reduce artificial light when natural light is available, which, in turn, can result in lower cooling requirements and lowered HVAC requirements to meet peak load. Smaller equipment size translates into reduced structural and electrical requirements, lower operating and maintenance costs, and lower construction costs, all of which lower the total cost of ownership.

The plumbing engineer, working with the structural and civil engineers, and the landscape architect can reduce the facility's potable water, sewer, and stormwater conveyance requirements. Some examples are waterless urinals or use of stormwater or graywater for irrigation of vegetation or for flushing toilets. Depending on the type of building, water from condensate can be used for graywater applications or for cooling tower makeup. The design engineer must weigh the benefits of water-cooled versus air-cooled condensers and the water versus electrical energy consumed by each. The engineer must examine the site climate, determine what alternatives and strategies can best be applied, and develop life-cycle analyses to guide the owner through the decision process posed by the maze of complex issues surrounding green design.

What Does Integrated Design Mean?

One of the key attributes of a well-designed, cost-effective green building is that it is designed in an integrated fashion, wherein all systems and components work together to produce overall functionality and environmental performance. This has a major impact on the design process for HVAC-related systems, as conceptual development must begin with HVAC system integration into the building form and into the approaches being taken to meet other green-building aspects. For example, consider the following:

- HVAC systems that use natural ventilation and underfloor air distribution, often used in green buildings, can have major impacts on building form.

- Other building energy innovations, such as daylighting, passive solar, exterior shading devices, and active double wall systems, often have significant impacts on the design of the HVAC system.
- On-site energy systems that produce waste heat, such as fuel cells, engine-driven generators, or microturbines, affect the design of HVAC systems so waste heat can be used most effectively.

Beyond these form-giving elements, there are many other specific features of a green building that affect (or are affected by) HVAC systems to achieve the best overall performance. Some of these strongly impact HVAC system conceptual design, and some require only minor adjustments to HVAC specifications. Such features might include the following:

- Using the commissioning process to document the owner's defined sustainable/green objectives and criteria and assist the project team to deliver
- Effective use of ventilation (and indoor air quality [IAQ] sensors linked to the ventilation system) to improve IAQ
- Provision of user controls for temperature and humidity control
- Reduced system capacities to reflect lower internal loads and building envelope loads
- Use of thermal energy storage systems to reduce the overall size of the chiller plant equipment, such as chillers, cooling towers, and pumps (to save capital costs, ongoing energy, and operational costs, and reduce outdoor noise levels during the daytime)
- Selection of non-ozone-depleting refrigerants
- Reduction and optimization of building energy usage below the levels of ANSI/ASHRAE/IES Standard 90.1-2016, *Energy Standard for Buildings Except Low-Rise Residential Buildings* (ASHRAE 2016) based codes or other applicable state and local energy codes (a reduction in overall energy consumption below that specified by the minimum energy code is becoming more common, either as local regulation or at least by strong encouragement)
- Use of reclaimed water for cooling tower makeup, and minimization of cooling tower blowdown discharge to the sanitary sewer system
- Testing of key systems, especially HVAC systems

Key Steps

The integrated design process (IDP) includes the following elements:

- Ensuring that as many of the interested parties as possible are represented on the design team as early as possible, including not only architects, engineers, and the owner (client), but also the CxA, construction specialist (contractor), cost estimator, operations/maintenance person, and other specialists (outlined below)

- Interdisciplinary work among architects, engineers, costing specialists, operations personnel, and other relevant persons from the beginning of the design process
- Discussion and documentation by the owners and the design team of the relative importance of various performance and cost issues, and the establishment of a consensus on these matters between client and designers, and among the designers themselves
- Provision of a design facilitator (or EDC) to suggest strategies for the team to consider, as well as a CxA to raise performance issues throughout the process and to bring specialized knowledge to the table
- Addition of an energy specialist to test various design assumptions through the use of energy and daylight simulations throughout the process and to provide relatively objective information on a key aspect of performance
- Addition of subject specialists (e.g., for daylighting or thermal storage) for short consultations with the design team
- Clear articulation of performance targets and strategies to be updated throughout the process by the owner and the design team

Iterative Design Refinement

The design process requires the development of design alternatives. To come up with the most effective combination, these alternatives must be evaluated, refined, evolved, and finally optimized. This is the concept of *iterative design*, wherein the design is progressively refined over time, as shown in Figure 3-3.

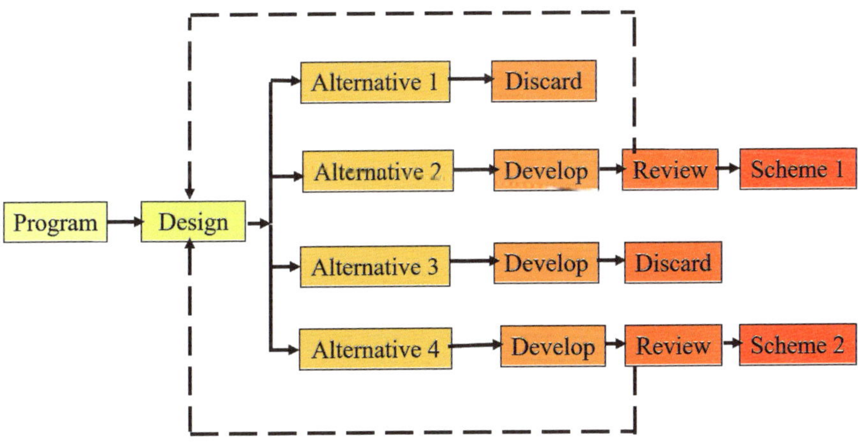

Image courtesy of Malcolm Lewis, CTG Energetics, Inc.

Figure 3-3 The iterative design process.

Fee and schedule pressures often lead the designer to want to lock in a single design concept at the beginning of the project and stick with it throughout. But this precludes the opportunity to come up with a better system that reflects the unique combination of loads and design integration opportunities for this specific building. This better design usually evolves during schematics and early design development in the iterative process.

CONCEPT DEVELOPMENT

The Big Picture

Designers should always keep in mind the three major steps for achieving a sustainable/green design:

- Reduce loads.
- Apply the most efficient systems.
- Look for synergies.

Reduce Loads. If you have a building with a normal 500 ton cooling load, is it possible to provide a more comfortable environment with, for example, only 300 tons? This may be difficult for a design team taking the typical approach to the problem, but it might be possible with a focused, integrated design team. Solar loads on the building would have to be reduced; lighting loads to the space would also have to be lowered; maybe the building could use daylight rather than electric light during the day, so the building shape would be influenced. The site for the building, its shape, its thermal mass, and its orientation could all work together to reduce the cooling load. Early and quick modeling can provide interesting information to assist decisions.

Similar things could be done with the heating load. Does the winter sun provide much of the building's heating needs during the day? With design changes, could the sun do more?

These considerations in a hierarchal design process are not typically brought to the attention of the HVAC&R engineer, who could significantly help improve the energy efficiency instead of just calculating the loads, and could apply energy-efficient systems. Significant reduction of utility consumption and environmental impact cannot occur by simply doing the same old job just a little bit better.

The HVAC&R or energy engineer can make a positive contribution to the success of a sustainable/green design. The value of the engineer to the project is significantly increased, and the results are reduced heating and cooling loads, and reduced overall utility consumption.

One attitudinal approach to building design that the designer should strive toward is making the building inherently work by itself. Building systems are there to fine-tune the operation and pick up extreme design conditions. In contrast, buildings are often traditionally designed like advanced fighter aircraft: if the flight computers are lost, the pilot cannot fly the plane.

Apply the Most Efficient Systems. This is the world of the energy-efficient engineer, and where ASHRAE generally operates. While efficient systems are very important, they are not enough by themselves. Many times, the most efficient system is not the most cost effective or the owner O&M personnel are not yet ready to operate and maintain such a system.

Look for Synergies. The preceding two major steps have the potential of increasing capital costs. Therefore, a well-designed, energy-efficient building might not be built because of high first cost, or the value-engineering knives might come out and cut the project back to a traditional, affordable project—proof perhaps that green design cannot work. Part of the solution to get around this syndrome is to look for synergies of how building elements can work together. This also relates to the cost transfer mentioned earlier.

If a building has a large amount of southern exposure, exterior-shading devices might significantly reduce the summer solar load while still admitting lower-angle winter sun. Daylight (but not direct sun) would allow for shutting off the electric lights on sunny days. The HVAC system for the south perimeter zones could be significantly reduced in size and cost as the simultaneous solar and electric lighting loads are reduced. Indeed, the very nature of the HVAC system might well be simplified because of the significant load reduction. Resulting cost savings can be used to pay for some or all of the additional treatments.

A major benefit of an on-budget integrated design is avoiding wasting a lot of time on elemental payback exercises and value engineering (and cost cutting) exercises. Many of the integrated solutions work so that if you save by cutting out an element, such as the exterior shades, there is an additional cost in another area, such as the size of the HVAC system.

EXPRESSING AND TESTING CONCEPTS

Expressing concepts is very important in green design, because that is how ideas and intentions are communicated to the owner and others on the design team. This is especially true because green design requires the close and active participation of many different parties.

There are three ways of expressing concepts in the design of buildings: the traditional verbal means, the diagrammatic or pictorial means, and the modeling means, which has come of age more recently alongside computers.

Verbal

Both the written and the spoken word play an especially important part in green design. Because there are many meetings or charrettes where ideas are explored and intentions voiced, getting across what is expressed accurately assumes significance. Then, succinctly and clearly putting down on paper what has been expressed (memorializing it) is also critical to the various team members, as they each go about filling their respective roles. To illustrate, see Chapter 4, Commissioning, to learn how important a written record of what happened during design (i.e., design intent, assumptions, etc.) is to the successful follow-through of a well-executed green design.

Diagrammatic/Pictorial

The use of diagrams, sketches, photos, and renderings, a tried and true method of communicating a lot of information, continues to be an essential part of green design (Figure 3-4). The old adage "one picture is worth one thousand words" most certainly applies here. But there is now a relatively new way of creating a picture (Figure 3-5) of a building, an energy system, or a year of operations—modeling.

Modeling

Modeling for energy performance is key to the design process. This can start from simply setting an energy benchmark or energy use intensity (EUI) target for the building and estimating the energy loads based on square footage. This is a great exercise to set expectations based on peer groups of buildings from ENERGY STAR Targetfinder, Labs for the 21st Century, or perhaps a portfolio of company buildings that

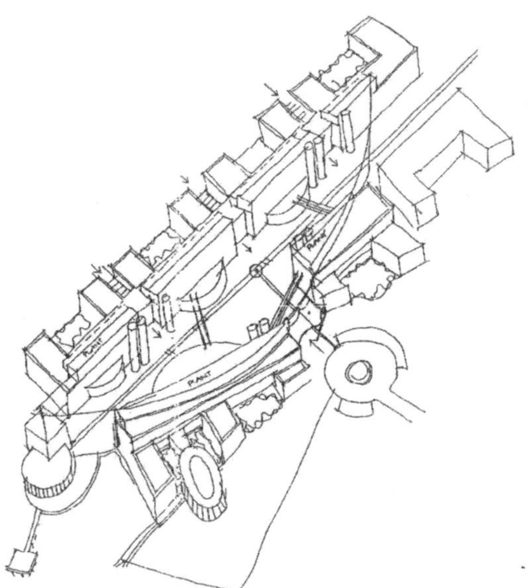

Image courtesy of Integrated Environmental Solutions (IES)

Figure 3-4 Example of diagrammatic building sketch.

exist already. As simple as a basic EUI model is, it should be the start of the more comprehensive modeling throughout the design process. Computer-based modeling plays such an important part in green design because of its speed, accuracy, and comprehensiveness.

Everyone is familiar with how speedy computers are, once the input data are entered. The slow part in this process is the human analyst, the one who converts intentions and ideas into computer modeling program input, which is why it is so important for that analyst to be very familiar with the modeling process. This is especially true for load and energy calculations that impact HVAC&R systems. The team has an idea, and they want an answer fast as to how well that idea would work.

Modeling programs have another advantage, especially the more sophisticated ones: they are comprehensive in what they can analyze simultaneously. The human mind can only accommodate so many ideas or concepts at once without getting confused and bogged down; a properly conceived model will not get confused and can provide answers that may be counterintuitive. As an example, a good modeling program can track heat gain from lights, plug loads, and solar energy, along with heat loss from the building envelope and infiltration, do it for every hour or every day of the year in whatever weather conditions are assumed, take into account mass effects of the structure, and still yield an accurate answer. It would be impossible for the human mind to do this in a reasonable time, unless it was very good at guessing.

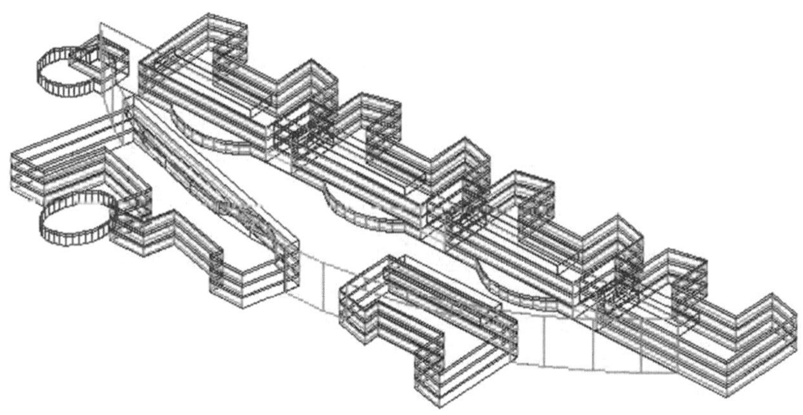

Image courtesy of Integrated Environmental Solutions (IES)

Figure 3-5 Example of computer simulation model derived from Figure 3-4.

BUILDING INFORMATION MODELING (BIM)/ BUILDING ENERGY MODELING (BEM)

Building design is moving away from conventional computer-aided design software to follow the way design software has evolved in the manufacturing sector. A building can now be a working digital prototype. Sustainable design is driving this solution to ensure that the buildings being erected are designed, constructed, and operated in a manner that minimizes their environmental impact and are as close to self sufficient as possible. This technology is known as *building information modeling* (BIM).

BIM is a software tool that uses a relational database together with a behavioral model that captures and presents building information dynamically. In the same way that a spreadsheet is automatically updated, a change in the parametric building modeler is immediately reflected everywhere. This means, for example, revising windows from one type to another not only produces a visually different graphic representation in all views of the building, but the insulation value of the glazing (R-value) is also revised. Because of the integration of BIM with existing tools of analysis, running energy calculations is greatly simplified. Visualization tools are sophisticated and allow three-dimensional building views and walk-throughs.

Multiple design options can therefore be developed and studied within a single model early in the design process to not only see the building and provide conventional documentation for construction, but also to interact with other software to perform energy analysis and lighting studies.

Using BIM helps with the demanding aspects of sustainable design, such as solar applications and daylight harvesting, and also automates routine tasks such as documentation. Not all BIM platforms are directly linked to building performance modeling simulation tools. Some BIM tools have built-in features that allow for building performance modeling within the BIM platform, but most energy-modeling tools are structured separately and can interact through the Green Building XML schema (gbXML). The gbXML schema has been around for more than 15 years and has been adopted by most platforms to translate information from the BIM model to other platforms that can perform specialty simulations for energy, daylighting, or thermal comfort analysis among others. The gbXML translation is not always perfect and requires designers in BIM to pay close attention to geometry creation, mechanical zoning, and, in some cases, zoning for daylight controls.

In the BIM platform, schedules of materials are generated directly from the model. If the model changes, so do the schedules. Architects are able to filter and sort material quantities automatically, bypassing the manual extraction/calculation process. Determining the percentages of material reuse, recycling, or salvage can be tracked and studied for various sustainability design options.

Engineers can perform year-round sun studies to understand when the building is provided natural shading and thus optimize the orientation of the building to maximize afternoon shading from the hot summer sun and properly size roof overhangs to minimize solar heat gain. Engineers can then reduce the capacity of cooling systems, demonstrating the building is exceeding the baseline building energy requirements.

In traditional energy analysis, tools using a two-dimensional computer-aided design file requires manually calculating the building values/areas from the floor plans and entering said data into an energy simulation application. As we move to BIM platforms, the data for supporting green design are captured during the design process in BIM and are able to be updated as the design progresses.

When comparing the design to energy standards such as ANSI/ASHRAE/IES Standard 90.1 (ASHRAE 2016), the energy performance for the baseline model averages the results of four simulations during one year of operation. One simulation is based on the actual orientation of the building on the site; the others rotate the entire building by 90°, 180°, and 270°, which enables the proposed design to receive credit for a well-sited building. This process is made easier when using BIM correctly to translate the data to the performance benchmarking tools.

The BIM software carries all the data required for building a structure. It quantifies building materials so designers can move walls or insert windows and translate the building's data on energy performance, daylight harvesting, and, perhaps most importantly, costs shift accordingly in real time. Designers can test the life-cycle performance of brick walls versus concrete walls. The digital representations behave like buildings and not just drawings.

BIM is transforming the way we work and will enable further endeavors in the practice of creating sustainable, cost-effective buildings.

Evaluating Alternate Designs

Modeling of alternate designs is made easier by the plethora of modeling tools available today. The chosen model should meet specific requirements depending on the level of accuracy needed. Error can be introduced into modeling if users forget the "garbage in/garbage out" rule. To reduce the chance of inappropriate or misunderstood input, modeling programs can use input and output formats that allow the user quick reality checks.

Various stages of design, requirements, and associated tools are shown in Table 3-1. While this is not a complete list of tools, it covers the general simulation tools and phases commonly used by the industry. A more comprehensive list of building simulation tools can be found at www.buildingenergysoftwaretools.com.

The following sections describe other approaches to the green-building design process that have been successful.

Table 3-1: Summary of Available Analysis/Modeling Tools

Stage	Requirements	Tools	Reality Checks
Scoping	Quick analysis Comparative results Reduce alternatives to consider Control strategy modeling	System analyzer Modified bin analysis (where load is not entirely dependent on ambient conditions) eQUEST IES VE	Operation cost/ft^2 (m^2) Payback or other financial measure
System design	Accurate output Industry-accepted methods	HAP TRACE 700 Elite Design IES VE	cfm/ft^2 (L/s/m^2) cfm/ton (L/s/kW$_R$)
Energy/ cost analysis	Accurate Complies with energy cost budget method requirements of Standard 90.1-2016 Industry-accepted methods Flexible Allows modeling of complex control strategies	EnergyPlus DOE HAP TRACE 700 SUNREL IES VE eQUEST	Btu/h·ft^2 (kWh/m^2)/yr Operation cost/ft^2 (m^2) Payback or other financial measure
Monitoring	Simplicity Intuitive interface Systemwide Interoperable	BACnet-compatible automation systems	Trended operating characteristics Benchmark comparisons (such as system kW/ton [kW/kW$_R$])

Zero Energy Design

A net zero building can be a net zero energy building (NZEB), a net zero water building, and/or a net zero wastewater building. Although there are various definitions and classifications for NZE in the literature, the best definition is provided in the 2006 NREL report "Zero energy buildings: A critical look at the definition," which defines four types of NZE buildings:

1. A site NZEB produces at least as much energy as it uses in a year, when accounted for at the site.

   For example, if a building consumes

- 3000 kWh of electricity, and
- 100 therms of natural gas (= 2930 kWh),

then the building's renewable energy system has to produce 5930 kWh for the building to be site NZE.

2. A source NZEB produces at least as much energy as it uses in a year, when accounted for at the source.
   For example, if a building consumes

   - 3000 kWh of electricity (site) = 10,020 kWh source (3000 × 3.34), and
   - 100 therms of natural gas site (or 2930 kWh site) = 3076 kWh source (2930 × 1.05),

   then the building's renewable energy system has to produce 13,096 kWh at source for the building to be source NZE. Assuming PV (electricity), 13,096 kWh/3.34 = 3920 kWh will need to be produced at site or 33% less than site NZEB. The site-to-source conversion factors used (3.34 for electricity and 1.05 for gas) are U.S. averages as reported by U.S. EPA. These factors vary throughout the United States and the world.
3. A net zero energy cost building is one where the amount of money the utility pays the building owner for the energy the building exports to the grid is at least equal to the amount the owner pays the utility for the energy used over the year.
   For example, if the building utility bill is

   - 3000 kWh of electricity at $0.12/kWh = $360, and
   - 100 therms of natural gas at $1/therm = $100,

   then the building's renewable energy system has to export energy in the cost of $460. Assuming PV (electricity) and the utility company buying at same rate of $0.12/kWh, the PV system needs to produce 3833 kWh.
4. A net zero emission building produces at least as much emissions-free renewable energy as it uses from emissions-producing energy sources.
   To perform the calculations, the emission factors need to be determined, which may be difficult to obtain in some regions (e.g., in the state of California, the emission factors are 0.524 lb CO_2/kWh electricity and 13.446 lb CO_2/therm [0.21 kg/kWh] of gas). In addition, although emissions encompass a variety of pollutants, this definition is still limited primarily to carbon dioxide.

An example of a NZEB is the Aldo Leopold Legacy Center in Baraboo, WI. The building is a carbon-neutral NZEB that uses 39.6 kW rooftop photovoltaic (PV) array to produce about 10% more electricity than that needed annually to operate the building. Specifically, the Legacy Center qualifies as a site NZEB, source NZEB, and emissions NZEB. Another recent example is the National Renewable

Energy Laboratory Research Support Facility (NREL RSF) that has demonstrated the production of a net zero energy building that includes plug loads and their data center while still maintaining a high degree of indoor air and environmental quality.

For additional information on the classification of net zero energy buildings, see the online resource listing for the net zero energy buildings database by the U.S. Department of Energy at the end of this chapter.

The following link provides information on incentive programs for all 50 states: www.dsireusa.org/.

REFERENCES AND RESOURCES

Published

Arthus-Bertrand, Y. 2002. *Earth From Above*. New York: Harry N. Abrams.

ASHRAE. 2013. ASHRAE Standard 202-2013, *Commissioning Process for Buildings and Systems*. Atlanta: ASHRAE.

ASHRAE. 2016. ANSI/ASHRAE/IES Standard 90.1-2016, *Energy Standard for Buildings Except Low-Rise Residential Buildings*. Atlanta: ASHRAE

CFR. 1995. Energy conservation voluntary performance standards for commercial and multi-family high rise residential buildings; mandatory for new federal buildings. Code of Federal Regulations 10 CFR 435, Office of the Federal Register, National Archives and Records Administration, Washington, DC.

Digging Deeper

ONE DESIGN FIRM'S TEN STEPS TO A NET ZERO ENERGY BUILDING

1. Define the zero you are seeking (site, source, carbon, or cost)
2. Establish the design criteria for system performance.
3. Understand the microclimate the of the project site (local climate surrounding the project site).
4. Design the siting, form, and thermal mass of the building to maximize the use of natural energy flows and reduce external loads to the absolute practical minimum.
5. Minimize lighting energy use through effective daylighting.
6. Reduce plug loads as far as possible
7. Employ extremely efficient HVAC systems to handle the remaining loads (including natural ventilation and mixed mode where feasible).
8. Develop accurate predictions of annual building energy consumption.
9. Provide renewable energy systems to offset resulting energy use
10. Commission your systems and ensure proper handover to O&M staff

JUSTIFICATIONS FOR GREEN DESIGN

Doing the Right Thing

The motivation and reasons for implementing green buildings are diverse but can be condensed into essentially wanting to do the right thing to protect the Earth's resources. For some, a wake-up call occurred in 1973 with the oil embargo—with the embargo came a realization that there may be a need to manage our planet's finite resources.

Regulations

Society has recognized that previous industrial and developmental actions caused long-term damage to our environment, resulting in loss of food sources and plant and animal species, and changes to the Earth's climate. As a result of learning from past mistakes and studying the environment, the international community identified certain actions that threaten our ecosystem's biodiversity. Consequently, it developed several governmental regulations designed to protect our environment. Thus, in this sense, the green design initiative began with the implementation of building regulations.

Lowering Ownership Costs

A third driver for green design is lowering the total cost of ownership in terms of construction costs, resource management and energy efficiency, and operational costs, including marketing and public relations. Ways to lower costs include providing a better set of construction documents that reduce or eliminate change orders, controlling site stormwater for use in irrigation, incorporating energy-efficiency measures in HVAC&R design, developing maintenance strategies to ensure continued high-level building performance and higher occupant satisfaction, and reducing marketing and administrative costs.

Case studies on commissioning show that construction and operating costs can be reduced from 1 to 70 times the initial cost of commissioning. A recent study by an international engineering firm indicated that treating stormwater on site cost one-third as much as having the state or local government treat stormwater at a central facility, significantly lowering the burden to the tax base.

A 123,000 ft^2 (11,427 m^2) higher education building constructed in 1997 in Atlanta, GA—a building that already had many sustainable principles applied during its design—provides an example of how commissioning, measurement, and verification play a critical role in ensuring that the sustainable attributes designed into the building are actually realized at \$1.00/ft^2 (\$10.80/m^2) savings when recommissioned, lowering the total cost

Digging Deeper

of ownership. Recommissioning identified several seemingly inconspicuous operational practices that were causing higher-than-needed consumption in the following areas: chilled-water cooling, by 40%; steam, by 59%; and electricity, by 15%. Implementing continuing measurement and verification ensures that the building will continue to perform as designed, again, lowering the total cost of ownership.

More esoteric is the cost of unsatisfied occupants, which includes administrative costs, marketing costs due to more frequent tenant turnover, or increased business cost because of absenteeism or reduced productivity, as well as the impact on marketplace image. Marketplace image is a significant driving force in promoting green/sustainable design. Green/sustainable projects provide owners an opportunity to distinguish themselves in public as well as promote their business or project in order to obtain the desired result. Promotion of green/sustainable attributes of a project can help with public relations and help overcome community resistance to a new project.

Increased Productivity

Another driver for green design is the recognition of increased productivity from a building that is comfortable and enjoyable and provides healthy indoor conditions. Comfortable occupants are less distracted, able to focus better on their tasks/activities, and appreciate the physiological benefits good green design provides.

A case study conducted by Pacific Northwest Laboratory points out many interesting observations about human response to daylight harvesting, outside views, and thermal comfort. The study, which compares worker productivity in two buildings owned by the same manufacturer, illustrates both the positive and negative impact the application of green design principles can have on human productivity (Heerwagen et al. 1995).

The first building is an older, smaller industrial facility, divided into offices, and a manufacturing area. It has high-ribbon windows around the perimeter walls in both office and manufacturing areas, providing only limited daylight harvesting. It has an employee lounge, a small outdoor seating area with picnic tables, and conference rooms.

The newer facility is 50% larger with energy-efficient features, such as large-scale use of daylight harvesting, energy-efficient fluorescent lamps, daylight harvesting controls, direct digital HVAC controls, environmentally sensitive building materials, and a fitness center at each end of the manufacturing area.

The study focuses on individual quality-of-work issues and the manufacturer's own production performance parameters. The study provides a mixed review of green design and gives insight on what conditions need to be avoided. While occupants perceived satisfaction and comfort stemming from daylighting and outside views, they also expressed complaints regarding glare and lack of thermal comfort. The green building studied did not appear to have controlled the quality of daylight through proper glazing selection, which may be the cause of the complaints about glare and thermal comfort.

Subsequent chapters discuss specific design parameters relative to building envelope design, including daylight harvesting, energy efficiency, and thermal comfort.

Filling A Design Need

There are increasing numbers of building owners and developers asking for green design services. As a result, there is considerable business for design professionals who can master the principles of green design and provide leadership in this arena.

Some publications that demonstrate the drivers of green design include *Economic Renewal Guide* (Kinsley 1997), *Natural Capitalism* (Hawken et al. 1999), and the *Earth From Above* books (Arthus-Bertrand 2002), and *Green to Gold: How Smart Companies Use Environmental Strategy to Innovate, Create Value, and Build Competitive Advantage* (Esty and Winston 2006).

Digging Deeper

Esty, D.C., and A.S. Winston. 2006. *Green to Gold: How Smart Companies Use Environmental Strategy to Innovate, Create Value, and Build Competitive Advantage*. New Haven, CT: Yale University Press.

Hawken, P., A. Lovins, and L.H. Lovins. 1999. *Natural Capitalism*. Little Brown.

Heerwagen, J.H., J.C. Montgomery, W.C. Weimer, and J.G. Heubach. 1995. Assessing the human and organizational impacts of green buildings. Pacific Northwest Laboratory, Richland, WA.

Kinsley, M. 1997. *Economic Renewal Guide*. Colorado: Rocky Mountain Institute.

Pierce, D.G. 2003. Public-private partnerships: Demystifying the process. Mondaq: http://www.mondaq.com/canada/article.asp?articleid=23737.

Torcellini, P., S. Pless, M. Deru, B. Griffith, N. Long, and R. Judkoff. 2006. Lessons learned from case studies of six high-performance buildings. Technical report, NREL/ TP-550-37542. National Renewable Energy Laboratory, Golden, CO.

NATIONAL RENEWABLE ENERGY LABORATORY'S NINE-STEP PROCESS FOR LOW-ENERGY BUILDING DESIGN

1. Create a base-case building model to quantify base-case energy use and costs. The base-case building is solar neutral (equal glazing areas on all wall orientations) and meets the requirements of applicable energy efficiency codes such as Standard 90.1 (ASHRAE 2016) and ANSI/ASHRAE Standard 90.2-2007, Energy Efficient Design of Low-Rise Residential Buildings (ASHRAE 2007).
2. Complete a parametric analysis to determine sensitivities to specific load components. Sequentially eliminate loads from the base-case building, such as conductive losses, lighting loads, solar gains, and plug loads.
3. Develop preliminary design solutions. The design team brainstorms possible solutions that may include strategies to reduce lighting and cooling loads by incorporating daylighting or to meet heating loads with passive solar heating.
4. Incorporate preliminary design solutions into a computer model of the proposed building design. Energy impact and cost effectiveness of each variant are determined by comparing the energy with the original base-case building and with the other variants. Those variants having the most favorable results should be incorporated into the building design.
5. Prepare preliminary set of construction drawings. These drawings are based on the decisions made in Step 3.
6. Identify an HVAC system that will meet the predicted loads. The HVAC system should work with the building envelope and exploit the specific climatic characteristics of the site for maximum efficiency. Often, the HVAC system is much smaller than in a typical building.
7. Finalize plans and specifications. Ensure that building plans are properly detailed and that specifications are accurate. The final design simulation should incorporate all cost-effective features. Savings exceeding 50% from a base-case building are often possible with this approach.
8. Rerun simulations before design changes are made during construction. Verify that changes will not adversely affect the building's energy performance.
9. Commission all equipment and controls. Educate building operators. A building that is not properly commissioned will not meet energy-efficiency design goals. Building operators must understand how to properly operate the building to maximize its performance.

Online

Building Performance Institute of Europe
www.bpie.eu/financial_instruments.html.

California Debt and Investment Advisory Commission. 2008. *Public-Private Partnerships: A Guide to Selecting a Private Partner*.
www.treasurer.ca.gov/cdiac/publications/p3.pdf.

Dena
www.dena.de/en.

European Commission, BUILD UP program
www.buildup.eu/.

High Performing Buildings magazine
www.hpbmagazine.org.

International Institute for Sustainable Laboratories, Labs 21 Tool Kit.
www.i2sl.org/resources/toolkit.html.

MURE (Mesures d'Utilisation Rationnelle de l'Energie), database on energy efficiency policies and measures
www.muredatabase.org.

Database of State Incentives for Renewables & Efficiency (DSIRE)
www.dsireusa.org.

U.S. Department of Energy, Energy Efficiency and Renewable Energy, Zero Energy Buildings
https://energy.gov/eere/buildings/zero-energy-buildings.

U.S. Environmental Protection Agency, Energy Star Target Finder.
www.energystar.gov/buildings/service-providers/design/step-step-process/evaluate-target/epa%E2%80%99s-target-finder-calculator.

CHAPTER FOUR

COMMISSIONING

Commissioning is defined by ANSI/ASHRAE/IES Standard 202-2013 (ASHRAE 2013a) as *a quality-focused process for enhancing the delivery of a project that provides continuous oversight from the predesign phase of the project throughout occupancy conducted under the administered by a commissioning authority (CxA)*. The commissioning process also provides an essential foundation for integrated design, delivery, and operation, and facilitates adapting to changing needs over the facility's lifetime. Commissioning establishes benchmarks for evaluating achievement of the owner's defined goals and objectives as they change over time. Commissioning is applied to new building projects and existing facilities to ensure a facility meets the needs of the occupants to effectively and efficiently deliver the owner's purpose for that building and their organization's and financial goals. The true measure of a high-performance building is how it performs over its lifetime; proper commissioning of the building and facilities help ensure that high performance. ASHRAE Standard 202 defines the specific details needed to establish and conduct a well-organized commissioning process for a project, and this chapter provides additional background and insight as it relates to green building design, and long term building performance on a high level.

Commissioning provides benefits to everyone: the owner, the designers and construction teams, building occupants, and building operators. It helps to reduce risk and, with fewer change orders, improve energy efficiency and lower operating costs. It also results in increased satisfaction of tenants and occupants. The commissioning process should begin early in the predesign phase to ensure maximum benefits from commissioning. Starting early improves designer and contractor quality control processes, makes the CxA part of the process, and identifies and helps resolve problems during design (when corrective action is the least expensive). During construction, commissioning can also provide benefit when the contractor has the materials and resources on site for efficient corrective action (minimizing postoccupancy rework and repairs).

How and to what extent an owner incorporates commissioning into their project generally depends on the owner's understanding the process and results of commissioning. Commonly, an owner might start with construction phase commissioning but may soon see how much more they would have benefited by starting in the predesign phase. Other factors that determine the extent of commissioning incorporated in any given project are how long the owner holds the property, the owner's staff capabilities and funding mechanisms for design, construction, and operation, the project schedule, and ownership experience.

The commissioning process focuses upon verifying and documenting that the building and all of the systems and assemblies are planned, designed, installed, tested, operated, and maintained to meet the owner's goals and objectives for the project. Commissioning is not only an important part of successful project delivery, but is an essential and, in many cases, required part of green-building design and construction. Commissioning is also now required in various energy codes and standards. Commissioning is not an exercise in blame. It is, rather, a collaborative effort to identify and reduce potential design, construction, and operational problems by resolving them early in the process, at the least cost to everyone. This chapter will include the selection and role of the commissioning provider, a discussion of various commissioning models, the choice of building systems for commissioning, and the long-term benefits afforded by implementing the commissioning process.

THE EVOLUTION OF THE COMMISSIONING CONCEPT

Initial guidance on commissioning was produced by ASHRAE in 2003 with ASHRAE Guideline 0, *The Commissioning Process*, which describes the overall process and provides guidance on implementation for commissioning, and ASHRAE Guideline 1.1, *HVAC&R Technical Requirements for the Commissioning Process* (2007), which contains additional information specifically focused on HVAC&R applications. However, these documents were guidelines, and there was a need for a "code language" standard that specifies the commissioning process. Thus, ASHRAE created Standard 202-2013, *Commissioning Process for Buildings and Systems*.

Commissioning of projects has been increasing in importance during the past decade, and is being adopted into building construction and energy codes. This is partly because of the LEED-mandatory requirement for commissioning and to the growing awareness of the benefits commissioning brings to a project. In addition, ANSI/ASHRAE/USGBC/IES Standard 189.1-2017, *Standard for the Design of High Performance Green Buildings*, includes a requirement for commissioning for all buildings with floor area greater than 5000 ft^2 (500 m^2), although changes to this threshold are being considered. There is some debate on whether all high-performance green buildings should be required to undergo a full commissioning process. Certainly, any building that includes complex systems or a larger number of energy and water consuming devices should be considered for commissioning.

COMMISSIONING PHASES AND PROCESS

Commissioning for new construction projects can be broken down into five phases: predesign, design, construction, acceptance, and warranty/ongoing commissioning. Commissioning of existing buildings also has four phases: investigation, analysis, recommendations, and implementation. Distinct commissioning activities occur during each of these phases. Achieving desired building performance starts with clearly defining the Owner's Project Requirements (OPR) at predesign (before development of the architectural program). During design, the CxA provides consulting and checklists to designers to assist them in their design quality control process and inform the designers about specifics the CxA will focus on during commissioning design reviews. Chapter 3 of this guide provides further information on commissioning activities in the design phase. During construction, the CxA conducts site visits and provides construction checklists to the contractors to assist them in their quality control process and, as in the design reviews, verifies that the contractor's quality control process is functioning well, and is documented. These efforts significantly improve the chances that the systems being commissioned will only need minor modifications during performance testing and will reduce the delivery team's efforts, and costs. The acceptance phase verifies through testing that the systems perform as intended and helps resolve issues prior to occupancy. During the warranty phase, the CxA monitors system performance and verifies that training provided was understood by operators, and maintainers. The CxA assists operators in better understanding their systems and maintaining maximum performance, which helps prevent inappropriate modifications by the operators (due to lack of understanding). During the new construction process, the CxA identifies specific deliverables required from the project team needed for the systems manual. The systems manual is the repository for the OPR, Basis of Design (BoD), commissioning reports, O&M manuals, equipment operation instructions, and record documentation, including any modifications made to the facilities systems and assemblies, along with the reasoning behind the modifications. Detailed information to be included in the systems manual is defined in ASHRAE Guideline 1.4, *Procedures for Preparing Facility Systems Manuals*, and Standard 202.

The Commissioning Role Implementation

Because of the nature of the delivery process, the further along the team is in the design and construction process, the more difficult and expensive changes become (see Figure 3-1). Waiting until after predesign to define project end goals, occupant requirements, and team roles and responsibilities can lead to increased project cost. The easiest time to evaluate goals, objectives, and criteria (and make changes) is during the schematic design phase, and this is also the best time to reduce the control cost of any changes. A steep decline in the feasibility to implement sustainability features occurs after the schematic design phase. The ability to make significant cost-efficient changes ends when the project moves into design development. The earlier a design team addresses and implements sustainable features into a design, the more likely these will be included without significant cost. This is a foundation of integrated design and critical to cost-effective green buildings. Historically, owners and contractors set up a contingency fund intended to cover the unpredictable cost of changes in a project. If the design does not meet the owner's needs, the owner may be forced to accept the project as is, because changes needed to meet his or her requirements would be too costly at that point. If the CxA is engaged as late as the construction phase, there is some, but very limited, opportunity to address potential design problems. The longer an owner waits to engage in the commissioning process, the less influence the CxA has on resolving problems cost effectively.

The CxA's Role

The role of the CxA varies according to the phase of the project. It is the CxA's role to lead the collaborative team effort required to balance competing interests so that the owner's needs and mission are not lost. To accomplish this task, a benchmark is needed. This benchmark is the OPR and the CxA should be involved with the OPR development.

Ideally, it is also in the predesign phase of a project, prior to engaging the design team, that the owner would engage a CxA to develop draft commissioning specifications and a commissioning plan. The draft commissioning plan defines the team's roles and responsibilities, suggested communication protocols, commissioning activities, and the schedule of the activities. Additional information on selecting the CxA is provided in following sections.

COMMISSIONING DOCUMENTS

The Owner's Project Requirements

The OPR is a written document that details the functional requirements of a project and expectations of how it will be used and operated. The OPR includes project and design goals, measurable performance criteria, budgets, schedules, success criteria, and supporting information (specific information that should be included can be found in ASHRAE Guideline 0 [ASHRAE 2013b] and ASHRAE Standard 202 [ASHRAE 2013a]). It also includes information necessary for all disciplines to properly plan, design, construct, operate, and maintain systems and assemblies.

Many confuse the OPR with an architectural program. The OPR is different from the typical architectural program, which focuses on project floor area needs, adjacencies, circulation, cost, and structural predesign test results. The OPR documents how the owner intends for this building to function and fulfill the needs of the owner and the occupants. For instance, an architectural program does not contain requirements for how the building will be operated. It also does not contain operation and maintenance (O&M) and training requirements or post-construction documentation requirements for the building, whereas the OPR does. Developing the OPR in conjunction with the architectural program reduces programming effort and provides valuable information to the designers that they typically do not have. In combination with the architectural program, the OPR provides a strong foundation for a successfully integrated design and delivery process and the building's operational criteria. The OPR, however, should not duplicate information contained in the architectural program. The OPR forms the basis from which the commissioning provider verifies that the developed project meets the needs and requirements of the owner. An effective commissioning process depends on a clear, concise, and comprehensive OPR with benchmarks for each of the objectives and criteria. This written document details the functional requirements of the building and the expectations of how it will be used and operated.

If no formal program exists, the OPR document can be used to assist with identifying the criteria the design team is tasked with meeting. However, the main purpose of the OPR is to document the owner's objectives and criteria. The designer's BoD documents, the assumptions the designers made to meet the OPR, and a summary of this information is provided to the operators of the project long after the design and construction team have left the project. As such, the OPR document provides the benchmark against which the design, construction, and project operating performance can be measured.

The OPR includes sustainable development goals that are in harmony with the owner's mission. Many teams focus on green rating system credits without knowing how these credit selections will affect the owner's needs. For example, on one project, the design team selected use of native and adaptive plants to achieve a green rating system credit but neglected to understand that this conflicted with that military facility's mission and requirements for no vegetation that was 4 in. (100 mm) in height above the ground.

Additional information on development on an OPR is given in Appendix D of ASHRAE Standard 202.

The Basis of Design (BoD)

This document should be submitted by the design team and outlines how they have designed the project to meet the goals of the OPR. It should be updated and reviewed as part of each design submission. Detailed examples for what information should be contained in the BoD are given in Appendix F of ASHRAE Standard 202.

The Commissioning Plan

The draft commissioning plan defines the team's roles and responsibilities, suggested communication protocols, commissioning activities, and the schedule of the activities. Project success is dependent on each team member's understanding of what is expected of them and obtaining their buy-in. The commissioning plan provides the owner with clearly defined roles and responsibilities for each team member for inclusion in contractual agreements and improved team efficiency. It is more economical to define these requirements early, before selecting and contracting the various project team members. After the project team selection, an owner will state the basic requirements of the project that form the starting point for the OPR and architectural program. This information will typically include justification for the project, programming needs, intended building use, basic construction materials and methods, proposed systems, project schedule, and general information (such as attaining LEED certification). Further discussion on the commissioning plan is included in the construction phase paragraphs below. Detailed examples for what information should be contained in the commissioning plan are given in Appendix G of ASHRAE Standard 202.

The Commissioning Specification

It is important that the contract documents include the requirements for the suppliers and contractors so they know of their duties and opportunities during the time of both bidding and construction. The design phase paragraph below includes further information on specifications. Good commissioning specifications will increase project efficiency and reduce conflicts. Detailed examples for what information should be contained in the commissioning specifications are given in Appendix L of ASHRAE Standard 202.

PROJECT PHASES

Predesign Phase

Commissioning is a team effort performed within a collaborative framework. The entire project team is part of the commissioning process. The commissioning plan defines the project team's roles and responsibilities, and execution of the commissioning process. It is essential that the owner clearly defines (contractually) the project team's commissioning roles and responsibilities as well as prepare the OPR. The value of the quality and quantity of information provided by owners is often related to their development experience. Institutional owners who have developed many buildings and who have held properties for extended periods have, over time, developed the information that design teams need to understand an owner's basic needs. One of the greatest values of involving a CxA in the predesign phase is to develop a comprehensive OPR document and a preliminary commissioning plan that serves as the project benchmark, guiding all project team members.

Design Phase

During design phase commissioning, the CxA often will develop design checklists and specifications that incorporate commissioning into the project. The design checklists reflect objectives and criteria that designers should check during their quality control process, the amount and type of information to be provided at each stage of the commissioning design review, and the designer's assertion that items in the commissioning checklists are complete. Typically, two or three reviews occur in most design phases. The specifications identify the roles and responsibilities of the construction phase project team, the systems that will be commissioned, and the criteria for acceptance of the commissioned systems.

Commissioning design reviews focus on assessment of the design meeting the OPR. Typically, the OPR provides the project vision; expected service life; energy and water efficiency goals; maintainability; training of operational and maintenance personnel; infrastructure for monitoring-based, ongoing commissioning; expectations of the building envelope's ability to resist weather, air, and water intrusion; and other functional requirements needed for occupants to effectively and efficiently deliver their daily mission. Commissioning design reviews can also randomly check the designer's quality control process by identifying systemic design issues that lead to potential change orders, including missing or misleading information and insufficient detail needed for accurate pricing and competitive bids. Commissioning design reviews can reduce owner and project team risk, allow corrections to be made at the lowest possible cost, and reduce requests for information, supplemental instructions, and schedule impacts. Third-party commissioning provides great benefits to the project team and owner. (See further discussion in the "Commissioning Models" section later in this chapter.)

Design phase commissioning promotes communication, identifies disconnects, questions design elements that appear incorrect, and shares experience to produce a better set of contract documents and a better high performance building.

Several engineering trade magazines and long-term owners, along with insurance companies who provide errors and omissions (E&O) coverage to the design community, have all voiced their concerns about the quality of construction documents and have charted how E&O premiums are affected as a result of judgments or settlements. Design phase commissioning reduces the risks of change orders, accompanying construction delays, and E&O claims, and helps clarify construction documents. Design phase commissioning, if correctly implemented, is a seamless process that provides benefits to the entire project team.

One role of the CxA is to assemble a review team experienced in the type of building being designed. Generally, the CxA has a team of reviewers with specific background and experience to review the disciplines selected for design phase commissioning. This process often requires the most senior individuals as part of the design phase commissioning provider team. (See "Selection of a CxA" later in this chapter.) A typical design review process is as follows:

- Written comments from the reviewers are provided to the design team and owner.
- Comments are reviewed, and the design team returns written responses.
- Meetings are scheduled with the review team and the designers to adjudicate comments as necessary, allowing the owner to understand the issues and have an opportunity to provide direction as needed.
- Design concerns, comments, and actions taken are recorded in the design review document. Changes are made as agreed and the commissioning review team verifies the change and closes the issue as appropriate.

Using a best-practices approach, the design phase commissioning process could occur four different times during the design: at 100% of schematic design, at 100% of design development, at 50% complete construction documents, and at approximately 95% of construction documents. (Combining the review of the first two phases can be done on smaller projects.) An advantage of four reviews, however, is that the design is evaluated based on the OPR document goals before the next phase starts.

Although four reviews is ideal, particularly for complicated projects, having at least two (as required for commissioning by Standard 189.1) provides at least some level of checks and balances.

Changes to optimize building performance, daylighting considerations, system selection, and stacking/massing synergies can best be addressed during schematic design review. Review during design development allows the team to identify potential problems and constructibility perspectives early enough to resolve many issues before the construction documents phase starts.

The quantity of design concerns typically increases during the 95% construction document phase because more detail is provided about each building system and component. The concerns identified at that point typically revolve around details—finishes, coordination conflicts, etc. Resolving these concerns provides clearer direction to the contractors, resulting in better cost and schedule predictions.

It is not uncommon that concerns identified at the 50% construction documents phase will go unaddressed by a design team. This is especially true in fast-track projects when designers, responding to owner and contractor demands for documents, struggle to finish and deliver their work product. This is why the later 95% review is so important.

Construction Phase

The CxA's role during the construction phase is to review the 95% complete construction documents and submittals; develop and integrate contractor construction checklists; identify and track issues to resolution; develop, direct, and verify functional and performance tests; observe construction of commissioned systems; review the O&M manuals provided under the contract; and oversee development of a systems manual. The purpose of these activities is to verify that the requirements of the OPR document have been met, commissioned systems are serviceable, commissioned systems perform as intended, and operational personnel receive the training and documentation necessary for operating and maintaining building performance.

The review of final construction documents is used to verify that the concerns and issues raised during the design reviews have been resolved, reducing the risk of contractors building flaws into the project. Sometimes, agreement between the designers and the commissioning design review team is not reached during the design phase. An example of this would be a disagreement over building pressurization control: the designer may feel that the design provides adequate control, and the CxA may disagree. The CxA must verify whether or not the building is correctly pressurized through performance tests. If the designer is correct, the CxA closes the issue after verification. If pressurization is an issue, the team then has the opportunity to correct the concern before more serious and costly ramifications can occur. Systems failing to perform are identified and the project team works to resolve the issue while the entire team is still engaged in the project.

The CxA reviews product submittals to look for potential performance problems and to verify that what was proposed is in fact delivered. An example of this would be contractor's ductwork shop drawings showing high-pressure loss fittings that would increase energy consumption. The CxA review activity does not take the place of the designer's review, nor is it meant to. The CxA review should reveal whether the contractors are following the designer's intent. Several purposes can be combined by the CxA in his or her review of the submittals, depending on the role the owner defines for the CxA. For instance, if the CxA is also assisting the team with LEED certification, the CxA can verify that the sustainable development goals identified in the OPR document are being met.

At the start of construction, a CxA may choose to be part of prebid conference meetings with the general contractor or construction manager to provide a description of commissioning activities, to describe the general roles and responsibilities the contractor will be asked to fulfill in the commissioning process, and to answer questions. Clear communication with the contractors during prebid has proven important in preventing high bids that reflect a fear factor from contractors who are unfamiliar with the commissioning process.

In the early stages of construction, the CxA updates the commissioning plan defining the commissioning process, the roles and responsibilities of the project team, lines of communication, systems being commissioned, and a schedule of commissioning activities. The CxA conducts an initial commissioning scoping meeting where the commissioning plan is reviewed by the project team and, based on this information, a final commissioning plan is developed and implemented.

Prior to the testing of systems and equipment, the CxA will ensure that the necessary prefunctional tests and observations have been completed. The prefunctional tests are those where the equipment installation is confirmed as ready to receive power. This includes checks such as correct phasing of electrical connections, correctly installed mounting fasteners, accessible valve operators, and secure pipe and duct connections. Often, the prefunctional test procedures are developed by the manufacturer and verified by the contractor and CxA. All prefunctional test reports should be retained and made part of the systems manual.

Throughout construction, the CxA observes the work to identify conditions that would impair preventive maintenance or repair, hinder operation of the system as intended and/or compromise useful service life, and to verify other sustainability goals (such as indoor air quality [IAQ] management) during construction. In recent years, there has been a growing recognition of the importance for building envelope commissioning, and it is during the construction process that envelope commissioning is most valuable. Typically, envelope commissioning is provided by third-party specialists and is often a separate contract from operating systems commissioning. The CxA may help develop activities to help perfect installation procedures at the start of the specific construction activity and coordinate activities to help ensure that the contractor's quality control process is working through verification of construction checklists, start-up procedures, and testing and balancing. When contractors have completed their construction checklist for a specific system being commissioned, they are stating that their systems will perform as intended. The CxA verifies this by directing and witnessing functional and performance testing.

Acceptance Phase

The functional testing phase of the commissioning process is often referred to as the *acceptance phase* and is the one phase that is most commonly associated with commissioning. With designer and contractor input, the CxA develops system tests (functional tests) to ensure that the systems perform as intended under a variety of conditions. Contractors under the CxA's direction execute the test procedures, while the CxA records results to verify performance. The tests should verify performance at the component level through inter- and intrasystem levels. Another practice is to have the contractors simulate failure conditions to verify alerts and alarms, as well as system reaction and interaction with associated systems. Problems identified are resolved while contractors and materials are still on site and the designers are engaged.

The functional tests are preestablished by the CxA and submitted to the contractor early in the construction phase. The contractor is aware at an early stage of all the tests that must be observed. Testing and balancing (TAB) of HVAC systems is typically provided by an independent third party (not the mechanical contractor) and a recent trend is to sometimes include this under the commissioning contract. The test and balance subcontractor and/or the installing contractor should include a dry run of the functional tests before the formal CxA witness test to ensure that the systems are performing as intended. Often, the systems are not checked in advance of the formal CxA-witnessed test and the systems fail, causing time delays and additional expenses for all parties involved. The contractor must correct the discrepancy, and the CxA must return for a re-witness. It is important that the CxA have clear contractual contingencies in place in order to adequately compensate for retests.

Warranty Phase

The first year of a project is critical to finding and resolving issues that arise, and the CxA plays an important role in helping ensure that a building performs at its optimum. The warranty period is also the period when the contractors and manufacturers are responsible for the materials and systems installed, and is the only time the owner has to identify warranty repairs without additional construction costs. As such, the CxA takes on a vital role to assist the owner and operational staff in identifying and resolving problems and assisting with resolution that gives the least interruption to the occupants during this critical period. This phase may also be coordinated with a postoccupancy evaluation, particularly if one is being performed as part of the LEED green-building rating program.

During the first several months, the CxA verifies that systems are performing as intended through monitoring of system operation. Many systems cannot be fully tested until the building is occupied or the systems are called to be operated during near-peak cooling or heating situations. There might be a small percentage of system components that pass functional testing but, under actual load, fail to perform as intended. These components must be identified in the warranty period and replaced or repaired as necessary. The CxA's role in conjunction with the operational staff is to search out problems that only become evident under actual load. To accomplish this, the commissioning provider performs several specific tasks.

The commissioning provider may also help identify system points to trend, verifying efficient system operation; installs independent data loggers to measure parameters beyond the capabilities of the building automation system (BAS); and monitors utility consumption. Using the trend system data from the selected BAS input points, the CxA can help in the analysis of the information, looking for operational sequences that consume resources unnecessarily and conditions that could compromise occupant satisfaction within the working environment. In addition, the CxA also looks for conditions that could result in building failure, such as high humidity in interstitial spaces in the building's interior, hot spots in the electrical distribution system, or analysis of electrical system harmonics where power quality is essential to the owner. These functions can only be tested after occupancy.

Some systems, such as the heating and cooling equipment, can only be fully tested when the season allows testing under design load conditions. The CxA works closely with the operational staff to identify and help resolve issues that become apparent in the warranty period and verifies that the operational staff fully understands and meets their warranty responsibilities. In addition, the CxA should provide operational staff with the specific functional test procedures developed for their use in maintaining building performance for the life of the building. By having the CxA work with the operational staff during the warranty period, the operators gain valuable insight into how the building should operate and what to look for to ensure continued performance. This helps to overcome a typical industry problem: the bypassing of system components and controls because of a lack of understanding of how the systems are intended to operate.

Ongoing Commissioning Phase

An important part of green design is verification that the goals defined by the owner are integrated by the design and construction team to achieve the owner's objectives for the lifetime of the building, as defined in the OPR. The OPR is a living document that changes over time as the owner requires change. Once the building is occupied, changes in the owner/occupant mission can occur that impact and modify the original project requirements. The original OPR is maintained for its historical value in marking the original project requirements. Using the original OPR as a starting point, the OPR is converted to the current facility requirements (CFR), which documents the changes the owner needs to match the changes in their daily mission. The CFR provides the foundation to guide the future project teams in modifications needed to meet the changing needs of the owner/occupant.

Ongoing commissioning in both new and existing projects provides the owner and operators with the tools necessary to efficiently manage financial and human resources to achieve desired returns on their investment, building performance, and reduction of environmental impact. Owners and operators cannot manage what is not measured. During occupancy, the CxA monitors the building's performance and identifies deterioration of performance. This deterioration typically falls into one of two categories: operator error or system malfunction or deterioration. Monitoring-based, ongoing commissioning identifies the cause of the deterioration and recommendations can be made for correction. Operator errors resulting in deterioration of building performance provide training opportunities to improve O&M staff understanding of correct building operation, which helps prevent future operational mistakes. Monitoring-based, ongoing commissioning identification of degradation of building performance due to system malfunction provides in-depth failure information, which lessens the O&M staff's troubleshooting efforts and can verify that the repairs performed resolve the problem. Monitoring-based commissioning is easily implemented by the CxA when integrated into the OPR or CFR. Implementation in existing buildings does require more effort and some additional cost, but it provides significant financial benefits immediately and over the lifetime of the building. Commissioning providers can also help integrate measurement and verification (M&V) plans and procedures that can be used to identify when a building begins operating outside of allowable tolerances, signaling the owner that corrective action is needed to maintain performance.

The operational staff's knowledge and understanding of how the building should be operated and their methods of operation in practice typically determine the actual building system's performance. The most common reason facilities fail to meet performance expectations or experience deterioration of performance is tied directly to how the building is operated and maintained. Changes in operational personnel often result in loss of the institutional knowledge needed to maintain the building's performance and meet occupants' needs. The system manual is the repository for essential information needed by the operator to maximize occupant satisfaction and building performance. Without this information, the operator is running blind, without reference points for correct system operation. The result is that it may be difficult to meet the owner's goals and objectives and operation costs may increase. Monitoring-based, ongoing commissioning provides regular monitoring and analysis of utility usage, system interaction, and operator performance. Performed on a regular basis, monitoring-based commissioning improves financial and operator performance by changing the culture from constantly responding to problems and replacing it with routine maintenance. It also helps to ensure that institutional knowledge is not lost through changes in personnel, occupant mission, and building modifications. For owners to get maximum performance from their facilities, they must know when systems fall outside of allowable performance tolerances or be aware of when operational personnel make changes that could negatively impact the facilities' performance. This is best done through monitoring-based, ongoing commissioning.

For ongoing commissioning activities, it is not necessarily required that a third-party CxA be retained to oversee the day-to-day activities. Ideally, the facilities staff can perform the function and integrate ongoing CxA into the management of the facility. Modern facility management is a highly skilled and high-tech enterprise and some major universities are now offering doctoral programs in this field.

ASHRAE Guideline 14, *Measurement of Energy, Demand and Water Savings* (ASHRAE 2014) provides several methods to establish operational benchmarks for energy consumption. Sustainable, healthy, high-performing facilities often measure other parameters including water consumption, waste generation, recycling, pesticide use, etc. The operational tracking of these parameters reduces total cost of ownership, lessens environmental impact, improves building occupants' quality of life, and can increase building value. An owner can obtain guidance on integrating sustainable operation practices by adopting USGBC's LEED for Existing Buildings Operation and Maintenance program at this stage of the building's life.

Additional information and insight into the application of these concepts in the operation of existing buildings can be found in Chapter 16, Operation, Maintenance, and Performance Evaluation.

DECISIONS TO MAKE DURING THE COMMISSIONING PROCESS

A good description of how to initiate a commissioning project is provided in Appendix C of ASHRAE Standard 202. This section outlines the series of decisions that must be made to get started.

Selection of the CxA

As with finding a doctor, lawyer, contractor, or other professional, the key element is that the commissioning provider should have experience in the types of systems an owner wants commissioned. In other words, an owner must match the experience with the job. A good CxA generally has a broad range of knowledge, including hands-on experience in O&M, design, construction, and investigation of building/system failures. CxAs must also be detail-oriented, good communicators, and able to provide a collaborative approach that engages the project team.

ASHRAE provides Commissioning Process Management Professional certification for owners implementing commissioning and engaging commissioning authorities. People who pass this certification understand how to apply the commissioning process and select commissioning providers, which helps ensure the owner gets what they paid for.

Selection of Systems to Commission

Commissioning was originally developed in areas where energy efficiency was a prime driver. The commissioning process has expanded beyond the original commissioning of HVAC systems to include building envelope, electrical, plumbing, lighting, security, etc. This expansion from HVAC is often referred to as *whole-building commissioning*.

Although numerous commissioning service models exist, commissioning should be performed by a third-party provider or the owner's own commissioning team whenever possible. Commissioning of all systems using the whole-building approach has proven to be beneficial. However, because of budget constraints, owners may want to look at commissioning systems that will yield the greatest benefit to them. In other cases, the systems selected for commissioning may be specified by code or standard, such as projects that are being conducted under the requirements of Standard 189.1. Long-term owners have an advantage and can apply their experience of where they have historically encountered problems and elect to commission only those systems. Others who do not have that depth of experience may wish to talk with long-term owners or insurance providers to gain perspective.

There are many factors that define which building systems should be commissioned, but there are no published standards yet to help guide owners through such selection (ASHRAE has produced guidelines, but not standards, in this area.) It often depends on the associated risk of not commissioning. Errors and omissions (E&O) insurance providers publish graphs of claims against design professionals by discipline. Interestingly, 80% of the claims against architects are for moisture intrusion, thus indicating a need for considering building envelope commissioning. In addition, some organizations have suggested that up to 30% of new and remodeled buildings worldwide may be the subject of excessive complaints related to IAQ. The reasons for these *sick buildings* include inadequate ventilation, chemical pollutants from both indoor and outdoor sources, and biological contaminates.

Commissioning provides several benefits, two of which are risk reduction and generally an overall lower total cost of ownership. Based on a specific climate such as Phoenix, AZ, the risk of not commissioning the building envelope is much less than in Miami, FL, or Bangkok (because of humidity concerns). Based on functional requirements, declining to commission the security systems in a conventional office building may have minimum risk compared to a federal courthouse. So, what should be commissioned?

The best time for determining what systems should be commissioned is during the development of the OPR document. Generally, there are three main system categories that, as a minimum, should be commissioned: building envelope, mechanical and plumbing systems, and electrical systems. If the building project includes an on-site renewable energy system, that should also be considered a definite candidate for commissioning. Depending on the functional requirements of a building and the complexity of systems, additional systems that may be commissioned include security, voice/data, selected elements of fire and life safety, irrigation and/or process water systems, energy or water monitoring systems, and daylighting controls.

In some cases, these systems may be commissioned by the manufacturer's installer, such as is the case with fire and life safety systems where specific licenses and certification is required. Standard 189.1 is more specific on the systems that should be commissioned and could be considered a good general guideline to follow. Some building systems are installed with oversight and field testing by the manufacturer (elevators, for example) and these systems do not normally require a third-party commissioning agent.

USGBC's LEED rating system recognizes the benefits of commissioning and its importance to green-building design, construction, and operation. As such, the LEED reference guide implies that building systems that affect energy consumption, water usage, and indoor environmental quality should be commissioned.

Each element of green design needs verification to ensure that the design, construction, and operation of the high-performance building meet the expectations of the team and realize the financial return envisioned by the owner.

Commissioning Models

Independent third-party commissioning is the preferred commissioning approach, because it reduces the potential for conflict of interest and puts a quality advocate squarely in the owner's corner. It also allows for integration of commissioning professionals specialized to meet a project's specific needs concerning the building envelope, security systems, labs, etc.

Having commissioning be part of the general contractor's responsibility has many of the same problems as the model using the design professional. Contractors, by the very nature of the construction business, are focused on schedule and budget. This focus is not always in the owner's best long-term interest. Most contractors are quality-minded and do their best to identify problems and assist with resolutions (though often to their detriment because they inadvertently take responsibility for the design in doing so). If a constructibility issue arises that will adversely affect the schedule or budget, the contractor may choose to fix the issue and hope that it does not create a warranty callback. The main problem is that many of the issues are discovered too late in the process, again resulting in change orders, construction delays, and additional costs, and building operational problems.

To truly be an owner's advocate, the CxA must owe allegiance to no one but the owner. A third-party CxA will verify that the goals defined by the owner and integrated by the design and construction team are achieved as intended, from the first day of occupancy. If the CxA is separate from the design professional or contractor, he/she will provide unbiased reporting of issues to the team and will guide them toward timely solutions without finger-pointing, delays, and liability.

Digging Deeper

ONE FIRM'S COMMISSIONING CHECKLIST

- Begin the commissioning process during the design phase; carry out a full commissioning process from lighting to energy systems to occupancy sensors.
- Verify and ensure that fundamental building elements and systems are designed, installed, and calibrated to operate as intended.
- Engage a CxA that is independent of both the design and construction team.
- Develop an OPR document and review designer's basis of design to verify requirements have been met.
- Incorporate commissioning requirements into project contract documents.
- Develop and use a commissioning plan.

- Verify installation, functional performance, training, and Operations and Maintenance (O&M) documentation.
- Complete a commissioning report.
- Perform additional commissioning:
    - Conduct a focused review of the design prior to the construction documents phase.
    - Conduct a detailed review of the construction documents when these are considered nearly complete and prior to their issuance for construction.
    - Conduct reviews of contractor submittals that are relevant to systems being commissioned.
    - Provide information required for recommissioning systems in a single document to the owner.
- Have a contract in place to review with operations staff the actual building operation and any outstanding issues identified during commissioning, and to provide assistance resolving these issues within the warranty period.
- Encourage long-term energy management strategies.
    - Provide for the ongoing accountability and optimization of building energy and water consumption performance.
    - Design and specify equipment to be installed in base building systems to allow for comparison, management, and optimization of actual vs. estimated energy and water performance.
    - Use M&V functions where applicable.
    - Tie contractor final payments to documented M&V system performance and include in the commissioning report.
    - Provide for an ongoing M&V system maintenance and operating plan in building O&M maintenance manuals.
- Operate the building ventilation system at maximum fresh air for at least several days (and ideally several weeks) after final finish materials have been installed before occupancy.
- Provide for the ongoing accountability of waste streams, including hazardous pollutants.
- Use environmentally safe cleaning materials.
- Train O&M workers.

REFERENCES AND RESOURCES

Published

ASHRAE. 2007. ASHRAE Guideline 1.1, *HVAC&R Technical Requirements for the Commissioning Process*. Atlanta: ASHRAE.

ASHRAE. 2012. ASHRAE Guideline 1.5, *The Commissioning Process for Smoke Control Systems*. Atlanta: ASHRAE.

ASHRAE. 2013a. ANSI/ASHRAE Standard 202, *Commissioning Process for Buildings and Systems*. Atlanta: ASHRAE.

ASHRAE. 2013b. Guideline 0, *The Commissioning Process*. Atlanta: ASHRAE.

ASHRAE. 2014a. Guideline 1.4-2014, *Systems Manual Preparation for the Commissioning Process*. Atlanta: ASHRAE.

ASHRAE. 2014b. ASHRAE Guideline 14-2002, *Measurement of Energy, Demand and Water Savings*. Atlanta: ASHRAE.

ASHRAE. 2015. Guideline 0.2-2015, *Commissioning Process for Existing Buildings and Systems*. Atlanta: ASHRAE.

ASHRAE. 2017. ANSI/ASHRAE/USGBC/IES Standard 189.1-2017, *Standard for the Design of High Performance Green Buildings*. Atlanta: ASHRAE.

The following guidelines have been or are being developed by ASHRAE to support the commissioning process:

ASHRAE. Guideline 1.2P, *Commissioning Process for Existing HVAC&R Systems*. Atlanta: ASHRAE.

ASHRAE. Guideline 1.3P, *Building Operation and Maintenance Training for the HVAC&R Commissioning Process*. Atlanta: ASHRAE.

CHAPTER FIVE

ARCHITECTURAL DESIGN AND PLANNING IMPACTS

OVERVIEW

The architect and landscape architect play particularly critical roles within any design team whose charge is to meet an owner's sustainability design intent. Careful considerations are important for site selection, building orientation and form, the structure's envelope, and the arrangement of spaces and zoning. These considerations are covered in this chapter, with the exception of site selection, which is covered in Chapter 7, Sustainable Sites.

Early site and architectural decisions can positively propel or alternatively severely hinder the ability of the rest of the team to optimize their roles in producing a high-performing building. Architectural design choices directly affect energy-related first costs for construction and the total embodied energy in construction materials. These choices further impact the operational energy use, resulting in carbon emissions as well as the well-being and comfort of the occupants. The importance of working as an integrated team with all design stakeholders throughout the entire design/construction process cannot be overemphasized.

This chapter is intended to help designers make design decisions that best effect positive sustainable/green project outcomes. A concise history of the design of buildings is provided, followed by a suggested design process and a listing of various sustainable design approaches.

CONCISE HISTORY OF THE DESIGN OF BUILDINGS

Before the industrial revolution, building efforts were often directed throughout design and construction by a single architect—the so-called *master builder model*. The master builder bore full responsibility for the design and construction of the building, including any required engineering. This model lent itself to a building designed as one system, with the means of providing heat, light, water, and other building services often closely integrated into the architectural elements.

Sustainability—semantically, if not conceptually—predates these eras, and some modern unsustainable practices had yet to arise. Sustainability in itself was not the goal of yesteryear's master builders. Yet some of the resulting structures achieved an admirable combination of great longevity and sustainability in construction, operation, and maintenance. It is interesting to compare the ecological footprint (a concept discussed in this book) of Roman structures from two millennia ago and heated by radiant floors to a 20th century structure of comparable size, site, and use. Also, some preindustrial building strategies, forgotten in the era of inexpensive energy, can produce energy savings when applied to modern buildings.

In the 19th century, as ever more complicated technologies and the scientific method developed, the discipline of design and engineering of building systems emerged separate from architecture. This change was not arbitrary or willful, but rather was caused by the increasing complexity of design tools and construction technologies; a burgeoning range of new materials, equipment and techniques; and the introduction of inexpensive energy. Complexity continued to increase throughout the 20th century and continues today. With the architect transformed from master builder to lead design consultant, most HVAC&R engineering practices performed work predominantly as a subcontract to the architect, whose firm, in turn, was retained by the client. Concurrent with these trends and inexpensive energy, the twentieth century doctrine of buildings over nature, emerged.

Under the buildings-over-nature paradigm, the architect, as the prime consultant, first conceives the shell and interior design concepts and then retains the other design professionals (structural engineers, HVAC&R engineers, electrical engineers, lighting designers, interior designers, and others) to fit designs for which they are responsible within the constraints developed by the architect. (Not coincidentally, this hierarchy and sequence of engineering involvement mirrors the relative expense of the subsystems being designed.) This linear sequence of design often further reinforces the buildings-over-nature trend of relying on the brute force of sizable HVAC systems to maintain acceptable occupancy conditions at the cost of wasted energy and excessive building materials and thus, opportunities to integrate and optimize architectural elements with engineered systems are missed.

Today's ideal of the integrated design team from the initiation of a project and emerging interoperability using new technologies (BIM, energy simulations, etc.) encourages the design and operation of sustainable green buildings.

DESIGN PROCESS FOR SUSTAINABLE ARCHITECTURE

The process in designing energy-conserving architecture involves the following:

- Analyzing climatic conditions at the macro- and micro-site levels
- Identifying passive and active strategies applicable for the climate
- Designing the building envelope, shape, and orientation to reduce energy loads while providing comfort

- Determining room placement and grouping of areas into HVAC zones to reduce energy loads while, again, providing desired comfort levels
- Researching case studies and building energy load data for similar building types in the same climate zone to identify typical resource-consuming systems and benchmarked energy demands
- Selecting potential optimal strategies
- Applying computer-based simulation studies and related data sources to evaluate energy consumption against accepted metrics and benchmarks to demonstrate the performance-based sustainable outcomes for alternative designs.

The identified steps are applied in an integrative fashion, as described below.

Climate Studies and the Determination of Most Applicable Passive Strategies and Active Systems

After selecting a site, the next step is researching the macro- and microclimatic conditions. Desired collected data includes ranges of design dry-bulb and wet-bulb temperatures (or humidity ratios), solar exposure, and wind velocity (predominant direction and intensity). Seasonal as well as daily ranges should be noted.

The climate dictates if the building will be either heating dominated or cooling dominated. The dominant need will prescribe the relative importance of HVAC strategies. Once temperature and humidity data is collected, plot their average minimum and maximum ranges on a psychrometric chart or use an available computer program to plot this data with its frequency of occurrence. Overall strategies and means to extend the comfort zone for both passive and active systems have historically been identified and plotted on psychrometric charts by several researchers as bioclimatic charts. This was originally done by Victor Olgyay in 1963, modified by Baruch Givoni in 1969, and then further developed in collaboration by Baruch Givoni and Murray Milne in 1979.

The strategies identified on the bioclimatic charts include sun shading, thermal high mass with or without night flushing, direct evaporative cooling, natural ventilation cooling, passive solar direct gain with low or high mass, humidification, wind protection, conventional cooling, and conventional heating. Free university-developed software is also available to plot a year's set of points and then evaluate these strategies by the percentages of time they can most effectively be applied. See Figure 5-1 for an example using this software for the city of Santa Rosa, California.

Once identified, key strategies can then be prioritized based on their percentage contribution to retaining comfort zone conditions. If an active system is the highest-rated strategy for restoring the comfort zone, then reducing the energy load for such equipment is a high priority. Decisions on design strategies should also include occupancy and seasonal use conditions, (e.g., a summer camp project in a heating-dominated climate should not be designed for heating loads, as a climate study might initially suggest.)

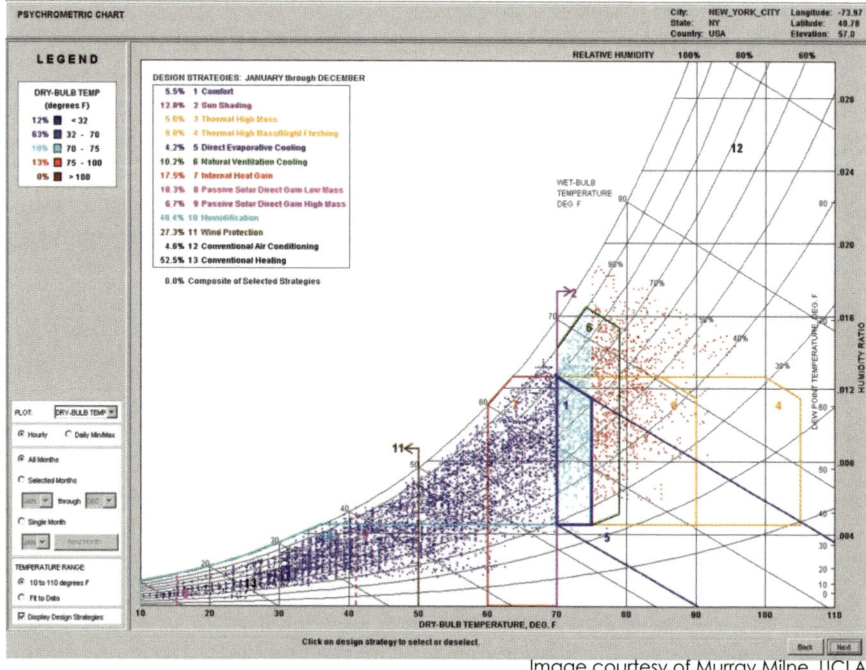

Image courtesy of Murray Milne, UCLA.

Figure 5-1 Climate Consultant Software (UCLA 2017) file report demonstrating the correlation of a location's year-long ambient air state points with applicable HVAC strategies.

Strictly speaking, bioclimatic analysis does not consider the building type, the building's intended use, nor projected internal load conditions. However some software developers are now making efforts in this area, (e.g., the UCLA-developed Climate Consultant Software includes internal heat gain as a design strategy.)

A complete climate analysis includes studies of the solar exposure, the seasonal effects of wind, and the availability of rainwater, which are briefly discussed in the next section and in more detail in Chapter 7, Sustainable Sites.

Building Form, Orientation, and Envelope Considerations

A challenge exists for architects to meet constraints in shape, orientation, and form, because these attributes are ones for which they have traditionally had great latitude and design freedom to explore varying options for aesthetic and building-use considerations. For sustainable design, architecture studies must be initiated with the knowledge of preferred shape/envelope/footprint ratios and orientation for optimization of energy use while providing comfort.

Decisions for designing a sustainable building's form, orientation, and envelope construction are informed by climate data, building site's existing landscape details, and potential optimization of natural on-site resources.

In predominately cold climates, a compact cube shape with a tight, well-insulated envelope is optimal to reduce heat loss. However, elongation in an east-west axis will increase exposure to the winter sun, thus increasing winter heat gain. Reducing overall glazing area, specifying high-performing windows and placing the bulk of windows on the equator facing exposure while minimizing glazing on other orientations is also recommended.

For a predominately hot climate, consider elevating the structure (for under airflow) and otherwise creating large shaded surface areas in contact with the outdoors. Create open courtyards, deep wraparound porches, and/or multiple building wings with several openings for ventilation paths aligned with prevailing wind patterns. For climates that contribute to overheating, architects also need to reduce direct solar gain on the equator-facing and west-facing surfaces. North- and east-facing surfaces rarely contribute to overheating from solar radiation.

The design of thermal mass and weight of building materials is a function of both climate and building use. "Elephant" building designs, those with high mass and heat absorbing materials, are suitable for buildings with high internal gains and in climates with large diurnal temperature swings. "Butterfly" buildings are those with lightweight materials suitable for building uses with low internal gains and moderate climates that would benefit from natural ventilation.

Charting the apparent sun paths for the year (e.g., on the 21st day of each month) and adjusting the irradiation values using the clear-sky radiation method described in the 2017 *ASHRAE Handbook—Fundamentals* is useful for determining the potential for passive building design and active solar energy collection. Sun angle data can be used for designing light shelves and other architectural strategies for daylighting to augment artificial light and also to design shading to reduce unwanted solar heat gain. See Chapter 10, Energy Sources for more information on using solar energy.

Wind rose charts can be consulted to find the seasonal direction and speed of the wind for a particular site. They are used to determine the feasibility for collecting wind energy and to analyze a site for protecting the building from unwanted wind exposure during the heating season or to determine options for channeling wind for increased ventilation for the cooling season. If wind rose charts are not directly available for a location, they can be developed with software or manually drawn using data from a local airport wind velocity-recording station. See Chapter 10, Energy Sources for more information on using wind energy.

Optionally, earth berms can be created or banks of trees or shrubs can be planted to help block wind to protect a building from wind-induced convective heat losses. Earth berms and shrubbery can also be used to redirect or focus wind for use in cooling or enhancing ventilation. Earth sheltering in contact with the building's envelope also diminishes both high and low temperature extremes. Sinking the building into the ground or designing the building to extend into a hill can provide cooling when outdoor temperatures are hot. Earth sheltering also acts as a buffer to the cold, when berming is located on the windward side of the building and particularly when the berms or hills are situated such that they allow the equator-facing side of the building to still be exposed (maximizing solar exposure). Use of a green roof provides multiple benefits for all climates: a tempering effect for cold or hot conditions, an offset for storm water which otherwise would be headed for a storm drain, and reduced heat island effect. See Chapter 7, Sustainable Sites, for more discussion on further optimizing site conditions.

Tight construction with high levels of insulation and careful placement of fenestration is central to energy conservation in cold climates. Glazing and daylighting should be considered in view of total building energy utilization. Based on studies with whole-building energy-analysis software in cold climates, providing an overuse of daylighting by increasing windows and skylights will result in an overall increase in the total energy use caused by the low thermal performance of glazing systems. The decisions regarding envelope design are central to reducing energy loads and take precedence over alternative energy-related power generation as a priority. Therefore, costly photovoltaic energy sources should not be used to power artificial lighting when daylighting can be provided to reduce electrical lighting loads.

Rainfall data can be used to determine rainwater collection potential for use in irrigating plants, toilet flushing, or for other uses. This is an opportunity for both saving both energy and conserving water. Rainfall data is also used for planning site water retention design schemes such as for rain gardens and bioswales. The collection, use, and maintaining of water on a site reduces or eliminates the need for storm water drain systems infrastructure.

For structures over three stories, design using ANSI/ASHRAE/USGBC/IES Standard 189.1, *Standard for the Design of High Performance Green Buildings.* This standard specifies minimum R-values, maximum U-factors, and F-factors for building-envelope insulation and windows based on climate zone.

Placement and Zoning of Building Spaces

Before placing and zoning internal spaces, consider minimizing the building size by designing spaces that are flexible or dual purposed (e.g., a lecture room also serving as a dining room). Considering future flexibility is also sustainable by extending a building's useful life. Room placement and zoning for heating and cooling goes hand in hand to provide a big impact on energy use and occupant comfort.

Zoning of the spaces is most effective when room placement and their relative positions are optimized so zoning can proceed based on exterior wall orientations, then on function, and finally on occupancy and equipment diversity.

Optimize room placement by considering thermal energy needs and applying the "thermal buffering" principle. Locate high thermal energy need areas (laboratories, natatoriums, etc.) in interior spaces surrounded by areas requiring less thermal energy (corridors, closets, etc.). This action reduces thermal losses from the more sensitive conditioned occupied spaces. Place rooms where people are sedentary on the equator-facing side of buildings. Place corridors, closets, storage rooms, Datacom, and utility rooms on non-equator-facing sides of buildings.

A formal thermal buffer zone can also be provided by constructing a double-skin façade system (DSF). A DSF, also called a *dual-skin wall system* or *envelope system,* is constructed by providing a glazed wall a short distance from a conditioned-space-enclosing wall. This is usually constructed on the equator-facing side of a building but may also be installed on the opposite facing outside wall. Air movement due to natural convection and stack effect between the dual wall skins can contribute to reducing both the cooling and heating load. In summer operation at the Yazaki World Headquarters in Canton, MI, air between the outer and inner walls rises by natural convection, pulling stale air from conditioned areas (through open windows in the inner wall) up and out of the building. In winter, air is introduced at a low level, is solar heated and rises between the walls (with inner windows closed), providing a thermal blanket. In some cases, tempered air is ducted into a conventional air handler or heat pump to eventually supply the conditioned space with heat.

There are many variations in DSF design. The interstitial space varies in dimension and design. Some DSFs have a solar tower at the top to increase the convective action. Movable blinds or photovoltaic panels may be added. Functioning similarly, a perforated, black-painted metal panel can be mounted on a sunny wall to provide tempered ventilation air. Many retrofitted commercial buildings in Europe and elsewhere have successfully used DSF principles.

Use of Case Studies and Benchmarking

Referenced buildings can be used as benchmarks to predict typical energy consumption for a building's various purposes (e.g., energy used for heating cooling, lighting, domestic hot water, and specialized equipment.) Case studies are readily available from professional organizations such as ASHRAE, AIA, and USGBC. When choosing benchmarks, the buildings selected should be similar in function, size, expected occupancy, and climate zone.

Using benchmarks along with microclimatic data should narrow and define the priorities and lead the designer to identifying the most viable design strategies for resource conservation. Benchmarking with high-performing reference buildings will provide the designer with exemplary targets. Once energy use targets are defined, the designer can examine alternative energy generation to meet them.

Selecting Optional Strategies Using Alternative-Energy Potential

Developing alternative-energy strategies begins with examining the magnitude and availability of solar and/or wind energy to the site. The use or application of alternative-energy technologies should be paired with high-load demand conditions, as well as cost considerations. If a building type has a high domestic hot-water demand load and is also located in a climate with strong solar radiation conditions, then evacuated tubes would be a prudent choice to reduce fossil fuel use for heating water. If a building has a need to be shaded from excessive solar radiation and high electrical loads, using photovoltaic panels as part of shading system on a southern wall would be a possible solution. Solar-air and water-based panels, evacuated tubes, photovoltaic panels and wind turbines are all alternative-energy systems that can be used to reduce fossil fuel use. Alternative-energy costs can be made affordable for clients when tax incentives, rebates, or low-interest loans are available and should be part of initial feasibility studies. See Chapter 10, Energy Sources for more on alternative-energy sources.

Although not an alternative-energy source, the use of geothermal heat pump systems is another way to reduce fossil fuel use.

Prioritizing Design Strategies Using Computer Simulations

Design strategies need prioritization based on analysis. Evidence-based design or performance-based outcomes require definable problems based on data. The increasing levels of adoption of performance-based metrics by municipalities, states, and corporations (e.g., those used for building labeling) will eventually place responsibility on designers to validate design proposals with appropriate data and whole building energy simulation.

Several computer programs are available to simulate alternatives and to assist in determining an optional building form, alternative footprints, orientations, shading, insulation levels, and glazing distributions for minimizing energy use. Building energy simulations are now possible with interoperable databases linking BIM 3D models with object-oriented software and embedded information where building data is exported directly into the simulation software. Unfortunately, not all building system types, use conditions, and environmental characteristics can be modeled using such software; for example computer programs simulating the effects of mass are more difficult to find. See the section titled, "The Role of Energy Modeling During Conceptual Design" in Chapter 6, Conceptual Engineering Design—Load Determination, for an in-depth discussion of this topic.

INTENTIONS IN ARCHITECTURE AND BUILDING SYSTEM DEVELOPMENT

Several philosophies and approaches may be applied by building designers for sustainable architecture. These vary from theoretical and abstract to those with more measurable, defined outcomes, and several may be combined to meet sustainability goals. Designers hold the safety and welfare of building occupants in their hands and, in striving for sustainability, the welfare of the environment with its effect on humans and humans' effect on the earth. The following philosophies and approaches which follow are offered for consideration by designers.

Ecosystem/Bioregionalism. Understand and design for local ecological systems above human development actions and prevent harmful effects by preserving ecologically sensitive areas and protecting endangered species.

Human Vitality. Design for human needs based on performance and productivity but constrained by a proenvironmental outlook. For example, when logging efforts in underdeveloped countries lead to soil erosion, flooding, and other negative effects on the ecosystem and human occupants, constraints to development are needed.

Circle of Life Principle. Minimize waste just as accomplished in natural processes; that is, waste in one process becomes food in another. The *cradle-to-cradle* concepts by William McDonough (2002) exemplify this approach.

Biomimicry. Consider nature as an example and model with its symbiotic relationships between organisms, protective skins, regenerative nature, and continuous nurturing processes. Solar thermal collectors imitate the polar bear. The dark- or selective-surfaced absorber plate absorbs solar energy just like the polar bear's black skin (yes, it is really black!). The bear's transparent and clear hollow-core hair is replaced by an insulating air space and glazing, which assists in allowing relatively short wave sunlight through while reducing long wave infrared radiative losses. The open-source "Ask Nature" website through the Biomimicry 3.8 Institute explores several ideas for applying biomimicry (www.asknature.org). In architecture, examples of applying biomimicry can be found in organic architecture initiated by Frank Lloyd Wright in the late 19th century and later by the Japanese Metabolists in the 1970s. A recent example is in the literal "organic growing of houses" by Mitchell Joachim. He has a molecular biology lab in his architecture office and prints three-dimensional, test-tube-grown products. Videos of his lectures can be found on the Internet by searching his name with "growing houses" (Joachim 2010).

Conservation and Renewable Resources. Energy is a critical resource. We must promote the use of clean, renewable, and non-carbon-emitting energy sources. The basic tenants of the environmental movement are reduce, reuse, and recycle materials, and use renewable rather than fossil energy.

Holistic Theory. Collaborate with other disciplines and promote integrated design practice to guide the process. Proponents attempt to challenge the rules, set aside conventional solutions, and use innovation in design.

Sustainable Design Intent. Use basic sustainable processes to conserve energy and water. This is best accomplished using a metrics-based approach that clearly defines sustainable criteria, measures, and outcomes. Implement sustainability standards (e.g., Standard 189.1) and apply building labels to measure success (e.g., ASHRAE Building EQ, LEED, Green Leaf, Green Globes, Go Green, GB Tools,

ENERGY STAR, etc.) Refer to Chapter 2, Green Building Rating Systems, Standards, and Other Guidance for more detailed descriptions of these resources.

Also consider using the 10 "Design Strategies for Sustainability" defined by the man who coined the term *sustainability*, John Tillman Lyle, in his book, *Regenerated Design for Sustainable Development* (Lyle 1996).

These strategies overlap those stated above and include the following:

- Let nature do the work.
- Consider nature as both model and context.
- Aggregate rather than isolate.
- Match technology to the need.
- Seek common solutions to disparate problems.
- Shape the form to guide the flow.
- Shape the form to manifest the process.
- Use information to replace power.
- Provide multiple pathways.
- Manage storage (as a key to sustainability).

REFERENCES AND RESOURCES

Published

Givoni, B. 1969. "Man, Climate and Architecture". Elsevier Architectural Science Series.

Joachim, M. 2010. Don't build your home, grow it! Presented at TED2010. www.ted.com/talks/mitchell_joachim_don_t_build_your_home_grow_it.

Lyle, J.T. 1996. *Regenerative design for sustainable development*. Hoboken, NJ: John Wiley and Sons.

McDonough, W. 2002. *Cradle to cradle: Remaking the way we make things*. New York: North Point Press.

Milne, M., and B. Givoni. 1979. *Energy conservation through building design*. Chapter 6, Architectural Design Based on Climate. Donald Watson, ed. New York: McGraw-Hill.

Olgyay, V. 2016. Design with Climate: Bioclimatic Approach to Architectural Realism. (New and Revised to the 1963 edition). Princeton, NJ: Princeton Press

Online

UCLA, Climate Consultant Software http://www.energy-design-tools.aud.ucla.edu/climate-consultant/request-climate-consultant.php.

CHAPTER SIX

Conceptual Engineering Design—Load Determination

Economic benefits can be realized by focusing efforts on reducing building loads. Reducing loads allow the designers to select smaller, less expensive HVAC equipment, which saves money and the reduced HVAC system's energy consumption lowers operational costs throughout the lifetime of the building.

Traditional load determination methods, such as the cooling load temperature difference method or rules of thumb, are rough, first approximations that are based on old correlations or simplified heat transfer calculations. Designing low-energy buildings requires the engineer to have a thorough understanding of the dynamic nature of the interactions of the building with the environment and the occupants. ASHRAE's *Load Calculation Applications Manual* (Spitler 2014) focuses on the most current load calculation methods: the heat balance method, and the radiant time series method. To optimize the design, detailed computer simulations that use these methods allow the engineer to accurately model the major loads and interactions.

Loads can be divided into those stemming from the envelope and those from internal sources. Envelope loads include the impact of the architectural features; heat and moisture transfer through the walls, roof, floor, and windows; and infiltration. Internal loads include lights, equipment, people, and process equipment. Examine all of the loads in two ways: separately, to determine their relative impact, and together, to determine their interactions.

The engineer must also understand the energy sources and flows in the building and their location, magnitude, and timing. Then they can be creative in coming up with solutions. The charts in Figure 6-1 show how average energy use breaks down in commercial and residential buildings. As designs are developed, breakdowns such as these should be kept in mind so that the energy-using areas that matter most are given priority in the design process.

When trying to minimize energy use in buildings, the first step is to identify which aspects of building operation offer the greatest energy-saving opportunities. For example, as shown in Figure 6-1, space heating and cooling, lighting, and water heating make up a combined 51% of energy use in the

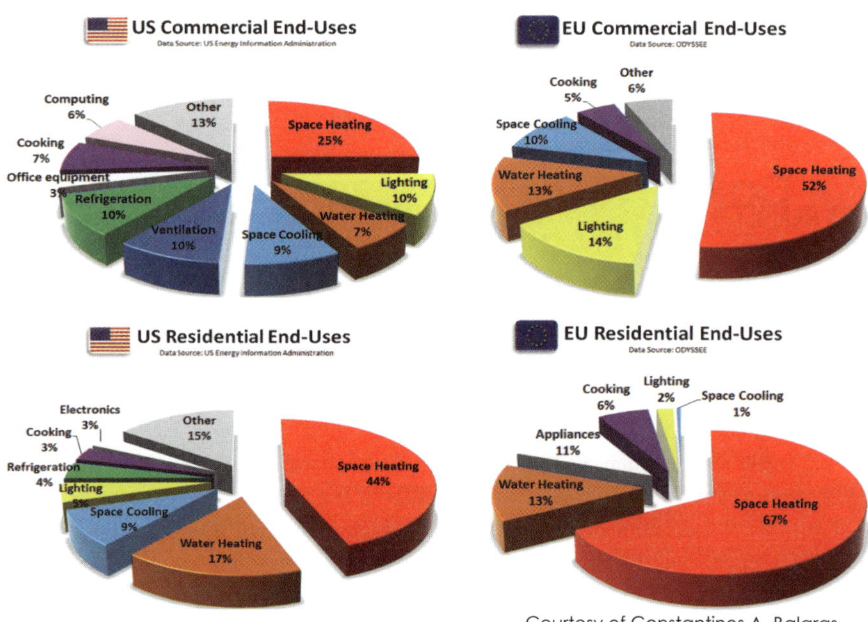

Courtesy of Constantinos A. Balaras

Figure 6-1 Average commercial and residential building energy use in the United States.

average commercial building in the United States. Similar breakdown is also provided for European buildings. Beyond the anticipated diversity of specific end-uses for different commercial building types, lighting and HVAC constitute the most important energy end uses; overall, space heating and hot water remain the main end uses.

Another important aspect of this exercise in the design process is that it allows the design team, along with the owner, to set the appropriate goal for energy usage. Many organizations and institutions have adopted the Architecture 2030 Challenge (see the "References and Resources" section at the end of the chapter for more information). Conceptual load calculations help to determine if a specific building design concept meets these requirements.

To find energy end-use statistics for many types of buildings (e.g., office, education, health care, lodging, retail, etc.), based on building location, age, size, and principal energy sources, consult the Commercial Building Energy Consumption Survey data published by the U.S. Energy Information Administration at www.eia.gov/consumption/commercial. In Europe, the ODYSSEE database (www.odyssee-mure.eu) contains detailed data on the energy consumption drivers by end-use (Figure 6-1) and CO_2-related indicators, and is regularly updated by national representatives, such as energy agencies or statistical organizations, from all 28 EU member states and Norway. Various data resources are also available from the Buildings Performance Institute Europe (BPIE) under the BPIE data hub for the energy performance of buildings (http://bpie.eu/focus-areas/buildings-data-and-tools/.).

THE ROLE OF ENERGY MODELING DURING CONCEPTUAL DESIGN

To achieve the greatest possible energy reductions, energy modeling should begin during the early conceptual stage of the design process. Decisions about the building's form, orientation, percentage of glazing on each façade, and construction materials are generally made during this stage and can have a profound impact on how the building will perform.

Often, the architectural designers will develop several design options to present to the client. Perhaps one option has an elongated shape, another may be L-shaped, and a third may be rectangular with an interior courtyard. The options give the client the opportunity to provide feedback on the proposed design and interior space planning. Of course, if an energy use assessment of each option can be provided at this stage, then that will be another very important factor for the team to consider.

Commercially available energy simulation software can now give design teams the capability to directly read geometry from 3-D design models. This is a great time saver, and allows the design team to get feedback quickly on the performance of proposed designs rather than having to re-create geometry each time the design changes. For designs that have very complex geometry, this feature becomes especially important.

Once the geometry is created, the energy model can be populated with information on the loads. At the conceptual design phase, exact information is not yet likely to be determined, so engineering judgment is important to provide a realistic assessment.

Suppose that the design team is working on a new office building. A massing model may be developed; however, it is unlikely that specific office locations have been determined, so at this stage it would be reasonable for the energy model to be divided into simple block zones (for each exterior exposure) and interior zones. ANSI/ASHRAE Standard 62.1, *Ventilation for Acceptable Indoor Air Quality.* (ASHRAE 2016a) provides data on typical occupancy densities and ventilation requirements, and ANSI/ASHRAE/IES Standard 90.1, *Energy Standard for Buildings Except Low-Rise Residential Buildings* (ASHRAE 2016b) provides data on typical lighting and equipment power densities for a wide range of different building types. Many energy simulation software tools have this information stored in a library (or allow creation of custom libraries) to allow users to quickly apply it to the energy model. Once this information is applied to the model, the design team can perform iterative simulations to determine the best performing massing scheme, the optimal orientation, the impact of different construction materials, and the impact of different window-to-wall ratios.

Because some assumptions must be made at this stage, the energy simulations may not predict energy consumption to the exact Btu (joule). However, if our assumptions are sound, we can get a good sense of the impacts of design decisions at an appropriate order of magnitude. Moving forward, we can be confident that the right concepts are in place and can then assess them in greater detail as the design progresses.

DETERMINING THE LOAD DRIVERS WITH PARAMETRIC SIMULATIONS

To gain an understanding of how the proposed design of the building affects energy use, the engineer should perform a series of parametric simulations. This process involves sequentially removing each load from the energy balance individually to determine the effect on overall energy use. The simulations that show the greatest energy reductions identify which loads are driving the building's energy consumption.

For instance, to effectively remove wall conductive heat transfer from the energy balance, set the wall thermal resistance to a very high value (such as R-100 [R-17.6]). Run the simulation model, and note the building energy requirements. Then, reset the wall insulation back to the actual R-value and proceed to the next parameter. Continue to do this individually with the floors, roofs, and windows. Then remove the window solar gain, daylighting, and other site-specific shading that may be involved. Assess the impact of infiltration and the effect of ventilation air by setting them sequentially to zero.

Parametric simulations should be completed for the internal loads in a process similar to that described in Chapter 5. Set the lighting load to zero, set the equipment load to zero, and then take all the people out of the building. In this manner, the engineer can understand what is driving the energy use in the building.

Plot the results for each case by annual energy use or by peak load to compare the impacts, as illustrated in Figure 6-1.

Look for creative solutions to minimize the impact of each load, starting with those with the largest impact, or those that are the easiest to implement. From the parametric analysis example in Figure 6-2, adding underfloor air distribution results in the largest impact on annual energy use, so it should be one of the first solutions considered. Alternatively, a solution such as high-efficiency lighting may not result in as large of an impact, but may be easier to implement.

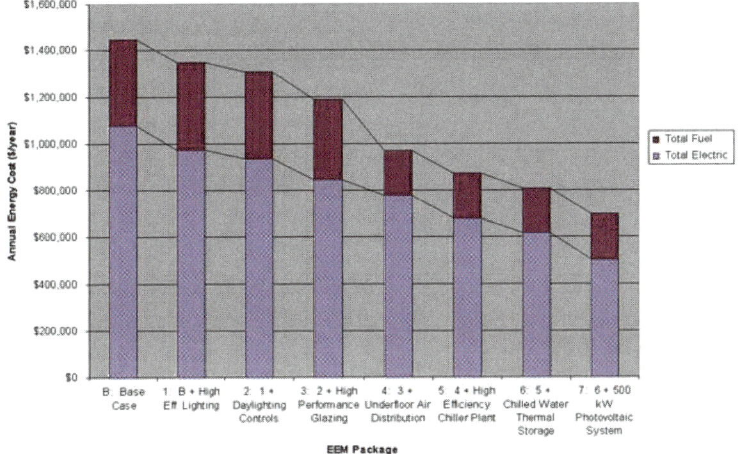

Image courtesy of CTG Energetics, Inc.

Figure 6-2 Parametric analysis for comparing various sustainability options.

ENERGY IMPACTS OF ARCHITECTURAL FEATURES

It is important to collaborate with the architects to help guide decisions that will improve the thermal envelope's performance. Determine the greatest sources of heat gains and losses through the building's skin, and look for opportunities to minimize the effect.

Solar radiation through the windows can be one of the largest gains during the summer cooling season. Providing external shading can effectively minimize these gains. Horizontal shading, or overhangs, on the south face of the building will block the high sun during the day for much of the year, and vertical shading on the east and west can reduce gains early and late in the day as the sun rises and sets. However, the microclimate around the building should be completely analyzed to ensure any external shading actually works as intended. For example, in areas where the summer months are mostly foggy (e.g., Santa Barbara, CA), providing external shading for high angle summer sun may not be very effective.

Another means to control solar gains is to specify high-performance glazing with a low solar heat gain coefficient. These glazing options can be either tinted glass, or spectrally selective glass. The benefit of the spectrally selective option is that they have a clear appearance and a high visible light transmittance, allowing for greater daylighting potential and views to the exterior. However, they are generally the more costly option.

In climates where heating the building is of primary concern, it can often be beneficial to allow solar heat gain into the building as it acts as a free heating source. In climates where heating and cooling are both significant concerns, use a building simulation software program to understand the trade-offs associated with various solar heat gain coefficient values.

Conduction gains through the building skin are also significant in the overall heat balance equation. Where possible, try to reduce the window-to-wall ratio. Even the best-performing glass selections cannot compare to the thermal characteristics of an insulated wall.

Glass selections with low U-factors will reduce the heat gains and losses. Also, try to minimize the U-factors of the walls and roof by using insulation with high R-values. Where thermal bridging may occur due to structural elements of the building, provide a continuous interior layer of insulation to minimize this effect.

It is very important for designers to look at the entire window assembly properties, including the frame and the spacers and not only for the glazing. Good glazing U-factors can be severely compromised by not having thermally broken frames.

Lastly, try to minimize infiltration gains from leakage of untreated air into the building. Positively pressurizing the building with the HVAC systems will minimize this impact; however, stack-driven and wind-driven pressure differentials will still cause exterior air to infiltrate the building. Opening and closing of doors to the exterior will also contribute to infiltration gains. Specify a continuous air barrier to minimize air infiltration through the building skin, and install entry vestibules or revolving doors to minimize airflow into the building through open doors.

THERMAL/MASS TRANSFER OF ENVELOPE

Basic, steady-state energy transfer through building envelopes is well known, well understood, and easy to calculate. Increased R-values are certainly beneficial in heating climates and can also help in cooling-load-dominated climates. Use as a minimum the recommended amounts spelled out in ASHRAE's energy standard (Standard 90.1 [ASHRAE 2016b], latest approved edition). While values above those recommended can be beneficial in simple structures, high values can be counterproductive over a heating/cooling season in more complex structures. It is always wise to evaluate R-value benefits through the application of load and energy simulation programs.

The effects of thermal mass are sometimes not as easy to gauge intuitively (see Figure 6-3). Therefore, the previously mentioned simulation programs, when properly applied, are very useful in this regard. If thermal mass is significant in the building being planned (or if increased mass would be easy to vary as optional design choices), then such programs are essential for evaluating the flywheel effects of thermal mass on both loads and longer-term energy use.

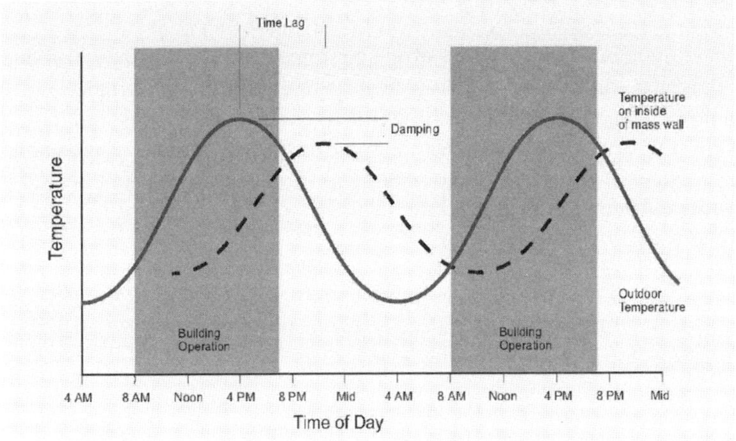

Image courtesy of U.S. Department of Energy Efficiency and Renewable Energy.

Figure 6-3 The effects of thermal mass.

GreenTip #6-1 describes a technique for combining nighttime ventilation with a building's thermal mass to achieve load and/or energy savings.

ENGINEERING INTERNAL LOAD-DETERMINING FACTORS

Internal loads are significant contributors to the energy balance, but can be effectively reduced during the design phase. For office buildings and other buildings in mild climates, lights are usually the major culprits. Therefore, optimize the daylighting and electric lighting design. Refer to Chapter 11 for a detailed discussion on lighting systems.

Likewise, evaluate office equipment loads. Make recommendations about the effects of the choices of computers, monitors, printers, and other types of equipment. In many offices, this equipment is left on all night. An office building should have very few loads when the building is unoccupied. Leaving an office full of equipment powered up all night and on weekends can easily add up to large energy consumption.

For example, assume one 34,000 ft^2 (3159 m^2) office building has nighttime plug loads of 10 kW, and assume the building is unoccupied for 14 h/day during the week and 24 h/day on the weekend. This adds up to around 6000 h/yr and 60,000 kWh—or \$4200 (at \$0.07/kWh)—of electricity that could easily be reduced. It is important for the engineer to bring up these issues during the design stage (and later, during the operation) of the building, because no one else may be paying attention to such details. Designing and building a great building is only half the job; operating it in the correct manner is the other—and often more important—half. The engineer can make better operation possible by designing piping, wiring, and controls capable of easily turning things off when not being used.

Educate owners about efficient office equipment and appliances to reduce plug loads, covering such things as flat screen computer monitors, laptops vs. desktop central processing units, copy machines, refrigerators, and process equipment (e.g., lab equipment, health care machinery, etc.). Consider measuring usage in one of the clients' existing buildings to get an accurate picture of load distribution and population profiles. Attempt to develop a total building electric load profile for every minute of one week.

ASHRAE GreenTip #6-1

Night Precooling

GENERAL DESCRIPTION

Night precooling involves the circulation of cool air within a building during nighttime hours with the intent of cooling the structure. The cooled structure is then able to serve as a heat sink during the daytime hours, reducing the mechanical cooling required. The naturally occurring thermal storage capacity of the building is thereby used to smooth the load curve and for potential energy savings. More details on the concept of thermal mass on building loads are included in Chapter 5, Architectural Design and Planning Impacts.

There are two variations on night precooling. One, termed *night ventilation precooling*, involves the circulation of outdoor air into the space during the naturally cooler nighttime hours. This can be considered a passive technique, except for any fan power requirement needed to circulate the outdoor air through the space. The night ventilation precooling system benefits the building IAQ through the cleansing effect of introducing more ventilation air. With the other variation, *mechanical precooling*, the building mechanical cooling system is operated during the nighttime hours to precool the building space to a setpoint that is usually lower than that of normal daytime hours.

Consider these key parameters when evaluating either concept:

- Local diurnal temperature variation
- Ambient humidity levels
- Thermal coupling of the circulated air to the building mass

The electric utility rate structure for peak and off-peak loads also is important when determining cost-effectiveness, in particular for a mechanical precooling scheme.

A number of published studies show significant reductions in overall operating costs by the proper precooling and discharge of building thermal storage. The lower overall costs result from load shifting from the day to the nighttime with its associated off-peak utility rates. For example, Braun (1990) showed significant energy cost savings of 10% to 50% and peak power requirements of 10% to 35% over a traditional nighttime setup control strategy. The percent savings were found to be most significant when lower ambient temperatures allowed night ventilation cooling to be performed.

For a system incorporating precooling to be considered a truly green design concept, the total energy used through the entire 24-hour day should be lower than without precooling. A system that uses outdoor air to do the precooling only requires the relatively lower power needed to drive the circulation fans, compared to a system that incorporates mechanical precooling. Electrical energy provided by the utility during peak demand periods also may be dirtier than that provided during normal periods, depending on the utility and circumstances.

The system designer needs to be aware of the introduction of additional humidity into the space with the use of night ventilation. Thus, the concept of night ventilation precooling is better suited for drier climates. A mechanical nighttime precooling system will prevent the introduction of additional humidity into the space by the natural dehumidification it provides, but at the expense of greater energy usage compared to night ventilation alone.

Both variations (i.e., night ventilation and night mechanical precooling) are not 100% efficient in the thermal energy storage in the building mass, particularly if the building is highly coupled (thermally) with the outside environment. Certain building concepts used in Europe are designed to increase the exposure of the air supply or return with the interior building mass (see, for example, Andersson et al. [1979]). This concept will increase the overall efficiency of the thermal storage mass.

For either type of system, the designer must carefully analyze the structure and interaction with the HVAC system air supply using transient simulations. A number of techniques and commercially available computer codes exist for this analysis (Balaras 1995).

WHEN/WHERE IT'S APPLICABLE

Night precooling would be applicable in the following circumstances:

- When the ambient nighttime temperatures are low enough to provide sufficient opportunity to cool the building structure through ventilation air. Ideally, a low ambient humidity level would also occur. A hot, dry environment, such as the southwestern United States, is an ideal potential area for this concept.

- When the building occupants would be more tolerant of the potential for slightly cooler temperatures during the morning hours.
- When the owner and design team are willing to include such a precooling system concept and to commit to (1) a proper analysis of the dynamics of the building's thermal performance and (2) the refinement of the control strategy (upon implementation) to fine-tune the system performance.
- More massive buildings, or those built with heavier construction materials such as concrete or stone, have a greater potential for benefits. Just as important, is the interaction of the building mass with the building internal and HVAC system circulating air. This interaction may allow for more efficient transfer of thermal energy between the structure and the air space.

PRO

- Night ventilation precooling has good potential for net energy savings because the power required to circulate the cooler nighttime air through the building is relatively low compared to the power required to mechanically cool the space during the daytime hours.
- Mechanical precooling could lead to net energy savings, although there will likely be a net increase in total energy use due to the less-than-100% thermal energy storage efficiency in the building mass.
- Both variations require only minor, if any, change to the overall building and system design. Any changes required are primarily in the control scheme.
- Night ventilation can provide a better IAQ environment, due to increased circulation of air during the night. A greater potential exists with the ventilation precooling concept. Both will be better than if the system was completely shut off during unoccupied hours.

CON

- Temperature control should be monitored carefully. The potential exists for the building environment to be too cool for the occupants' comfort during the early hours of the occupied period. This will result in increased service calls or complaints and may end with the night precooling being bypassed or deactivated.
- The increased run time on the equipment could lead to lower equipment life expectancy or increased frequency in maintenance. Careful attention should be given to the resulting temperature profile through the day during the commissioning process. Adjustments to the control schedule may be necessary to keep the building within the thermal comfort zone.
- Proper orientation must be given to the building operator so he or she can understand how the control concept affects the overall system operation throughout the day.
- Future turnovers in building ownership or operating personnel could negatively affect how successfully the system performs.
- Occupants would probably need at least some orientation so that they would understand and be tolerant of the differences in conditions that may prevail with such a system. Future occupants may not have the benefit of such orientation.

KEY ELEMENTS OF COST

The following provides a possible breakdown of the various cost elements that might differentiate a nighttime precooling scheme from a conventional one and gives an indication of whether the net cost for the precooling option is likely to be lower (L), higher (H), or the same (S). This assessment is only a perception of what might be likely, but it may not be correct in all situations. There is no substitute for a detailed cost analysis as part of the design process. The listings below may also provide some assistance in identifying the cost elements involved.

First Cost

- Mechanical ventilation system elements S
- Architectural design features S
- System controls H
- Analysis and design fees H

Recurring Cost

- Energy for mechanical portion of system
    - Ventilation precooling L
    - Mechanical precooling S/H
- Total cost to operate cooling system L
- Maintenance of mechanical ventilation and cooling system S/H
- Training of building operators H
- Orientation of building occupants H
- Commissioning cost H
- Occupant productivity S

SOURCES OF FURTHER INFORMATION

Andersson, L.O., K.G. Bernander, E. Isfält, and A.H. Rosenfeld. 1979. Storage of heat and cool in hollow-core concrete slabs. Swedish experience and application to large, American style buildings. Second International Conference on Energy Use and Management, Lawrence Berkeley National Laboratory, LBL-8913.

Balaras, C.A. 1995. The role of thermal mass on the cooling load of buildings. An overview of computational methods. *Energy and Buildings* 24(1):1–10.

Braun, J.E. 1990. Reducing energy costs and peak electrical demand through optimal control of building thermal storage. *ASHRAE Transactions* 96(2):876–88.

Keeney, K.R., and J.E. Braun. 1997. Application of building precooling to reduce peak cooling requirements. *ASHRAE Transactions* 103(1):463–69.

Kintner-Meyer, M., and A.F. Emery. 1995. Optimal control of an HVAC system using cold storage and building thermal capacitance. *Energy and Buildings* 23:19–31.

Ruud, M.D., J.W. Mitchell, and S.A. Klein. 1990. Use of building thermal mass to offset cooling loads. *ASHRAE Transactions* 96(2):820–29.

ASHRAE GreenTip #6-2

Night-Sky Cooling

GENERAL DESCRIPTION

When using chilled-water-based cooling systems, peak load reduction opportunities exist through the use of thermal storage systems. Many thermal storage systems use installed chiller capacity to charge the storage system during off-peak periods. However, while this takes advantage of cheaper energy during off-peak periods, it can actually increase overall energy consumption. Yet, as nighttime utility generation capacity is generally from cleaner sources than during peak periods, thermal storage systems can be an effective way to reduce greenhouse gas emissions as well as save operating costs.

Another good way to create a source of cooling is to use a roof spray cooling system. This system would spray water on a roof that is cooled by evaporation and radiation heat transfer. Such a system would work best in areas with limited cloud cover at night and with lower humidity, thus limiting where it would be effective. The consumption of water is also a concern that needs to be balanced with the energy effectiveness.

The cooled water from the roof spray system is collected from the roof surface, filtered and stored for use the following day. Researchers at UC Davis' Western Cooling Efficiency Center have reported Annual Energy Efficiency Ratios (EERs) of 40 to 150 for night-sky radiant cooling roof spray systems, versus EERs of 13 to 20 for ground-source heat pump systems (generally considered a highly efficient system). One manufacturer of these systems claims EERs of 50 to 100.

The roof spray cooling system consists of a piping grid and conventional water spray nozzles located on the roof surface and connected to a water pumping and filtration system. During the cooling season, water is sprayed over the roof surface at night. This water is first cooled by evaporation during the spray process and then further cooled by radiation to the night sky. The water is recirculated through this spray-cooling process until it is cooled and then stored for later use in providing cooling to the building. Studies have shown that the minimum water temperature reached during operation can be 5°F to 10°F (3°C to 5°C) less than the minimum dry-bulb air temperature reached at night.

WHEN/WHERE IT'S APPLICABLE

Research data suggests that, where summer dry-bulb temperatures fall below 65°F (18°C) at night, system capacities can be expected to be approximately 25 ton-hours per 1000 ft^2 of roof area (300 Btu/ft^2 [0.95 kWh/m^2]). Another report places capacities at approximately 290 Btu/ft^2 (0.9 kWh/m^2).

Cooling via night-sky radiation relies on the fact that the effective temperature of the night sky can be significantly cooler than the ambient air temperature. Thus, an object set outside at night will radiate more heat to the night sky than it will absorb from the night sky and the net loss of heat will cause the object to cool below the surrounding air temperature.

Most of the cooling via night-sky radiation occurs as the water film left on the roof from the spray process exchanges radiant heat with the night sky. The typically large roof areas available allow for substantial radiative cooling at relatively little additional building cost. The effective nighttime sky temperature is a function of both the dry-bulb air temperature and the humidity content of the air, with higher humidity reducing the differential between the effective sky temperature and dry-bulb air temperature. Although not an easily measured parameter, sky temperature data are available in typical meteorological year (TMY) weather data files or can be estimated from other weather parameters using established algorithms.

PRO

- Very high efficiency cooling system.
- Enhances redundancy of systems.

CON

- Higher first cost, unless chiller capacity can be reduced by the peak tons available from the chilled-water storage system.
- Sensitive to quality of the potable water used for makeup (e.g., clogging of spray nozzles).
- Requires good design of stratified thermal storage tank.
- Available capacity subject to weather patterns.

KEY ELEMENTS OF COST

The following provides a possible breakdown of the various cost elements that might differentiate a night sky radiant cooling system from a conventional one and an indication of whether the net cost for an option is likely to be lower (L), higher (H), or the same (S). This assessment is only a perception of what might be likely and it obviously may not be correct in all situations. There is no substitute for a detailed cost analysis as part of the design process. The listings below may also provide some assistance in identifying the cost elements involved.

First Cost

- Chilled-water storage system — H
- Interface of chilled-water storage to cooling system — H
- Roof spray system — H
- Controls and instrumentation — H

Recurring Cost

- Energy consumption — L
- Testing and balancing (TAB) — S
- Maintenance — S/H
- Commissioning — H

SOURCES OF FURTHER INFORMATION

Allen, C.P. 1977. *The estimation of atmospheric radiation for clear and cloudy skies*. Master's thesis. San Antonio, TX: Trinity University.

Berdahl, P., and R. Fromberg. 1982. The thermal radiance of clear skies. *Solar Energy* 29(4):299–314.

Bourne, R., and C. Carew. 1996. Design and Implementation of a Night Roof-Spray Storage Cooling System. Proceedings of the 1996 ACEEE Summer Study on Energy Efficiency in Buildings, Washington, D.C.

Bourne, R. and B. Chen. 1989. Full year performance simulation of a direct-coupled thermal storage roof (DCTSR), Proceedings of the annual meeting of the American Solar Energy Society, Denver, CO.

CEC. 1992. Cool Storage Roofs ETAP Project, Consultant Report. Prepared by the Davis Energy Group. CEC P500-92-014, Davis, CA: California Energy Commission.

Chen, B., J. Kasher, J. Maloney, and G.A. Girgis, and D. Clark. n.d. *Determination of the clear sky emissivity for use in cool storage roof and roof pond applications*. Omaha, NE: Passive Solar Research Group, University of Nebraska.

Martin, M., and P. Berdahl. 1984. Characteristics of infrared sky radiation in the United States, *Solar Energy*, 33:321–326.

PNNL. 1997. Technology installation review. *WhiteCap™ roof spray cooling system: cooling technology for warm, dry climates*. Richland, WA: Pacific Northwest National Laboratory.

ASHRAE GreenTip #6-3

Plug Loads

GENERAL DESCRIPTION

As the efficiency of various energy-using systems (i.e., HVAC, lighting, or building envelopes) improves, the impact of plug loads on the total annual building energy usage increases. This is accelerated by the rapidly growing use of computers and other electronic equipment in most building types. The purpose of this GreenTip is to identify some key issues to watch for that are relative to plug loads.

TYPICAL PLUG LOAD ENERGY BUDGETS

Historically, plug loads in office buildings have been designed to support connected loads in the 2 to 5 W/ft^2 (20 to 50 W/m^2) range. For laboratories or computing-intensive buildings, this can go up to 10 to 15 W/ft^2 (100 to 150 W/m^2). The electrical distribution system must be designed to support this level of load, in compliance with the *National Electrical Code*© (NFPA 2014). Yet the actual diversified load exerted on the building electrical demand and upon the HVAC system will more likely be 1 to 1.5 W/ft^2 (10 to 15 W/m^2) or even less for highly efficient installations, as described below.

DIFFERENCE BETWEEN CONNECTED LOAD AND EFFECTIVE LOAD

The reason for this disparity is the fundamental difference between connected load for electrical equipment and the effective load imposed on the HVAC systems and electrical systems by the actual equipment when diversity of use is taken into account. The nameplate ratings of equipment are not representative of the actual energy usage of that equipment, which is typically a fraction of nameplate on average. Thus, if energy usage and cooling loads from office equipment and other plug loads are based upon nameplate, they can be overestimated by 200% to 500%.

There is debate about how much further plug loads are likely to drop in the future, as ever-improving efficiencies of equipment are being offset by increasing amounts of equipment being used. Programs like the U.S. EPA's ENERGY STAR program will continue to improve the energy efficiency of individual pieces of equipment.

CONTROLS OF PLUG LOADS

One of the key factors in reducing effective plug loads is the impact of controls on the equipment loads. The primary issue is the power management controls now built into all computers and most other office equipment—putting the equipment in standby mode or completely shutting it off after preset intervals of no activity. Another way to accomplish this is through the use of occupancy sensors connected to plug strips that control equipment at a workstation, so it can be shut off when not occupied. This also works well for task lighting, which is not otherwise equipped with power management.

SOURCES OF FURTHER INFORMATION

NFPA. 2014. National Electrical Code®. Standard 70-2014. Quincy, MA: National Fire Protection Association.

Piette, M., et al. June 1995 and subsequent updates. *Office technology energy use and savings potential in New York*. Berkeley, CA: Lawrence Berkeley National Laboratory.

REFERENCES AND RESOURCES

Published

ASHRAE. 1998. *Fundamentals of Water System Design*. Atlanta: ASHRAE.

ASHRAE. 2015. *ASHRAE Handbook—HVAC Applications*. Atlanta: ASHRAE.

ASHRAE. 2016a. ANSI/ASHRAE Standard 62.1-2016, *Ventilation for Acceptable Indoor Air Quality*. Atlanta: ASHRAE.

ASHRAE. 2016b. ANSI/ASHRAE/IES Standard 90.1-2016, *Energy Standard for Buildings Except Low-Rise Residential Buildings*. Atlanta: ASHRAE.

NFPA. 2014. *National Electrical Code®*. Standard 70-2014. Quincy, MA: National Fire Protection Association.

Spitler, J. 2014. *Load Calculation Applications Manual*. Atlanta: ASHRAE.

Trane. 2000. *Multiple-Chiller-System Design and Control Manual*. Lacrosse, WI: Trane Company.

Online

Air-Conditioning, Heating and Refrigeration Institute
www.ahrinet.org.

Alliance to Save Energy
http://ase.org.

American Council for an Energy-Efficient Economy
www.aceee.org.

Architecture 2030 Challenge
www.architecture2030.org.

Buildings Performance Institute Europe (BPIE)
http://bpie.eu/focus-areas/buildings-data-and-tools/.

California Energy Commission
www.energy.ca.gov.

CoolTools Chilled Water Plant Design Guide
https://energydesignresources.com/resources/publications/design-guidelines/design-guidelines-cooltools-chilled-water-plant.aspx.

ENERGY STAR®
www.energystar.gov.

Geoexchange
www.geoexchange.org.

Heschong Mahone Group
www.h-m-g.com/projects/daylighting/projects-PIER.htm.

ODYSSEE MURE European energy efficiency & CO_2 indicators, and energy efficiency policy measures
www.odyssee-mure.eu/.

Savings by Design
www.savingsbydesign.com.
U.S. Department of Energy, Federal Energy Management Program
www.eere.energy.gov/femp.
U.S. Energy Information Administration,
Commercial Building Energy Consumption Survey
www.eia.gov/consumption/commercial.
U.S. Environmental Protection Agency
www.epa.gov.

CHAPTER SEVEN

SUSTAINABLE SITES

One interesting phenomenon that has come out of the encompassing of the sustainable design and integrated design processes is that design professionals are becoming more aware of the work done by disciplines outside their own. The *ASHRAE GreenGuide* is primarily focused for the needs of its members and related disciplines, but for those involved with green-building design projects it is important to have at least a basic knowledge of all aspects of green design. In 2012, ASHRAE announced a rebranding of the society to becoming a sustainability resource for the industry, including changing the Society's tagline to "Shaping Tomorrow's Built Environment Today."

Sustainability will be part of the design in all features of the future built environment, and this includes the site development as well. For example, in addition to the LEED rating system, ANSI/ASHRAE/USGBC/IES Standard 189.1 and the *International Green Construction Code* (IgCC) both include whole sections on "Sustainable Sites," and much of the material in this chapter relates to how those issues are treated.

This chapter provides a summary of the key issues in the following topical areas:

- Where to locate the building project
- Landscaping
- Urban heat island effect
- Exterior lighting/light as a pollution source
- Stormwater management

Each of these is briefly summarized in the following sections.

LOCATION OF THE BUILDING PROJECT

For most locations, at least in most developed countries, local or regional regulations and agencies control many decisions as to exactly where development would be allowed and where it would be not allowed. These rules and regulations are designed to, as a minimum, prevent building development from taking place in environmentally sensitive areas (such as near a wetland). Other local regulations govern where land-use planning is a primary method for determining where new development is allowed. Depending on the region, these regulations can be fairly strict or more lenient. However, it is safe to say that this is primarily a local or regionally controlled decision.

The building's location may be allowable under minimum codes or local regulations, but not desirable for a building project that was intended to being "green." For example, consider a highly efficient office building with many sustainability features incorporated but where the project developer has decided to pass up a more urban location for a rural setting. Would that building be considered truly a green building if all the occupants had to drive private vehicles to the site each day? There is no right or wrong answer to that question as all projects are unique, but certainly transportation to the site is a consideration. Thus, urban sprawl is another topic for consideration to moving toward a more sustainable built environment. Urban sprawl affects more than just environmental impact considerations. It affects issues such as quality of life, but these issues are beyond the scope of this guide.

The decisions and issues associated with this topic go well beyond just the design of the physical building structure and systems and thus are somewhat outside the scope of this book. However, the design professional of the future should be at least aware of these issues and, when possible, contribute to the overall societal discussion on these topics.

URBAN HEAT ISLAND EFFECT

The urban heat island effect is the physical effect of urban areas to have ambient air temperatures higher than the surrounding rural areas in the same region. This is a complex topic, and the contributors to the urban heat are many. Those who have participated in a building project that was working for LEED certification, or compliance with one of the other green-building standards or programs, are most likely aware of the credit points offered that are intended to address the urban heat island effect. Those credit points address one of the main contributors to the urban heat island effect, that is, the tendency of buildings and the site hardscape areas to absorb and retain heat from the sun.

There are many other contributors, or potential contributors, to the urban heat island effect that are beyond the control of the building designers. These include items like industrial activity and motor vehicle transportation. And of course, outdoor condensing units play a big role in the development of the urban heat island effect as all the thermal energy from inside the building (plus the energy used to power the refrigeration system that provided that cooling) is transported to the ambient air.

Why Be Concerned About the Urban Heat Island Effect?

In some circumstances, and from the thinking of some people, the urban heat island effect may be considered a good thing. For example, if the urban heat island effect in a major metropolitan area means that precipitation falls as rain rather than snow, many motorists in that area might be thankful. However, in summer that same urban heat island effect, and the resulting increase in ambient air temperature beyond what it normally would be, means that every air-cooled condensing unit is working that much harder to provide the required cooling. The additional load placed on the compressor means additional thermal heat load to the ambient air, thus increasing the heat island effect even more. For example, a simple, air-cooled air-conditioning unit operating with a 100°F (38°C) ambient air temperature will have approximately 10% lower compressor power input than for a comparable unit providing the same tons of cooling and operating in an ambient air temperature of 105°F (40°C).

In that regard, the heat island effect should therefore be looked at as a societal issue and not just evaluated from the cost effectiveness of reductions in building heat gain from the installation of roof with a lower solar absorption rate on the building. In fact, much of the benefit from measures that reduce the urban heat island effect, such as a cool roof or lower solar reflective index (SRI) hardscape, is experienced by all in the local area and not by just that particular building's owner. Therefore, this is a subject that can be considered to be a benefit to society as a whole.

Heat Island Mitigation Methods Associated with a Building Project

Users of this guide are already likely familiar with many methods that the building designer has available to help minimize the contribution of their building project to the heat island in the area where this building is to be built. One of the primary methods includes the use of materials that have a higher SRI. The SRI is a metric that indicates the ability of a material to absorb the sun's heat (in terms of the absorptivity of the material) as well as the ability to lose heat by thermal radiation, which is determined by the thermal emissivity of the material. Other methods include the use of shading to prevent a building wall or the hardscape from absorbing the sun's heat, green-roof systems, or porous paving materials.

The Interrelationship of Urban Heat Island and Building Heat Gain

There is a complicated relationship between the use of materials that have a higher SRI value and their impact on the building heat gain (and hence cooling load) and the urban heat island effect. The use of higher-SRI materials will have varying impact on the building heat gain, depending on how thermally coupled the building interior is with the roof and walls. In some structures, the reduction in heat gain is practically insignificant, while in other cases it will help reduce the overall cooling load to a noticeable degree. However, green-building standards and programs are concerned not only with the building heat gain but also with the contribution of the building structure and associated hardscape with the heat island effect. The heat island effect is a sites issue, rather than strictly a building energy issue, although criteria such as a high SRI roof can affect both. That is why criteria are included in the sites section in green-building rating systems and standards. This also helps explain the differences in why the heat island requirements, specifically those for the roof materials, differ between ANSI/ASHRAE/IES Standard 90.1 and ANSI/ASHRAE/USGBC/IES Standard 189.1. Standard 90.1 is solely focused on the impact on the building heat gain and cooling load, while Standard 189.1 is intended to provide a balanced approach to treating environmental issues. Thus, Standard 189.1 contains more stringent requirements in the heat island section because it is intended to consider both building energy as well as the heat island effect.

A second consideration that is starting to gain attention is the reverse: the impact of anthropogenic waste heat on the urban heat island. Heat rejected by building air-conditioning systems, transportation, and industrial processes does not just disappear into thin air, especially in dense urban environments. Thus, whatever can be done to reduce energy consumption and improve efficiency of HVAC&R systems can also have the positive benefit of reducing the addition to the overall thermal energy balance of urban areas.

Potential Negative Impacts

One recent study has pointed out a potential negative impact of high-SRI building surfaces on the local urban air quality (Landgraf 2015). This study found that brighter-colored buildings, which are a common strategy used to combat the urban heat island effect in cities, can be detrimental to the air quality at ground level. White or high-SRI surfaces can cool down the urban area quicker in the evening and result in decreased vertical air mixing and can, in some circumstances, result in higher levels of pollutants concentrated at the more occupied ground level. This is just one more example of the complex interactions that must be considered in sustainable design.

EXTERIOR LIGHTING

Similar to the heat island issue, the design of exterior lighting levels and lighting fixture selection for a green building involves both consideration of lighting power (for total energy consumption) and the impact of the building and its systems on the surrounding locality. In the case of exterior lighting and the building site, it is also a matter of how the lighting is directed to the area intended to be lit.

To prevent light from escaping the building project property boundary to areas that would be adversely affected by that light, the lighting design should specify fixtures that meet certain standards. Exterior lighting fixtures are rated on how well they perform in terms of backlight, uplight, and glare (the "BUG" ratings). It is also important to consider the locality where the building is proposed to be built. In some areas, there already is a light pollution problem that exists from all the other surrounding developments, and, thus, one more building contributing to that problem will have minor impact on the total problem. In other localities away from city lights and near areas where additional light in the surrounding environment would cause unwanted consequences, special concern should be given to the lighting fixture design to prevent unwanted light from escaping the property boundary.

A number of references are available from organizations that are concerned with exterior lighting. For sustainable site considerations (above and beyond just the energy consumption aspects), the International Dark Sky Association provides a series of resources. See the references listed at the end of this chapter for a link to this valuable resource.

STORMWATER MANAGEMENT

Development of properties for buildings and the associated infrastructure has led to the creation of impervious surfaces that contribute to increased surface runoff into surface waterways. That increased runoff is the cause, or at least a significant contributor, to flooding problems downstream. In most localities, limited requirements are in place for addressing this issue. This is one area where the design and features of the building project can be included that provide a measurable positive contribution to the environment. While many of these measures are outside the normal responsibility of the building systems design engineers and are more the responsibility of the landscape architect, some of the measures can be used to also provide an alternative source of water that can be used to reduce the overall demand for potable water.

Prescriptive Versus Performance Focus

Existing green-building rating systems, codes, and standards approach the stormwater management issue from various perspectives. One approach is to specify or require the building project achieve a particular performance basis. This approach will seek to address the amount and quality of stormwater runoff based on comparison to a set criterion. For example, there could be a target for the project design team to end up with the postdevelopment runoff amount to not exceed that which would have occurred from the same site if left in an undisturbed, predevelopment state. A performance approach is what is used by U.S. Green Building Council LEED rating systems. Performance is what ultimately counts, but this is achieved by the actual design and selection of specific measures.

A different approach could be taken, one prescriptive in nature. Instead of specifying specific performance criteria to be achieved, this alternative approach would require the inclusion of best management practices that have been proven to address the problems caused by unmitigated stormwater from the built environment. This approach is described as being prescriptive in nature, as it prescribes specific technologies or practices to be implemented. Some of these practices are briefly outlined in the section that follows.

Stormwater Mitigation Techniques

The best designs for minimizing stormwater impact will result in no more surface runoff from that property than would have occurred from a natural, undisturbed landscape at that same location. This can be achieved by a combination of one or more of the following design features.

Maintaining a Larger Vegetated Percentage of the Property. Obviously, maintaining as much as possible of the property with some form of vegetated surface is a prime consideration. However, some properties offer limited, if any, potential for this, so additional measures are needed.

Minimizing Unnecessary Impervious Surfaces and Compaction of the Soil. Impervious surfaces, such as roofs, and concrete or asphalt pavements, do not allow precipitation to infiltrate into the soil. In addition, during the site preparation and construction process, compaction of the soil may occur. Compacted soil allows for less water infiltration and more surface runoff. By careful planning and construction management, the amount of unnecessary compaction can be minimized. The site development planning should also consider the use of porous concrete or pavers, or other methods to increase the rainfall absorption on-site and thus minimize runoff.

Rain Gardens. Rain gardens are low-lying regions on the property located to collect surface runoff from impervious surfaces (such as parking areas) slowing down that runoff enough to allow for a large portion of it to naturally infiltrate into the soil or leave by evaporation or transpiration from plants in the rain garden. In this regard, rain gardens provide a similar function as a traditional retention pond on site, but with the additional benefits of providing a more natural landscape look and avoiding some of the negative aspects of retention ponds such as liability concerns and the need to fence that area off. Rain gardens are the subject of GreenTip # 7-1 at the end of this section.

Rainwater Harvesting. Rainwater collection for later use in the building or on the site as irrigation water has two sustainability-related benefits. First, this water would typically be collected from impervious surfaces, such as a rooftop (Figure 7-1), and thus all water collected is that much water that is prevented from quickly running off the property and contributing to stormwater problems downstream. (This is thus a sustainable site issue.) If the collected water is used to displace potable water, for example, as makeup water to a cooling tower, then rainwater harvest-

Courtesy Tom Lawrence

Figure 7-1 Example of rain garden used to manage stormwater runoff from a building roof.

ing also contributes toward water efficiency. Rainwater harvesting is a subject of GreenTip #12-3 in this guide.

Green Roofs. Green provide offer the ability to absorb at least a portion of the precipitation that falls on the roof. The amount collected and stored is a function of soil depth, condition of the soil before the storm event (whether dry or already containing some water), and the type of soil in the green-roof system. Green roofs have been recognized by green building rating systems and standards (LEED, Standard 189.1) as having a positive effect of reducing the local urban heat island, but the amount of this benefit depends on the moisture content of the soil and any shading from vegetation on the green roof. Green roofs do provide additional environmental benefits, but can also be a source of additional water consumption if not properly designed and irrigation is used. To minimize these negative aspects, the design team should select native plants to minimize water consumption and, depending on the green roof type, may need to consider the load-bearing capacity of the roof (especially for existing buildings) along with the need for the installation of waterproofing material, drainage, and other necessary improvements. An example green roof over an underground car parking area with native or adapted species planted can be seen in Figure 7-2.

The synergy of integrating green roofs with solar thermal collectors or photovoltaic systems usually mandates the use of extensive-type roofs (i.e., thin growing medium, low weight and capital cost of minimal maintenance plants). They may even improve the efficiency of photovoltaics because of the resulting cooling effect from the green roof during hot days.

Courtesy Tom Lawrence

Figure 7-2 Example green roof planted with native or adapted species.

Regulations and guidelines for green roofs are available in some countries, for example, the first one published in the mid-1990s in Germany (FLL 2002); the UK guidelines on best practices for the design, installation, and maintenance of green roofs (NFRC 2014); and ASTM. In North America, the top five ranked cities with the most amount of green roofs installed are Washington, D.C.; Toronto, ON; Philadelphia, PA; Chicago, IL; and New York, NY with over 4.3 million ft^2 (0.4 million m^2). Most of these municipalities support green roofs directly through policy, rebate programs, tax credits or other incentives.

Green roofs are more common in Europe. In Germany, the market boomed in the 1980s and has reached about 100 million ft^2 (10 million m^2) of new green roofs added each year. More than 75 European municipalities currently provide incentives or requirements for green roof installation. Copenhagen, Denmark, which has set an ambitious target to become the world's first carbon-neutral capital by 2025, mandates all new flat roofs at or under a 30° pitch have a green roof. Other leading European cities include Berlin and Stuttgart, Germany along with Linz and Vienna, Austria.

REFERENCES AND RESOURCES

Published

ASTM. 2015. ASTM E2400/E2400M-06(2015)e1, *Standard Guide for Selection, Installation, and Maintenance of Plants for Green Roof Systems*, West Conshohocken, PA: ASTM International.

U.S. Environmental Protection Agency. 2007. *Developing your Stormwater Pollution Prevention Plan: A Guide for Construction Sites.* www3.epa.gov/npdes/pubs/sw_swppp_guide.pdf

Online

Cool Roof Rating Council
www.coolroofs.org.

FLL 2002. Bonn: The Landscaping and Landscape Development Research Society
www.fll.de/shop/english-publications/green-roofing-guideline-2008-file-download.html.

Green Roofs for Healthy Cities
www.greenroofs.org.

International Dark Sky Association
www.darksky.org.

International Greenroof and Greenwall Projects Database
www.greenroofs.com/projects/plist.php.

International Green Roof Association
www.igra-world.com.

Landgraf, M. 2015. Brightening Buildings to Fight Urban Heat Islands Has Negative Side Effect on Air Quality
http://phys.org/news/2015-05-bright-facades-trees-smog.html.

National Federation of Roofing Contractors. 2014. The GRO Green Roof Code, London: National Federation of Roofing Contractors Limited.
www.nfrc.co.uk/green-roof-installations.

U.S. Environmental Protection Agency, Heat Island Effect
www.epa.gov/hiri.

U.S. Environmental Protection Agency, Stormwater Management
https://www.epa.gov/npdes/npdes-stormwater-program

ASHRAE GreenTip #7-1

Rain Gardens

GENERAL DESCRIPTION

In 1990, in response to a demand for better stormwater management in Prince George's County, Maryland, bioretention systems, or rain gardens, were used as a financially and aesthetically pleasing solution (Prince George's County 2014). Since their utilization in Maryland, case studies and research have taken place to determine the potential impacts that rain gardens can have on decreasing the amount of stormwater runoff in local areas, as well as the amount of pollutants that can be absorbed by the vegetation incorporated in rain garden designs. The results of these studies have been positive.

The concepts of a rain gardens are basic. Where there are impervious structures that rain water encounters, there will be runoff. This runoff is often directed to infrastructure designed to carry runoff (and the associated contaminants) to nearby waterways. Rain gardens intercept runoff before reaching those systems and filter a portion of the runoff using soil, mulch, and appropriate vegetation. In order for runoff to be captured, rain gardens must be on a shallow slope away from the aforementioned impervious structures in the direction runoff flows toward stormwater management systems (EPA 2013b).

Rain gardens most often include a variety of native plants which are drought tolerant and wet tolerant that are found in low areas (Cleanwater Campaign 2012). While rain garden designs will vary from owner to owner, there are some specifics. Rain gardens can be installed in different soils, including sandy soils and clay soils. Their size can vary depending on the catchment area, and they can be installed in lawns, roadsides, or parking lot islands (North Carolina State University 2001). Most are dug approximately 4 to 6 in. (100 to 150 mm) deep with mulch and specific vegetation used to absorb water and filter pollutants. During a typical rain event, the first 1 in. (25 mm) of rain washes away the majority of pollutants from roof tops, roads, parking lots and other structures. Rain gardens are generally designed to hold water from a 1 in. (25 mm) rainstorm event.

During that time, pollutants are absorbed and filtered through the mulch and vegetation instead of entering the stormwater drains and ultimately the local waterways (University of Rhode Island 2013). The design is shallow enough that rain will not collect and stay for more than 24 hours so that mosquitoes and other insects are unable to breed, nest, and lay eggs (Cleanwater Campaign 2012). These designs will provide more absorption and filtration than common grass lawns while providing a more diverse landscape (University of Rhode Island 2013). The absorption rate of rain gardens will decrease over time, so application of fresh mulch will help replenish the soil's capacity to filter pollutants. An overflow with outlet pipe should be installed to divert excess water out of the rain garden, especially in clay soils. The top of the pipe can be set at the desired maximum water depth (ranges from 6 to 12 in. [150 to 300 mm], with 9 in. [225 mm] considered standard depth).

WHEN AND WHERE THEY ARE APPLICABLE

While some alterations may need to be made in colder climates (snow and salt content must be considered), rain gardens are applicable almost anywhere in the United States (EPA 2013b). In most cases, the existing soil type will not prohibit designs, since the areas are dug out first and a specific soil mixture is added onto the cleared ground. An ideal soil mixture suitable for rain gardens is a mix of 50% to 60% sand, 20% to 30% topsoil, and 20% to 30% compost (Cleanwater Campaign 2012). Slightly depressed areas are ideal locations, but a flat site can work if depressed areas do not exist on site.

PRO

- Development and design is similar to that of common landscape projects which require no major technology or infrastructure.
- Due to the specific plant type requirements, rain gardens require little maintenance over time other than observation for erosion and clogging.
- With adequate planning and implementation, they can be built to serve most regions and soil types.

- While providing water management and pollutant filtration, rain gardens are most often aesthetically pleasing to the owner.
- Rain gardens can be designed on new developments or retrofitted to existing developments (Oklahoma Farm to School 2013).

CONS

- If not designed properly, standing water or increased erosion can occur (Oklahoma Farm to School 2013). These issues should be avoidable by following guidelines provided by local, state, or national entities such as the Environmental Protection Agency.

KEY ELEMENTS OF COST

First Cost

- Soil amendment mix (specific mix dependent upon local soil conditions)
- Possible excavation and hauling away existing soil
- Pipe and filter fabric for the overflow
- Mulch
- Desired vegetation
- Initial watering (Daily for approximately two weeks) (Engineering Technologies Associates and Biohabitats 1993)

Recurring Cost

- Remulching void areas and periodic application of fresh mulch to maintain spoil function
- Treating diseased vegetation
- Replacement of tree stakes and wiring as needed (Engineering Technologies Associates and Biohabitats 1993)
- Removal of invasive plants or weeds as needed

REFERENCES USED IN THIS GREENTIP

Cleanwater Campaign: Solutions to Stormwater Pollution. 2012 http://northgeorgiawater.org/education-awareness/clean-water-campaign/.

Engineering Technologies Associates and Biohabitats. 1993. *Design Manual for Use of Bioretention in Stormwater Man-*

agement. Prepared for Prince George's County Government, Watershed Protection Branch, Landover, MD.

Environmental Protection Agency—Bioretention (Rain Gardens). 2013. www.epa.gov/soakuptherain/soak-rain-rain-gardens.

Oklahoma Farm to School. www.okfarmtoschool.com/

Prince George's County. 2014. Stormwater Management Design Manual www.princegeorgescountymd.gov/DocumentCenter/Home/View/4627.

University of Rhode Island, Healthy Landscapes http://web.uri.edu/riss/take-action/simple-steps-at-home/rain-gardens/rain-garden-resources/.

SOURCES OF FURTHER INFORMATION

Environmental Protection Agency—Experimental Rain Gardens www.epa.gov/water-research/experimental-permeable-pavement-parking-lot-and-rain-garden-stormwater-management.

State University of New York College of Environmental Science and Forestry www.esf.edu/sustainability/action/raingarden.htm.

University of Rhode Island Cooperative Extension, Rain Gardens: A Design Guide for Homeowners in Rhode Island—Helping to Improve Water Quality in Your Community http://cels.uri.edu/rinemo/publications/SW.RGBrochure_text.pdf.

ASHRAE GreenTip #7-2

Green-Roof Systems

GENERAL DESCRIPTION

Living vegetative surfaces on a building rooftop, green roofs are classified in two categories based upon soil depth:

Intensive:

- Minimum soil depth: 12 in. (300 mm)
- Capable of housing large trees, shrubs and recreational facilities
- Heavy; requiring structural reinforcement, 80 to 150 lb/ft^2 (400 to 750 kg/m^2)

Extensive:

- Average soil depth: 1 to 6 in. (25 to 150 mm)
- Often installed using modular plots
- Lighter; less of everything, 10 to 15 lb/ft^2 (50 to 75 kg/m^2)

Green-roof systems are composed of a series of layers used to house the vegetation, filter and collect water, protect the building from water damage, and insulate the building. The uppermost layer is the vegetation layer. This layer varies the most in application due to climatic factors and depth of soil. Maintenance required is determined by the types of plant life used; an intensive system will require more attention than an extensive system. The second layer is the growth media, the soil mix layer in which the vegetation grows. This layer will also vary with location as different soil types will require a variation in the soil mix ratio.

In extensive systems, under the growth media is typically a layer of filter fabric. This filter allows for water penetration, but must be tough to withstand pressure. Typical materials for filter fabric systems are geosynthetic fabrics or geotextiles.

In intensive systems, the filtration system is also a part of the drainage system. A layer of gravel beneath the fabric-protected growth media allows tree roots to grow down and increase the stability of the plant layer above. A root barrier is layered beneath the gravel. Water can be stored in the gravel, drained to a cistern, or, in the case of extensive systems, be contained in a molded modular storage unit system. This is also the layer in which irrigation can take place to access and encourage deep roots.

Beneath the drainage layers of green-roof systems is a waterproofing layer. Watertight materials can be applied as a liquid-applied membrane, a single-ply specialty sheet, or a three-layer laminated roof system. One or more layers of insulation provide reduced heat transfer through the roof as well as lower noise transmission.

WHEN AND WHERE THEY ARE APPLICABLE

Green-roof systems are particularly well-suited for dense urban application. Unused rooftops can become productive spaces. Structural constraints make retrofit projects more difficult for the implementation of a green-roof system, but a new building can handle the structural details in the design phase.

Climate plays a crucial role in the actual return on investment figures. For example, stormwater retention may be less of a benefit for a green roof located in low rainfall areas or regions where rainfall tends to be concentrated in the winter months.

KEY ELEMENTS OF COST

What climate region a green-roof system is implemented in will determine actual cost savings as the same system will not perform identically in different locations. Using L to represent lower cost, H to represent higher cost, and S to represent the same cost; here is a brief comparison of several expenditure types:

First Cost

- Installation — H/S (H in retrofits)
- Design fees — S

Recurring Cost

- Overall energy costs — L
    - Water costs — L (if rainwater harvesting is done)
    - Training for operations — H
    - Overall maintenance costs — H

Table 7-1: Potential Benefits and Drawbacks of Green Roofs

Advantages	Disadvantages
Stormwater runoff reduction	Additional structural load
Reduced heat gains (in summer) and heat loss (in winter) to building structure	Cost
Longer life for the base roofing system (may not apply to an intensive green roof)	Additional maintenance, ranging from limited for an extensive green roof with low-maintenance plants to high for a manicured landscape intensive roof
Reduced noise transmission from outside	Optimal roof type, plant materials, and soil depths will vary depending on climate
Aesthetic benefits to people in or around the building with the additional green space	Documentation of benefits such as reduction in heat island effect has not been proven
Other general environmental benefits, such as reduced nitrogen runoff (source: bird droppings), air pollutant absorption, potential carbon sink, bird habitat	—

SOURCES OF FURTHER INFORMATION

The City of London. 2008. Greater London Authority, Design for London, London Climate Change Partnership. *Living Roofs and Walls—Technical Report: Supporting London Plan Policy.* London: Greater London Authority.

"Greenroofs 101" 2011. www.greenroofs.com/Greenroofs101/.

CHAPTER EIGHT

INDOOR ENVIRONMENTAL QUALITY

INTRODUCTION

Indoor environmental quality (IEQ) is inarguably one of the most important characteristics of green buildings intended for human occupancy. While it is challenging as well as important to provide good IEQ in an energy efficient manner, no sacrifice of IEQ can be justified to obtain energy use reductions. After all, the purpose of such buildings is to support the activities for which the building exists and to do so in a manner that does the least harm to the environment while enhancing the health and well-being of the occupants.

The four major aspects of indoor environmental quality are indoor air quality (IAQ), thermal conditions, illumination, and acoustics. Perhaps even more important are the interactions within and among these factors and between the building, the occupants, and the indoor and outdoor environment.

To provide for good IEQ, the process must begin at the design phase and be carried through to the operation and maintenance of the building. The most effective way to accomplish integrated design is to assemble the entire design team at the beginning of the project and to brainstorm siting, overall building configuration and zoning for intended building use, ventilation, thermal control, and illumination goals as a group. These goals should also be considered in conjunction with energy supply, water utilization, material selection, and aesthetics.

Once the initial design concept is agreed upon, then the evolution of the design through its various stages can occur with a shared concept and the potential for direct interaction among team members as challenges arise later in the process. The integrated design process is discussed in more detail in Chapter 3 of this book. ASHRAE Guideline 10-2016, *Interactions Affecting the Achievement of Acceptable Indoor Environments*, is also devoted to discussion of the interactions among indoor environmental conditions and the considerations important for designers as well as building operators (ASHRAE 2016a).

INDOOR AIR QUALITY (IAQ)

In the context of *ASHRAE GreenGuide*, IAQ refers to the quality of the air in buildings intended for nonindustrial use. Good IAQ is defined by the absence of contaminants at concentrations likely to be harmful and the absence of conditions that are likely to be associated with occupant health or comfort complaints. The major means for achieving good IAQ are eliminating or reducing the sources of pollution in building materials and furnishings and from appliances, equipment, and consumer products. After vigorous efforts to eliminate pollutant sources, ventilation is applied to remove unavoidable pollutants, including the metabolic products from occupants and those from materials and furnishings used. Finally, if possible, measurements of indoor contaminant concentrations should be made and compared with design goals or available benchmarks to verify that good IAQ is achieved as intended.

From Design to Completion: Meeting IAQ Objectives

During predesign, the commissioning authority (CxA) should help the owner articulate the key functional requirements for the project. For IAQ, this could include such things as ensuring that the building enclosure keeps out rain and humid air that may cause mold and moisture problems or ensuring that outdoor air intakes are located to minimize introduction of outdoor contaminants and that materials selected during the design phase are actually installed. Failure to meet such requirements has been responsible for serious IAQ problems in many buildings. During predesign, the CxA should also review and provide input to the project schedule to ensure that the key steps required to achieve the owner's objectives are incorporated. For example, commissioning of the building enclosure may require that mock-ups be built and tested or that assemblies be inspected while they are still open and visible. Additional discussion on commissioning is found in Chapter 4 of this book.

Compressed scheduling, improper sequencing of work, and poor construction management can cause IAQ problems. For example, an unrealistic timeline or a bonus for early completion can create pressure to complete work while materials or equipment may still be exposed to moisture. Ductwork stored outdoors can accumulate dirt that facilitates future growth and circulation of microorganisms, even though high-efficiency filtration is used in the finished project. Operating HVAC equipment for temporary heating or cooling during construction or for occupied areas in a partially completed building can cause widespread system contamination, transfer of construction dust and volatile organic compounds (VOCs) to the occupied area. Schedule and construction management issues that affect IAQ are discussed in more depth in the *Indoor Air Quality Guide* (ASHRAE 2009a).

HVAC System Selection for Optimizing IAQ and the Energy Impacts of Ventilation

When selecting HVAC systems, their ability to meet IAQ objectives should be carefully considered. All too often, the type of HVAC system is decided by the architect before the engineering firm is even retained. In other cases, several types of HVAC systems may be compared, but their implications for IAQ are not fully considered. Therefore, optimal IAQ requires a fully integrated design process. For example, buildings that use natural ventilation and passive cooling require close coordination of the architectural and mechanical design to meet IAQ and comfort objectives. In mechanically cooled buildings, it is essential to reduce the risk of condensation and mold growth in the building enclosure. Some types of HVAC systems are less able, or even unable, to maintain positive pressurization. Similarly, HVAC systems vary in their ability to control indoor humidity levels, the flexibility they allow in the location of outdoor air intakes, their ability to provide higher efficiency particle filtration, time and effort required to perform IAQ-related maintenance, and other factors. HVAC systems also differ in the amount of energy they consume in delivering the minimum ventilation set by building codes and standards. The following sections describe a few ventilation systems that can also save energy. For more details, see Strategy 1.3 and Strategy 8 in the *Indoor Air Quality Guide* (ASHRAE 2009a).

Dedicated Outdoor Air Systems (DOASs). DOASs are 100% outdoor air systems. Having the ventilation system decoupled from the heating and air-conditioning system can provide many advantages (ASHRAE 2015b). DOASs must address latent loads, primarily from the outdoor air and, in some cases, DOASs may be designed to remove latent load from both outdoor air and the building. If the exhaust airstream is located near the ventilation airstream, both sensible and latent energy can be recovered, For very humid outdoor air conditions, an active desiccant wheel or a passive dehumidification component may be cost effective. Figure 8-1 shows the components of a typical DOAS. For additional information on DOAS, please see *ASHRAE Design Guide for Dedicated Outdoor Air Systems* (ASHRAE 2017a).

Energy Recovery Ventilation. Energy recovery ventilation is required by energy standards for certain applications (ASHRAE 2016c). Even when not required by the standard, energy recovery systems can provide sufficient payback in overall system sizing and reduced operating costs over the life of the system, making them attractive despite an increase in system fan power caused by increased flow resistance. Their use could also yield credit towards green rating systems, such as LEED. Figure 8-2 shows different types of heat recovery ventilation systems.

Energy recovery ventilation devices can (1) transfer heat and moisture between incoming and exhaust air (called *total energy recovery ventilators [ERVs]*) or (2) transfer only heat, not moisture (called *heat recovery ventilators*

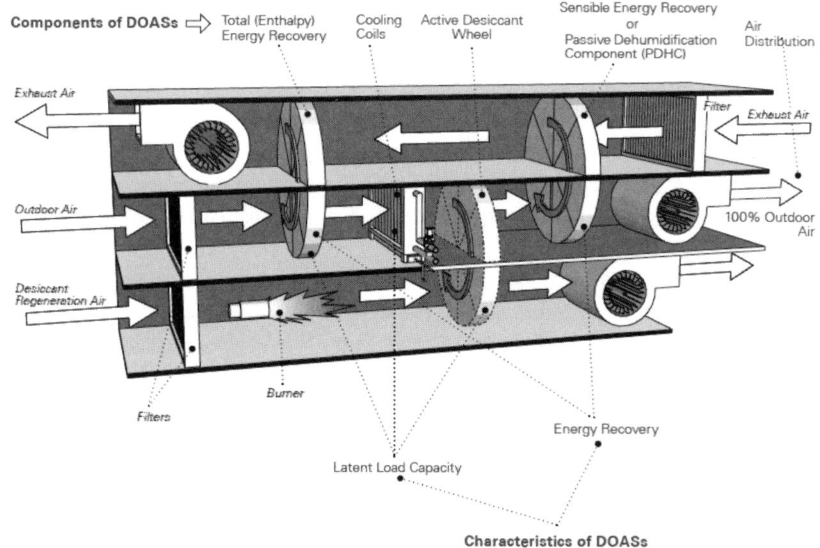

Source: *Indoor Air Quality Guide*.

Figure 8-1 Characteristics of dedicated outdoor air systems.

[HRVs]) (ASHRAE 2015b). Improved humidity control can be an added benefit of the ERV, critical for controlling condensation, mold growth, and thermal comfort. General design considerations include those associated with filtration, controls, sizing, condensation, and sensible heat ratio. Control systems, possibly including bypass dampers, should be incorporated to minimize fan power use, particularly during conditions of low heat recovery (i.e., economizer mode).

Demand-Controlled Ventilation (DCV). DCV is "a ventilation system capability that provides for the automatic reduction of outdoor air intake below design rates when the actual occupancy of spaces served by the system is less than design occupancy" (ASHRAE 2010) offering a means to ensure adequate outdoor airflow while reducing energy consumption at part-load conditions. DCV is often considered (or required) for densely occupied zones and zones with large occupancy fluctuations. Standards 90.1 and 189.1 set requirements for when DCV is required. Readers may also refer to ASHRAE research project RP-1547, *Carbon Dioxide Based Demand Control Ventilation for Multiple Zone HVAC Systems*, which studied VAV system control logic that effectively reduced energy use without underventilating (Lau et al. 2014).

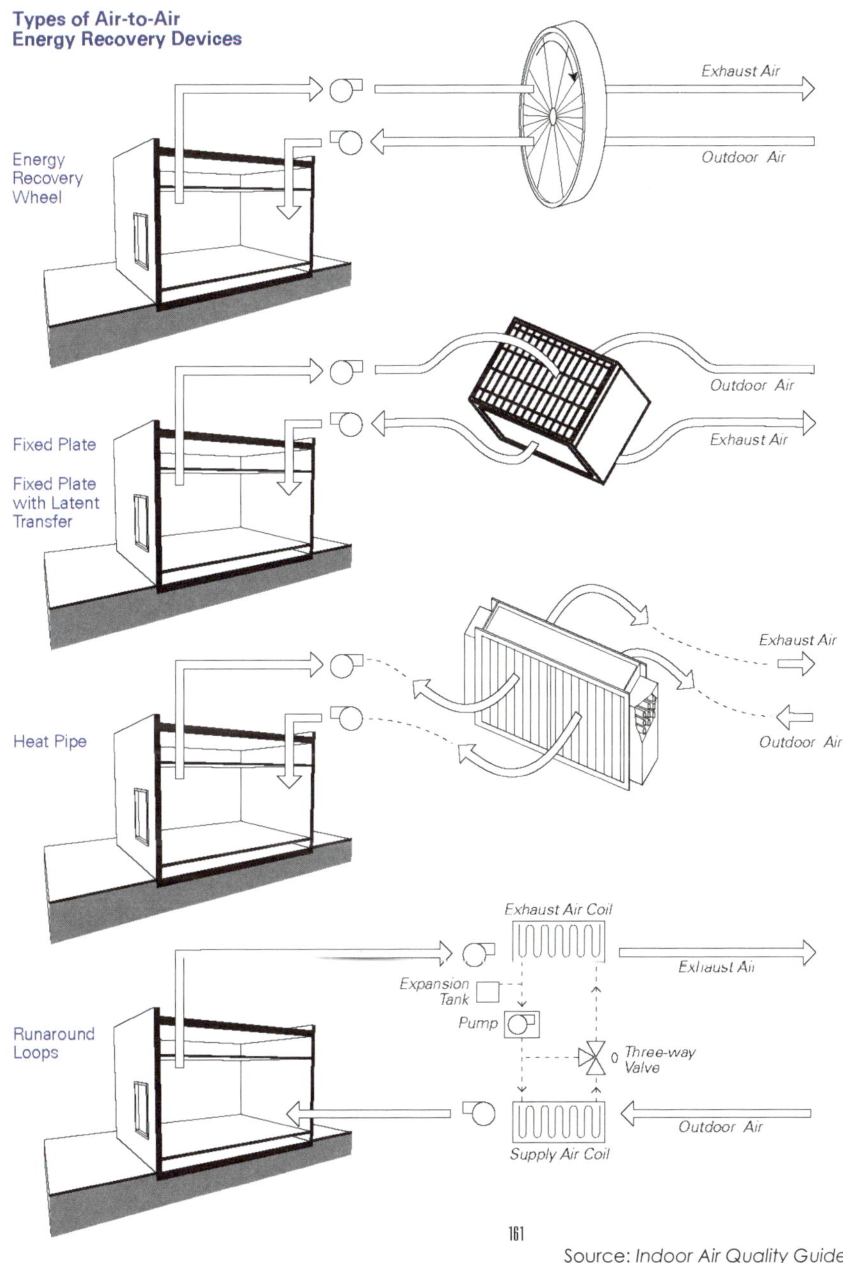

Source: *Indoor Air Quality Guide*.

Figure 8-2 Types of heat recovery ventilation systems.

Natural or Mixed-Mode Ventilation. Naturally ventilated buildings take advantage of and adapt to dynamic ambient conditions to provide outdoor ventilation. Buildings in temperate climate zones can take advantage of natural ventilation strategies for a significant portion of the year. When considering the use of natural ventilation, the prevailing climate (the ambient air temperatures, humidity, cleanliness of the outdoor air, and wind airflow patterns) must be considered. Tools such as the Climate Suitability Tool, developed by and available for free from the National Institute of Standards and Technology (NIST), can estimate the number of hours per year a building may be able to take advantage of natural ventilation based on a handful of building inputs (internal heat gain, ventilation rate required, and heating/cooling set points).

ASHRAE Standard 62.1-2016 requires that spaces using the natural ventilation procedure also be provided with a mechanical ventilation system for use when natural ventilation is not adequate or appropriate. Sophisticated systems using pressure sensors and motor-driven dampers can be used to control pressures in various parts of buildings and to take advantage of stack effect or wind pressure to deliver ventilation where and when it is needed. Nevertheless, the high potential return in energy performance and IAQ can often justify the extra effort. Strategy 8.4 in the *Indoor Air Quality Guide* (ASHRAE 2009a) discusses other key design issues, such as ensuring a satisfactory acoustic environment, security of the building, and smoke control.

Personalized Ventilation. The personalized ventilation system (as provided in many automobiles and airplanes, for example) is fundamentally aimed at improving ventilation in the immediate breathing zones of occupants. Personalized ventilation systems can greatly improve ventilation effectiveness, leading to improved contaminant control, which then offers the possibility of reducing the quantity of outdoor air provided by a central ventilation system, as well as its delivered temperature (warmer temperatures in cooling months or cooler temperatures in heating months), both of which can lead to energy savings.

Use of the ASHRAE Standard 62.1 IAQ Procedure. Standard 62.1's IAQ Procedure (IAQP) provides designers with an important option or adjunct to the prescriptive Ventilation Rate Procedure (VRP) in ASHRAE Standard 62.1, thereby increasing the potential for good IAQ control.

In general, the attainment of acceptable IAQ can be achieved through the removal or control of irritating, harmful, and unpleasant indoor contaminants. Source control approaches should always be explored and applied first, because they are usually more cost effective than either ventilation (dilution) or filtration and air cleaning (FAC) (removal). Application of the IAQP typically uses a combination of source control, dilution, and FAC to reduce the minimum outdoor air intake required by the prescriptive VRP. More details on the IAQP and its potential benefits can be found in Strategy 8.4 of the *Indoor Air Quality Guide*.

Limit Entry of Outdoor Contaminants

Because epidemiological studies linking ambient fine particle, ozone levels, and other contaminants with morbidity and premature mortality have generated increased concern, attention should be paid to limit the entry of outdoor contaminants through the building enclosure and ventilation equipment.

Make an Airtight Enclosure. Among the benefits of airtight enclosures are reduced chances of thermal comfort complaints, amount of energy used to condition the building, chances of condensation inside enclosure assemblies in both heating and air-conditioning mode, and chances of ice dam problems in snowy climates. Tight enclosures also form the basis for effective pest exclusion. Lastly, airtight enclosures benefit by making it easier to manage the building pressurization relative to the outdoors and between spaces and floors of multistory buildings, allowing ventilation systems to perform more effectively. Keep in mind that as enclosures are made more airtight, the need for mechanical ventilation systems will increase. Target airtightness levels developed by three building programs as follows:

- 0.40 cfm75/ft^2 (2.0 L/s·m^2 at 75 Pa) enclosure used by the Government Services Administration (GSA 2014) and the 2015 International Energy Conservation Code (ICC 2015), where "cfm75" refers to the amount of air required to maintain an indoor-pressure difference of 75 Pa.
- 0.25 cfm75/ft^2 (1.27 L/s·m^2 at 75 Pa) enclosure used by the Army Corps of Engineers (USACE 2012)
- 0.60 ach50 used by the Passive House Institute U.S. (PHIUS 2015) (Note: 0.60 ach50 roughly translates to 0.09 cfm75/ft^2 (0.46 L/s·m^2 at 75 Pa) enclosure for a three-story building with 30,000 ft^2 [2800 m^2] of gross floor area.)

For data on the measured airtightness of commercial buildings in the United States, the reader is referred to Emmerich and Persily (2014).

Locate Outdoor Air Intakes to Minimize Introduction of Contaminants. Whenever possible, the outdoor air entry locations should be separated from known external pollutant sources such as loading docks, exhaust stacks from the building itself and others in its vicinity, and regions of heavy vehicular traffic. Industrial processes in the vicinity can also be important pollutant sources that should be shielded from building air intakes. Pollutant sources and prevailing wind conditions should be evaluated together. ASHRAE Standard 62.1 and Strategy 3.2 of the *Indoor Air Quality Guide* discuss specifics on minimum distances between outdoor air intakes and contamination sources (Figure 8-3).

Use Air Filtration/Cleaning to Remove Contaminants in Outdoor Supply Air. Following the requirements in ASHRAE Standard 62.1-2016 (ASHRAE 2016b), the ventilation design team should to determine compliance with outdoor

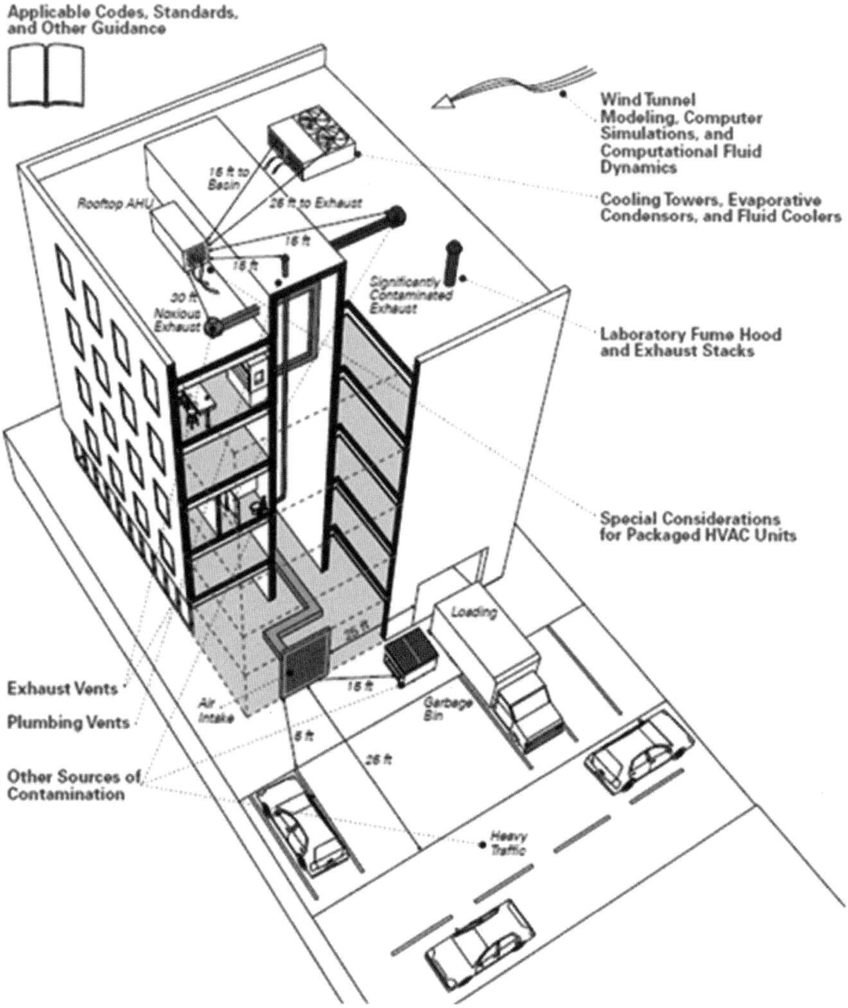

Image source: *Indoor Air Quality Guide*.

Figure 8-3 Examples of problem pollutants and air intake locations for good IAQ.

air quality standards in the region where the building will be located. A primary resource for information on outdoor air pollution is in the Green Book on the U. S. Environmental Protection Agency (EPA) website, where maps show areas that are not in compliance (nonattainment) with the National Ambient Air Quality Standards (NAAQS) (EPA 2017). The NAAQS addresses particles, ozone, lead, carbon monoxide (CO), nitrogen dioxide (NO_2), and sulfur dioxide (SO_2). The next step is to determine if there are any local sources of outdoor air pollution, such as traffic or industrial and commercial operations. Filtration, air cleaning, and location of outdoor air intakes can then be considered as a means of reducing the entry of outdoor contaminants into the building.

Typically, filtration of outdoor particles includes the use of media filters and electrostatic precipitators, while air cleaning of outdoor ozone, nitrogen dioxide, and volatile organic compounds uses adsorption and chemisorption filtration. The incorporation of filtration and/or air-cleaning systems for particle and gas contaminant removal in any building requires investment in design, labor, equipment, and maintenance, and, as for all building systems, must be evaluated in economic terms. There are a few published studies that have reported the cost/benefits of indoor air cleaning with regards to airborne particle removal (e.g., Fisk et al. 2011). These studies have reported insignificant filtration costs relative to salaries, rent, health insurance costs, loss of productivity, morbidity, and mortality.

Particle Filtration Technologies. Filters tested by ANSI/ASHRAE Standard 52.2-2012, *Method of Testing General Ventilation Air-Cleaning Devices for Removal Efficiency by Particle Size* (ASHRAE 2012a) are assigned minimum efficiency reporting value (MERV) ratings. Filters need to be at least MERV 8 to have any effective removal efficiency on the smaller particles (PM_{10}), and filters with MERV ≥11 are much more effective at reducing $PM_{2.5}$. These designations mean particles less than 10 microns or less than 2.5 microns, respectively. The high-efficiency particulate air (HEPA) filter is the best-known air filter. To qualify as a true HEPA filter, it must be able to capture at least 99.9997% of all particles 0.3 m or larger in diameter. The breakdown of particle distribution by size, count, and weight in typical outdoor air systems is found in Figure 8-4 and the particle capture efficiencies listed in Table 8-1 for the various filter ratings.

Electrostatic precipitators use electrical charges to attract and deposit particles on their collection plates. Ion generators use corona discharge to generate ions with the intention of charging and clumping particles until they are settled to the ground or other room surfaces due to gravity, which have to be cleared away. A disadvantage of electrostatic precipitators and ion generators is that almost all of them create small amounts of ozone or ultrafine particles (Morrison et al. 2014).

Gas-Phase Air Cleaning. Gas-phase air-cleaning technologies include adsorption filters (e.g., activated carbon), chemisorption filters, and photocatalytic oxidation. The main purpose of these technologies is to remove VOCs, ozone, nitrogen dioxide odors, and other nonparticulate gases from the indoor air. Disadvantages of adsorption filters are that they become ineffective over time, requiring regular replacement, and photocatalytic oxidation technology produces ozone.

Control Entry of Radon and Vapor Intrusion of Other Subsurface Contaminants. Radon is a radioactive gas formed from the decay of uranium in rock, soil, and groundwater. Exposure to radon is the second leading cause of lung cancer in

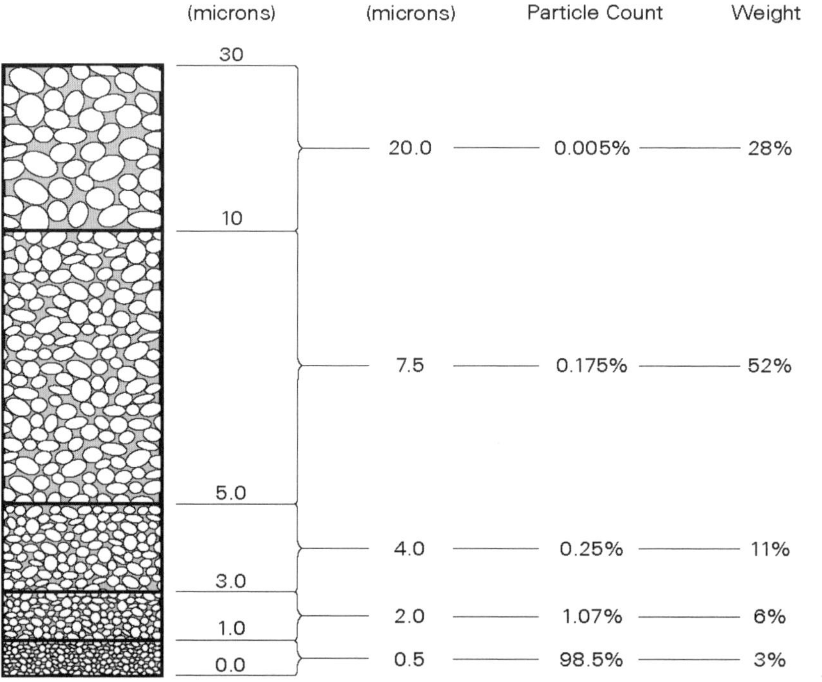

Figure 8-4 Particle distribution by size, count, and weight in typical outdoor air samples.

the United States after cigarette smoking. Radon most commonly enters buildings in soil gas that is drawn though joints, cracks, penetrations, or porous regions of building foundations when the building is at a negative pressure relative to the ground. The potential for high radon levels varies regionally, seasonally, and with discrete weather events (Figure 8-5). There is also variation from building to building in the same region.

Design for control of radon entry has three major elements: (1) active soil depressurization (ASD), (2) sealing of radon entry routes, and (3) use of HVAC systems to maintain a positive pressure in ground-contact rooms (Figure 8-6). Details on these control strategies can be found in Strategy 3.3 of the *Indoor Air Quality Guide.*

Vapor intrusion of other organic vapors from subsurface soils or groundwater is also important to control because contaminants can travel in groundwater plumes for

Table 8-1: Comparison of MERV Level, Prior Rating, and Efficiency by Particle Size

MERV Level	Dust Spot %	Typical Particulate Filter Type	% 0.3–1 µm	% 1–3 µm	% 3–10 µm
1	N/A	Low-efficiency fiberglass and synthetic media disposable panels, cleanable filters, and electrostatic charged media panels	Too low efficiency to be applicable to ASHRAE Standard 52.2 (ASHRAE 2007) determination		
2	N/A				
3	N/A				
4	N/A				
5	N/A	Pleated filters, cartridge/cube filters, and disposable multi-density synthetic link panels			20–35
6*	N/A				36–50
7	25%–30%				50–70
8	30%–35%				>70
9	35%–40%	Enhanced media pleated filters, bag filters of either fiberglass or synthetic media, rigid box filters using lofted or paper media		>50	>85
10	50%–55%			50–65	>85
11	60%–65%			65–85	>85
12	70%–75%			>80	>90
13	80%–85%	Bag filters, rigid box filters, minipleat cartridge filters	>75	>90	>90
14	90%–95%		75–85	>90	>90
15	>95%		85–95	>90	>90
16	98%		>95	>95	>95
The following classes are determined by a different methodology than that of ASHRAE Standard 52.2 (ASHRAE 2007).					
NA	N/A	HEPA/ULPA filters evaluated using IEST Recommended Practice CC001.3 (IEST 1993). Types A through D yield efficiencies at 0.3 mm and Type F at 0.1 mm	99.97% IEST Type A		
NA	N/A		99.99% IEST Type C		
NA	N/A		99.999% IEST Type D		
NA	N/A		>99.999% IEST Type F		

* MERV 6 is prescribed by ASHRAE Standard 62-2001 (ASHRAE 2001) for minimum protection of HVAC systems.

Source: *Indoor Air Quality Guide*

several miles (ASHRAE 2010b) from accidental spills, improper disposal, leaking landfills, or leaking storage tanks (both below and above grade). Vapor intrusion is a concern because of chronic health effects caused by long-term exposure to low-level contaminants. ASTM International developed a standard for the assessment of vapor intrusion on properties involved in real estate transactions (ASTM 2012a).

Provide Effective Track-Off System at Entryways. Tracked-in dirt contains contaminants that can result in health effects for the occupants, contains abrasives that contribute physical damage to floor finishes, and is reported to be the source of the majority of the dirt and dust that must be cleaned from buildings. The simplest and most effective way to reduce this impact is to design the landscape and entryways to prevent dirt entry. Suggestions for the design of entryways can be found in Strategy 3.5 of the *Indoor Air Quality Guide* (ASHRAE 2009a) and requirements for them in ANSI/ASHRAE/USGBC/IES Standard 189.1-2017 (ASHRAE 2017b).

Limit Entry of Contaminants from Indoor Sources

Understanding the emission characteristics of various indoor sources is an important step in devising strategies to limit contaminants emissions. In this sec-

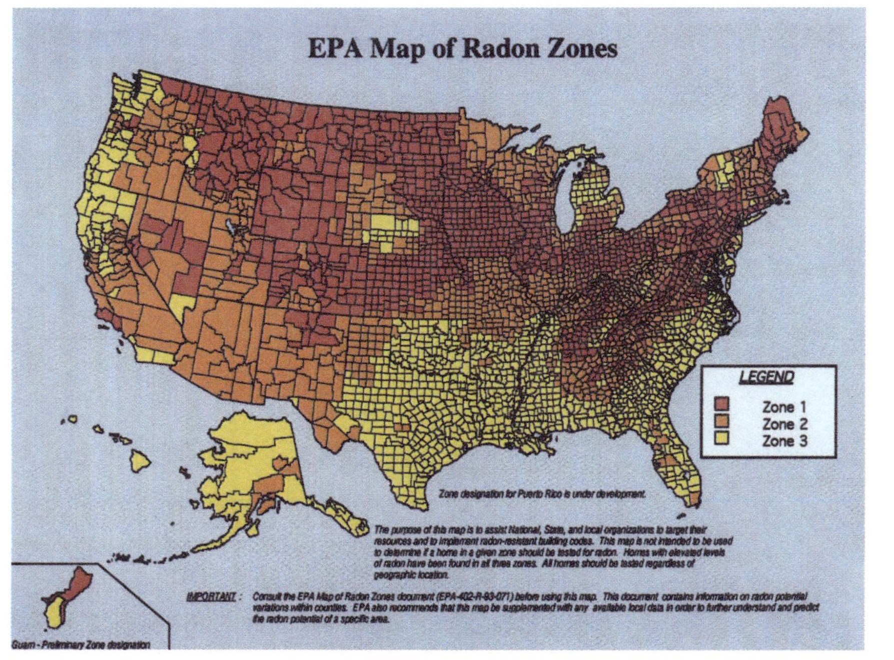

Source: U.S. EPA website.

Figure 8-5 U.S. EPA map of radon zones in the United States.

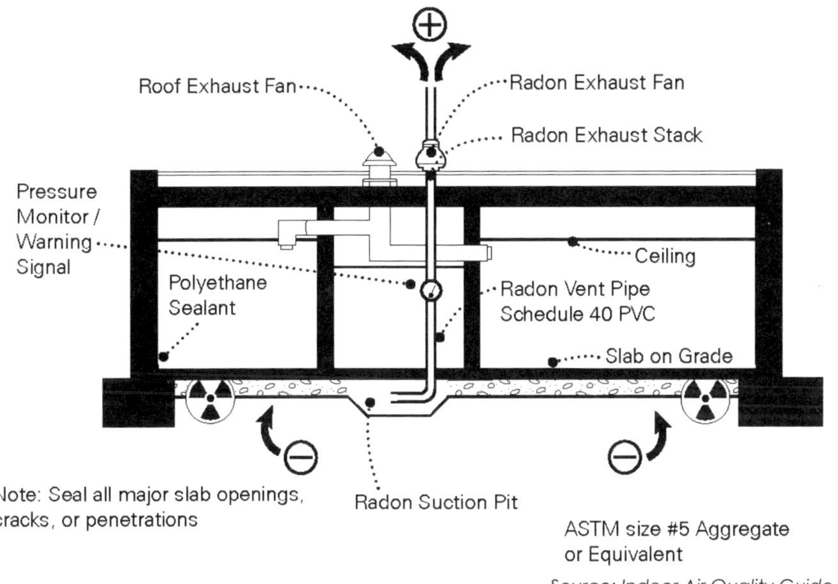

Figure 8-6 Schematic of an active soil depressurization scheme for control of radon entry into buildings.

tion, we first classify the typical indoor contaminant sources in terms of their emission behavior and discuss the major influencing factors and general control strategies. We then provide more specific guides for selecting low-emitting materials, and methods for limiting the impact of the emissions. In addition, the following discussion is intended as an example rather than an extensive list, though the compounds listed are of importance in various IAQ guides or emission testing protocols. It should be emphasized that there is still more to learn in understanding indoor emissions, especially secondary emissions.

Passive Sources. These are the building materials, finishes, and furnishings that do not involve occupants and exist in buildings regardless of the activities of the building occupants or operation of the ventilation system. Major contaminants of concern from these sources include VOCs and semivolatile organic compounds (SVOCs). The emission rates are generally highest following installation or application and then decrease over time throughout a building's service life. Cleaning products also contain VOCs. Actual emission patterns depend strongly on the requirements, material and its position in a material assembly, the chemical compound of interest, and the environmental conditions (temperature, relative humidity, and surrounding air velocity and concentration level). For example, architectural coatings such as paint, wood stain, and varnishes that are applied wet can initially have emission rates that are several orders of magnitude higher than dry materials such as engineered wood products, but then decrease quickly. After several weeks, the emission rate of the dry material may be an order of magnitude higher than the wet material. The emission rates of dry materials start at a relatively low level, but can persist over a long period of time (months to years). The types of emitted compounds also vary dramatically among different material types (ASHRAE 2009a). Oil-based wet coating materials using petroleum products as raw materials emit a significantly higher number of VOCs, while water-based coating materials generally have fewer compounds emitted as well as a lower emission rate selection, FAC options, and implementation and documentation of the IAQP.

Selecting low-emitting materials/products in building design can result in reduction of indoor contaminant concentrations by a factor of two or more (ASHRAE 201b). Standard test methods and guideline emission limits have been established for many building materials, furnishings, furniture, copiers and printers, and computers (e.g., California Department of Public Health, the Collaborative for High Performance Schools, GreenGuard, ASTM, Carpet and Rug Institute, and Business and Institutional Furniture Manufacturers Association [BIFMA]). Emission test methods are used to measure the emission rate per unit of materials or products tested (called emission factors). Green-building rating programs such as LEED have assigned credits for using materials and products that have measured emission factors less than the established limits to encourage the use of low-emitting materials and products.

There are also green-product labeling programs based on the chemical contents in the materials (e.g., GreenSeal, EcoLogo). At best, this type of labeling program can be used for relative comparison among the same type of materials or products, because the actual chemical emission rates can differ significantly from the chemical content. A manufacturer's materials safety data sheet (MSDS) may only be useful for initial screening, because only chemicals with more than 0.1% of the material mass are listed, and requirements for listing are limited to chemicals identified by the U.S. Occupational Safety and Health Administration (OSHA) as hazardous. Chemicals of IAQ concern often occur in much smaller mass amount in the materials. For those compounds that are listed, listing the actual amount in the product is not required.

To reduce the impact of passive sources, allowing a period for the emission rates to decline immediately following the production or installation of materials is an effective strategy. Ventilating the whole building with the highest practical outdoor air ventilation rate for a period of time before occupancy will also help reduce the exposure of occupants to the indoor source contaminants. The "bakeout procedure," airing the building at elevated temperature, is not recommended because of the potential of increasing emissions.

Active Sources. These source types also generate heat. Examples include computers, office equipment, and room air cleaners. They can emit particles, especially ultrafine particulate matter, as well as emit VOCs and SVOCs. These types of sources are associated with occupant activities or use of the building, and their emission rates are therefore highly dependent on the timing of their respective activities or operation. The emission rates of the sources are typically dramatically higher during the operation or activity than during idle or power-off conditions. While proper selection of materials or equipment can reduce emissions from the sources, using local exhaust for these sources or increased dilution flow during the operation can limit the contamination from these sources. Consider also careful placement of these sources in so that their emissions have minimal impacts to the occupants.

Occupants. Contaminants from occupants are mainly bioeffluents (including exhaled breath, sweat, and shed skin cells) and depend on personal hygiene, health conditions, and the immediate environment they were in before entering the space or building of interest. The can include VOCs, SVOCs, and both viable and nonviable particles. There are also emissions from clothes such as dry-cleaning chemicals and chemicals adsorbed from air during previous exposure situations. Emissions from smokers even when they are not smoking can still be significant. For emissions from occupants, adequate personal hygiene, proper selection of personal deodorants, cosmetics, fragrances, and laundry detergents are perhaps a good starting point.

Secondary Sources. Ozone-initiated gas-phase and surface reactions, producing ultrafine particles, formaldehyde, and other aldehydes, is an example of a secondary source. The most effective approach of reducing secondary sources is to limit sources of ozone, both outdoor and indoor.

Limiting Humidity/Moisture Issues

One of the primary functions of the building envelope is to protect building materials, finishes, and occupants and their possessions from the elements. In this section, "moisture" refers to liquid water and water vapor.

The first line of defense is drainage so that water can exit the envelope system and is kept away from materials that might absorb water or otherwise be damaged. Use gravity to drain roofs, roof penetrations, walls, window and door openings, surrounding landscapes, and subgrade foundations away from the building. Sometimes sumps and pumps are needed to drain subgrade foundations, but precautions should be in place so that they operate when needed even when no one is available to physically turn them on. Use flashing, overhangs, and drip edges to direct water away from vulnerable joints, intersections, and penetrations.

The second line of defense is to specify materials that form the outer skin of a building that are able to tolerate getting wet and, in some climates, wet and frozen. Stone, concrete, brick, decay-resistant woods, paints and coatings, synthetic membranes and boards, glass, corrosion resistant metals, composite wood, and fiber- and cement-based materials have all been successfully used as exterior materials on buildings. It is also important to have an effective, continuously sealed air barrier covering all six sides of the building envelope to address infiltration/exfiltration issues and to avoid interstitial moisture accumulation in building materials as well as on cold interior surfaces (ASHRAE 2015b, Chapter 44; ASHRAE 2017b). These measures are primarily aimed at avoiding water damage and long-term wetting that may result in microbial growth. These measures should also help in avoiding thermal bridges. For more information, see Strategy 2.1 in the *Indoor Air Quality Guide* (ASHRAE 2009a) and ASHRAE Position Document, "Limiting Indoor Mold and Dampness in Buildings" (ASHRAE 2012b).

Control Indoor Humidity. It is critical to keep the indoor dew point low enough to ensure that there is no condensation on the exposed surfaces of cool HVAC components or persistent, excessive moisture accumulation on sensitive building materials or furnishings (ASHRAE 2009b). If humidity is too high, condensation on cool, indoor surfaces is more likely. If such condensation occurs frequently or for extended periods of time, microbial growth will occur, which can then cause indoor air quality problems and damage to building materials. There is also some evidence that airborne pathogens may survive longer in low-humidity indoor environments, possibly increasing disease transmissions.

Humidity control is less of an issue in cold climates where humidification is used. In comparison, hot, humid climates pose a greater challenge where high humidity can lead to potential risks of mold growth. The challenge in such climates is not so much in being able to dehumidify but to do so without having to overcool. Based on considerations provided in ASHRAE Standards 55-2017, 62.1-2016, and 90.1-2016, it is imperative that the efficiency dimension associated with the cooling and dehumidification process involve an absolute improvement in component/system efficiency as well as some means of energy recovery (ASHRAE 2017c, 2016b, 2016d). Design considerations can help to control indoor humidity include the following:

- Tight envelopes promote substantially better control over indoor humidity. Though, if the sensible load is reduced, dedicated dehumidification may be needed because thermal set points are met but not humidity set points.
- Buildings in hot, humid climates should be designed with the control of relative humidity as a priority without overcooling.
- Systems should have humidity control at design- and part-load conditions.
- Thermal comfort acceptability of tropically acclimatized subjects may be different than their temperate climate counterparts. Warmer and drier thermal environments or a cooler microenvironment (immediate breathing zone) coupled with warmer ambient conditions may be more acceptable.
- Take into account the humidity control in energy recovery systems.

Select Suitable Materials, Equipment, and Assemblies for Unavoidably Wet Areas. Bathrooms, showers, spas, indoor pools, kitchens, entryways, custodial closets, and conditioned garages are examples of spaces that are likely to get wet during ordinary operation. Use materials that can tolerate regular wetting without damage and that can dry quickly enough to avoid moisture-related problems; these include tile, glass, resins, cement-based products, and moisture-resistant forms of oriented strand board and gypsum board. Lastly, specify mold-resistance testing criteria that are appropriate for materials (e.g., a score of 10 when tested using ASTM D3273, *Standard Test Method for Resistance to Growth of Mold on the Surface of Interior Coatings in an Environmental Chamber*) (ASTM 2012b).

Control Impacts of Landscaping and Indoor Plants on Moisture. There are potential advantages and disadvantages associated with the presence of plants as a component of the building envelope (e.g., green roofs or vertical gardens) or on other locations in the interior space (e.g., atrium gardens and living walls). Benefits of green roofs are thought to include reduction in stress (i.e., thermal stress) on the waterproofing membrane, reduction in heat island effects in urban areas, and reduction in stormwater runoff. However, the building architect and the landscape (garden) architect need to work together to ensure that the water-proofing membrane is installed with excellent workmanship and that penetrations through the membrane are avoided. Inside buildings, the water released from plants can be significant, so it must be well-managed to avoid humidity and moisture-related issues.

Control of Moisture and Dirt in Air-Handling Systems. Moisture and dirt in air handlers can lead to several problems. Moisture can build up at several locations within the air handlers, including at the outdoor air intake, in the condensate pans, and downstream from the air humidification unit. (See Figure 8-7 for example of good drain pan design.) Dirt and debris can reduce the airflow below the design levels, particularly when dirt buildup on the fan blades is sufficient to alter the fan curve. Also, dirt buildup on cooling coils can restrict airflow and increase the fan power needed and, coupled with the insulating properties of the dirt, can lead to energy waste. Finally, moisture and dirt can facilitate microbial growth in the air-

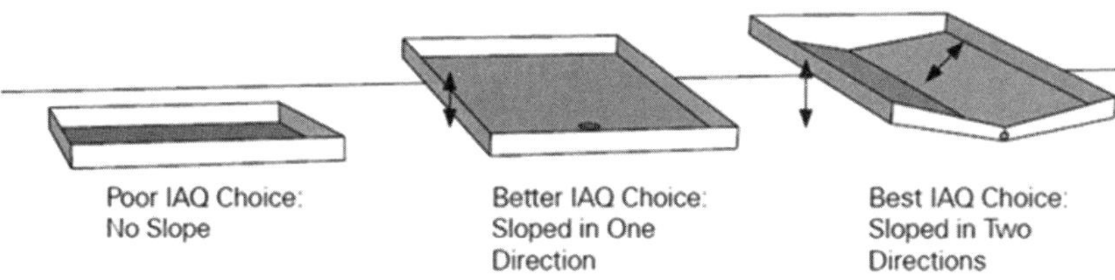

Source: *Indoor Air Quality Guide, Adapted from Stanke et al. (1998).*

Figure 8-7 Drain pan examples with and without slope.

handling unit, in turn leading to the release of potentially irritating or disease-causing biological and chemical agents into the occupied spaces. ASHRAE/ACCA/ANSI Standard 180 (ASHRAE 2012c) identifies 10 inspection/maintenance tasks that relate to control of dirt, microbial growth, and/or excess moisture in air handlers.

Control Moisture Associated with Piping, Plumbing Fixtures, and Ductwork. Mold growth can occur on cold water pipes or cold air supply ducts with inadequate thermal insulation or a failed vapor retarder and result in material damage, IAQ problems, and adverse health impacts on occupants. Liquid water from condensation can damage materials nearby such as ceiling tiles, wood materials, and paper-faced wallboard. Leaks from poorly designed plumbing within walls or risers may go unnoticed until damage, including mold growth, becomes evident in occupied spaces. Implementation of design strategies that limit condensation on cold-water piping and ducts and that reduce the likelihood of piping leaks hidden in building infrastructure will lessen the likelihood of these potential problems.

Operation for IAQ

All building systems will in some way contribute the contaminant emissions in a building, whether they are passive or active sources, particulate or gaseous, or moisture-related. Therefore, maintaining proper operation of building systems is key to maintaining IAQ.

Maintain Proper Building Pressurization. Maintaining proper pressure relationships ensures that air flows in desirable directions. For instance, it is critical is to ensure that air in a space with high contaminant levels does not escape into adjacent spaces that are expected or required to be cleaner. Facilities built for health care, laboratory, food service, swimming (natatoriums), and many industrial uses have spaces that require negative pressurization. Pressurization is also used to minimize the flow of warm, humid air through the building envelope toward cooler surfaces, which may result in condensation inside the envelope and microbial growth. ASHRAE Guideline 10-2016 (ASHRAE 2016) provides guidance on pressure relationships.

Pressurization is achieved by maintaining supply airflow higher than exhaust and return airflow in positively pressurized spaces and higher exhaust flows than supply flow in negatively pressurized spaces. Stack effects and wind pressures, however, must also be considered and overcome when determining system flow rates (Persily 2004). Thus, the effectiveness of a building pressurization strategy will be greatly enhanced by building an airtight enclosure, as discussed earlier in this chapter.

Facilitate Access to HVAC Systems for Inspection, Cleaning, and Maintenance. Access to the HVAC systems at critical locations such as after supply air silencers, at junctions, duct interiors, and near duct ends should be made available for regular inspection, cleaning, and maintenance. Specific locations, size, and type of access panels are described in the HVCA and NAIMA (North American Insulation Manufacturers Association) guidelines (HVCA 2002, NAIMA 2007). In existing ductwork systems without access panels, it is possible to cut through duct walls or use existing openings such as grilles, registers, and diffusers (NAIMA 2007, NADCA 2006). Prior to cutting, considerations should be given to the potential presence of asbestos-containing wrapping and lead paint on the ductwork. Regardless of the access means used, the openings should be able to be closed airtight. There should be no obstruction or alteration to the airflows and no degradation to the thermal or functional integrity of the HVAC system.

Control *Legionella* in Water Systems. Legionnaires' disease, also called *Legionellosis*, is a bacterial disease commonly associated with water-based aerosols that have originated from warm water sources (Figure 8-8). It is often associated with poorly maintained cooling towers and potable water systems, including from fountains and whirlpool spas. Legionnaires' disease can develop when a susceptible individual inhales or aspirates aerosolized water droplets containing *Legionella* bacteria. The Centers for Disease Control and Prevention (CDC) estimates 5000 cases of Legionnaires' disease are reported in the United States annually,

Photo courtesy of Hal Levin.

Figure 8-8 Bellevue Stratford Hotel, Philadelphia. Site of 1976 Legionnaires' disease outbreak in which 29 people died from exposure in the hotel and 5 people died from exposure in the vicinity.

though many cases may go undiagnosed. Approximately 9% of cases are fatal, though the rate may approach 40% in health-care-acquired cases because of underlying health issues that exist in that population (Garrison et al. 2016).

Thus, preventive maintenance should be carried out for effectively operating cooling towers and other evaporative equipment. These maintenance activities include using effective drift eliminators, periodical cleaning, and employing a water treatment program with a proper biocide. Regular monitoring of *Legionella* levels in cooling-tower water is recommended by many experts. ANSI/ASHRAE Standard 188-2015, *Legionellosis: Risk Management for Building Water Systems* (ASHRAE 2015a) and Guideline 12-2000, *Minimizing the Risk of Legionellosis Associated with Building Water Systems* (ASHRAE 2000) share knowledge about monitoring potential sources of *Legionella* as well as providing technical guidelines for investigation, control and prevention of Legionnaires' disease. A toolkit, *Developing a Water Management Program to Reduce Legionella Growth and Spread in Buildings: A Practical Guide to Implementing Industry Standards*, is available from the CDC website and provides a checklist to help identify if a water management program is needed, and to help identify where *Legionella* could grow and spread in a building and ways to reduce risk of contamination.

Consider Ultraviolet Germicidal Irradiation. Ultraviolet germicidal irradiation (UVGI) has been used successfully to control airborne pathogens such as *Mycobacterium tuberculosis*, the bacterium that causes tuberculosis. Commercially available lamps tend to be low-pressure mercury vapor, which emit UV primarily at 254 nm and damages the DNA of irradiated microorganisms, rendering them unable to reproduce.

In addition to inactivating airborne microorganisms, UVGI directed at environmental surfaces can damage microorganisms present or growing on the surface. Lower-intensity UVGI is effective for surface inactivation because irradiation is applied continuously. UVGI from lamps in air-handling unit (AHU) plenums has been used successfully to inactivate microorganisms present on airstream surfaces such as cooling coils and drain pans. Recent field studies (including ASHRAE RP-1738 [Bahnfleth and Firrantello 2017]) have identified the energy savings potential of UVGI when used to eliminate microorganism growth (biofouling) on cooling coils (Firrantello et al. 2016; Wang et al. 2016).

The designer must be aware that the use of UVGI lamps in AHUs, ductwork, and upper air spaces (areas of the room that are located above eye level where their UVGI fixtures are shielded from direct view from below but still provide additional protection). requires careful attention to safety considerations to prevent in-advertent exposure of people to UV light. For example, lockout/tagout procedures are necessary to prevent accidental turning on of UVGI lamps when facility maintenance personnel are working in AHUs. Refer to Chapter 17 of the *ASHRAE Handbook—HVAC Systems and Equipment* (ASHRAE 2016d) and Chapter 60 of the *ASHRAE Handbook—HVAC Applications* (ASHRAE 2015b) for a comprehensive review of other safety considerations For more complete information, refer to *Indoor Air Quality Guide*, Section 4.5 (ASHRAE 2009a).

Properly Vented and Unvented Combustion Equipment. For vented combustion equipment, it is important that the venting system be designed, installed, and maintained. Otherwise, potentially dangerous pollutants and combustion gases may be released into occupied indoor spaces. All vented combustion appliances will be supplied with information describing the requirements of the venting system that must be used with that appliance. For example, natural draft equipment usually requires a draft hood that allows room air and combustion products to flow up the venting system.

When natural- or induced-draft-vented combustion appliances are located within the building envelope and installed with venting systems that comply with manufacturer instructions, they will be able to operate in the presences of negative pressures. However, excessive negative pressures may cause backdrafting. This can best be avoided by providing makeup air that is interlocked with the exhaust or otherwise ensuring that extreme negative pressures are not created.

Unvented combustion appliances are designed such that the combustion products can be released into the occupied space. The combustion products of properly operating equipment will be primarily carbon dioxide and water, but small amounts of carbon monoxide and oxides of nitrogen, particularly nitrogen dioxide, will also be present. It is important to follow the manufacturer's installation instructions, which will include information on acceptable room size and provision of adequate ventilation air. Unvented combustion appliances need regularly scheduled maintenance to ensure that emission rates for contaminants remain within allowable tolerances. See ASHRAE's position document "Unvented Combustion Devices and Indoor Air Quality" (ASHRAE 2012c) for additional information on these appliances.

Provide Local Capture and Exhaust for Point Sources of Contaminants. It is much more efficient to capture contaminants at the source and directly exhaust to the outdoors than it is to allow them to be released into the space and then remove them with dilution ventilation. It is important to ensure that pollutants that have been captured, and are being transported through an exhaust system, do not leak back into occupied space or air distribution systems. Exhaust systems should be at a negative pressure relative to surrounding space such that any leaks in the system result in airflow into the exhaust. Alternately, Standard 62.1 requires exhaust systems that are not negatively pressurized relative to their surroundings be sealed in accordance with Sheet Metal and Air Conditioning Contractors' National Association (SMACNA) Seal Class A.

Continuously Monitor and Control Outdoor Air Delivery. Accurate monitoring and control of outdoor air intake at the air handler is important for providing the correct amount of outdoor airflow to a building (not overventilating or underventilating). In particular, it has been common practice for designers to use fixed minimum outdoor air dampers. However, the approach does not necessarily provide good control of outdoor air intake rates, particularly in VAV systems (ASHRAE 2010b).

Continuous monitoring of the outdoor air rates at the air handler does not guarantee that the proper amount of ventilation is delivered locally within the building. Poor air mixing both in the ductwork and in the occupied space, especially in larger and more complex air distribution systems, can result in parts of a building receiving less than the design minimum amount of ventilation.

Effectively Distribute Ventilation Air to the Breathing Zone. Ventilation only works when the air is effectively distributed to the breathing zone. Different methods of air distribution have different efficiencies. For an inefficient system, the quantity of outdoor air at the air handler needs to be increased in order to provide the minimum quantities in the breathing zone that are required by code and by ASHRAE Standard 62.1 (ASHRAE 2016b).

The airflow rate that needs to be distributed to a zone varies by the effectiveness E_z of the distribution within the room. Values of $E_z > 1.2$ are attributed to systems with superior distribution, such as low-velocity displacement ventilation (DV). Values of $E_z = 1.0$ are attributed to systems with effective distribution, such as mixing ventilation (MV) air distribution systems (ceiling supply of cool air and ceiling return, ceiling supply of warm air and floor return and UFAD systems).

THERMAL COMFORT AND CONTROL

Thermal conditions indoors, combined with occupant activity and clothing, determine occupant comfort (Figure 8-10). Achieving and maintaining thermal

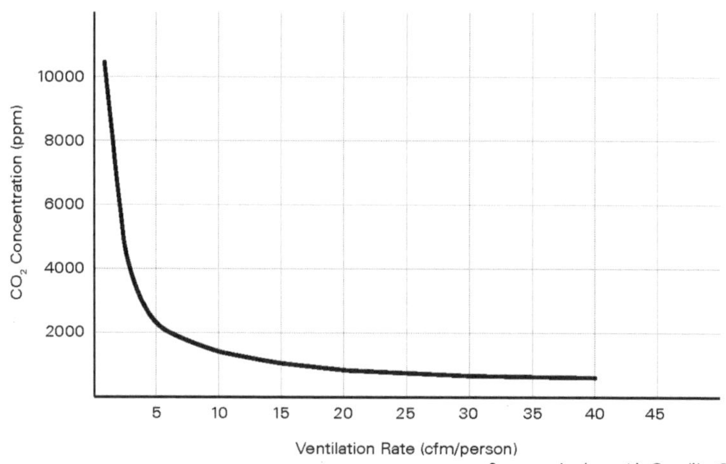

Source: *Indoor Air Quality Guide*

Figure 8-9 Relationship between ventilation rate and pollutant concentration from an indoor source with no outdoor source.

comfort for the occupants of buildings drives much of the energy consumption by buildings. Thermal comfort is an essential requirement for building users, is a key metric of a building's overall IEQ, and has an influence on the health and productive performance of building occupants. The challenge facing HVAC&R engineers and designers of green buildings is to provide healthy, thermally comfortable, and productive environments as energy efficiently as possible. The thermal environment within a building is also important for the long-term durability of the building fabric and its contents.

What Is Thermal Comfort, and What Factors Are Involved?

ANSI/ASHRAE Standard 55-2017, *Thermal Environmental Conditions for Human Occupancy* (ASHRAE 2017c) defines thermal comfort as "that condition of mind that expresses satisfaction with the thermal environment." The standard acknowledges that the sensation of thermal comfort experienced is specific to the individual and that it is difficult to satisfy all occupants. Therefore, the standard specifies the combination of environmental conditions, in relation to a given mix of meta-

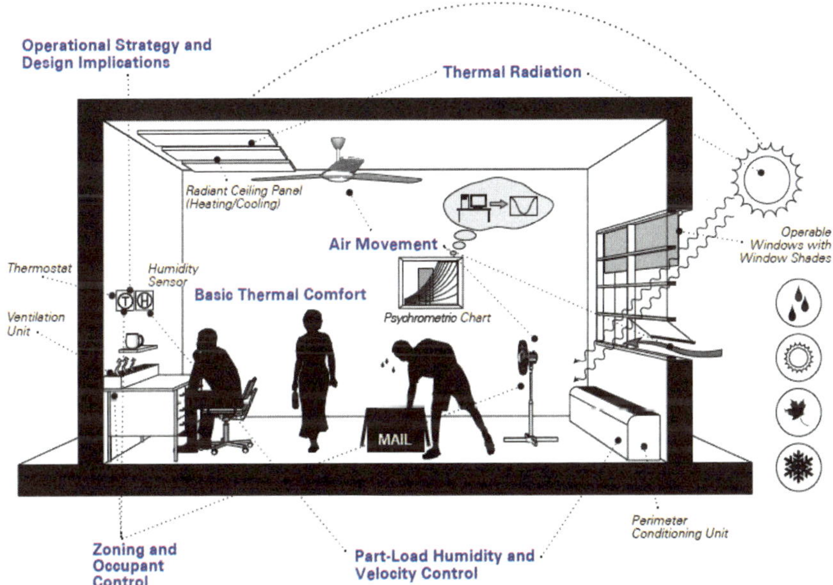

Source: *Indoor Air Quality Guide*, based on a sketch by Larry Schoen.

Figure 8-10 Thermal comfort is maintained by correctly understanding all of the relationships in the figure and applying effective tools, as illustrated in Strategy 7.6 in the *Indoor Air Quality Guide*.

bolic rates and clothing ensembles, that will satisfy a majority of occupants within the space.

There are six primary factors that affect the conditions for thermal comfort of occupants. The environmental factors are dry-bulb air temperature, radiant temperature, air speed, and relative humidity. The personal factors are metabolic rate and thermal insulation of clothing.

Evaluating Thermal Comfort

Metrics such as Predicted Mean Vote and Predicted Percentage Dissatisfied are documented in ASHRAE 55-2017 (ASHRAE 2017c) for mechanically conditioned spaces. However, in environments where thermal conditions are regulated primarily by the occupants, thermal comfort is found in practice to be achieved over a wider band of environmental conditions. This is the result of an individual's expectations and prior experience of thermal environments (including the environment outdoors), availability of control, as well as other effects where such adaptive opportunities are available to individuals. This suggests that occupants take a more active role in achieving thermal comfort, rather than simply being passive recipients of their thermal environment, which is termed *active thermal comfort.*

Role of HVAC and Control for Thermal Comfort

The five primary HVAC operation factors that impact thermal comfort, and thus imply the need for adequate control are air movement (e.g., from diffusers, avoidance of drafts), air temperature and humidity conditions, vertical air temperature gradients (floor to ceiling air temperatures), and radiant surface temperatures (e.g., warm or cold floors, ceilings, or other surfaces).

Localized and personalized control of conditioned air meets individual preferences, as well as potentially reduces heating and cooling energy demands. Shutdown of localized delivery to unoccupied spaces can also improve energy efficiency.

LIGHT AND ILLUMINATION

Maintaining proper light levels is essential for virtually all human occupancy and, although not commonly thought of in terms of the health consequences (except eyesight), insufficient or inappropriate illumination can result in other health and safety hazards as well as poor performance. Energy efficiency in illumination is an increasingly important part of the effort to reduce buildings' environmental impacts, including those associated with energy use. As buildings become less energy intensive overall through increased use of more efficient HVAC approaches (both passive and active), much of the additional energy savings will be achieved through more effective use of daylight and electricity for illumination (Figure 8-11).

Illumination Sources. There is some evidence that higher daily light doses benefit all individuals' wellbeing. Increasing use of daylighting together with changing to more energy-efficient light sources (e.g., light-emitting diodes [LEDs]) might provide the opportunity to increase light levels during daytime, without increasing energy use. This all depends on the spectrum, intensity, duration, pattern, and timing of light exposure (see Figure 8-12 for the light spectrum). However, guidelines on the best combinations of these factors do not yet exist.

Automatic Control. Illumination will increasingly be controlled by occupancy sensors, timers, and, in some cases, by timer-limited switches. This is common in Europe and Japan already and will become increasingly common in green buildings in the United States.

Individual Control. There is strong evidence that individuals differ in their preferences for illumination levels (Despenic et al. 2017). Having the opportunity to choose one's own light levels creates a positive mood that carries over into other

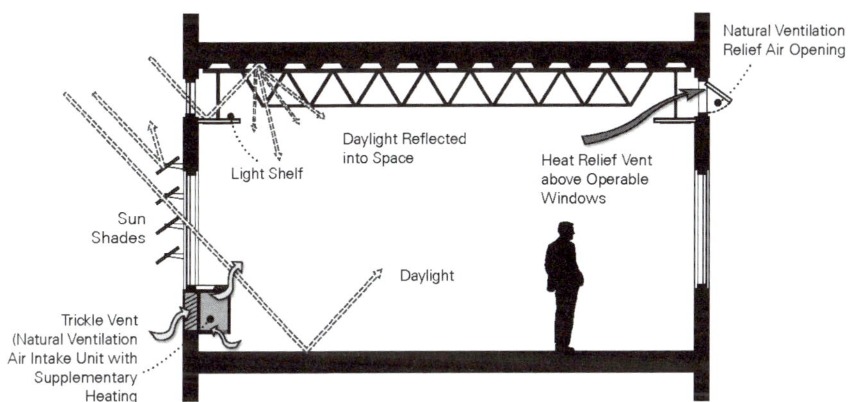

Source: *Indoor Air Quality Guide*

Figure 8-11 Daylight used in combination with shading and mixed-mode ventilation.

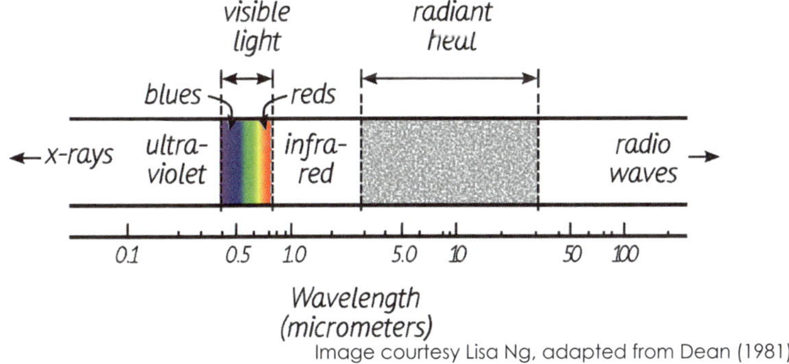

Image courtesy Lisa Ng, adapted from Dean (1981)

Figure 8-12 Light spectrum.

dimensions of well-being. On average, individual (personal) control also reduces energy use by 10%–20% over typical fixed light levels. There is some research interest in the possibility that individual control over light source spectrum, which is possible with LED lighting, might add to these benefits

Sophisticated Glazing. A well-designed window glazing system should reduce indoor and/or solar heat gain or loss. “Smart” windows include those that use electrochromic devices, suspended particle devices, microblinds, and liquid crystal devices to adjust the amount of incoming light according to the designer or occupant’s intent.

Exterior Shading. Increased shading of windows subject to direct insolation is important, as well as using more sophisticated light guides and interior shade to increase useful daylight while preventing both visual and thermal discomfort.

Light Tubes. These roof-mounted tubes can carry daylight into the core of a building, supplying a large fraction of a building’s lighting requirements.

Design Considerations. Lighting design for green buildings could consider the following:

- Electrical illumination should be accomplished with user-controlled task lighting wherever practical.
- Lighting controls in buildings should be localized, allow variable light levels, and be accessible to users wherever practical.
- A variety of lighting conditions through diverse illumination sources should be provided to allow maximum user choice in control and regulation of interior lighting.
- Design of lighting and performance standards for lighting should be based on light delivered to actual task stations in buildings, not on illumination of general spaces. Standards should discourage delivery of overillumination to nontask areas.
- Public health authorities and building operators and occupants should consider the possible connection between occupant discomfort or illness and environmental lighting, including spectral characteristics, illumination levels, electromagnetic and chemical emissions from fixtures, and user access to effective lighting control.
- The use of newer light sources should be applied cautiously until the health effects research indicates that the impacts are within acceptable limits.

Additional discussion of lighting systems can be found in Chapter 11 of this book.

ACOUSTICS

Noise and vibration are commonly combined into one term, *acoustics*. Acoustics involves far more than the control of noise emitted from mechanical equipment. Sound intensity, often characterized in decibels, can be too strong or too weak for building occupants to successfully use buildings for their intended purpose. Individuals have different sensitivities to different noise frequencies, so control of noise spectrum also matters. Speech recognition requires sufficient volume without competing background noise, while concentration during activity or sleep and relaxation may require significantly quieter conditions.

Poor acoustic conditions are cited as the top source of dissatisfaction in many postoccupancy evaluations of building occupants, particularly in green buildings. It is incumbent on architects and designers to consider the resulting acoustic quality during design to improve indoor environmental quality and occupant satisfaction.

Good acoustic environments are possible in green buildings. Some suggestions include the following:

- Limit the amount of glass in a building to the amount just needed for daylighting requirements.
- For natural ventilation designs, place ventilation inlets on the building envelope so that they do not face external noise sources (e.g., traffic), and use louvered openings designed for acoustic attenuation purposes on the exterior and interior of the building.
- Consider using a sound-masking system in low-background-noise conditions to improve speech privacy.
- Place acoustical absorption on allowable surfaces within spaces and/or consider using hanging acoustical elements such as banners, drapes, etc. Also explore acoustically absorptive materials that may be more green (i.e., made from recycled content and/or recyclable).

More information to help with designing appropriate acoustics in green buildings may be found in Field (2008) and Chapter 48 of *ASHRAE Handbook—HVAC Applications* (ASHRAE 2015b).

REFERENCES AND RESOURCES

Published

ASHRAE 2000. Guideline 12-2000, *Minimizing the Risk of Legionellosis Associated with Building Water Systems*. Atlanta: ASHRAE.

ASHRAE 2009a. *Indoor Air Quality Guide*. Atlanta: ASHRAE. Available for free download at www.ashrae.org/FreeIAQGuidance.

ASHRAE 2009b. ASHRAE Standard 160, *Criteria for Moisture-Control Design Analysis in Buildings*. Atlanta: ASHRAE.

ASHRAE 2010. *Standard 62.1-2010 User's Manual*. Atlanta: ASHRAE.

ASHRAE 2012a. ANSI/ASHRAE Standard 52.2, *Method of Testing General Ventilation Air Cleaning Devices for Removal Efficiency by Particle Size*. Atlanta: ASHRAE.

ASHRAE 2012b. ASHRAE Position Document. *Limiting Indoor Mold and Dampness in Buildings*. Atlanta: ASHRAE. www.ashrae.org/about-ashrae/position-documents.

ASHRAE 2012c. ANSI/ASHRAE/ACCA Standard 180-2012, *Standard Practice for Inspection and Maintenance of Commercial-Building HVAC Systems*. Atlanta: ASHRAE.

ASHRAE. 2012d. ASHRAE Guideline 1.5, *The Commissioning Process for Smoke Control Systems*. Atlanta: ASHRAE.

ASHRAE. 2013. ANSI/ASHRAE Standard 202, *Commissioning Process for Buildings and Systems*. Atlanta: ASHRAE.

ASHRAE 2015a. ANSI/ASHRAE Standard 188-2015, *Legionellosis: Risk Management for Building Water Systems*. Atlanta: ASHRAE.

ASHRAE 2015b. *ASHRAE Handbook—HVAC Applications*. Atlanta: ASHRAE.

ASHRAE 2016a. Guideline 10-2016, *Interactions Affecting the Achievement of Acceptable Indoor Environments*. Atlanta: ASHRAE.

ASHRAE 2016b. ANSI/ASHRAE Standard 62.1-2016, V*entilation for Acceptable Indoor Air Quality*. Atlanta: ASHRAE.

ASHRAE 2016c. ANSI/ASHRAE Standard 62.2-2016, *Ventilation and Acceptable Indoor Air Quality in Low-Rise Residential Buildings*. Atlanta: ASHRAE.

ASHRAE 2016d. ASHRAE/ANSI/IES Standard 90.1-2016, *Energy Standard for Buildings except Low-Rise Residential Buildings*. Atlanta: ASHRAE.

ASHRAE. 2017a. *ASHRAE Design Guide for Dedicated Outdoor Air Systems*. Atlanta: ASHRAE.

ASHRAE. 2017b. ANSI/ASHRAE/USGBC/IES Standard 189.1-2017, *Standard for the Design of High-Performance Green Buildings Except Low-Rise Residential Buildings*. Atlanta: ASHRAE. (Will be incorporated into the 2018 version of the *International Green Construction Code* [IgCC].)

ASHRAE. 2017c. ANSI/ASHRAE Standard 55-2017, *Thermal Environmental Conditions for Human Occupancy*. Atlanta: ASHRAE.

ASHRAE 2017d. *ASHRAE Handbook—Fundamentals*. Atlanta: ASHRAE.

ASTM 2010. ASTM E779-10, *Standard Test Method for Determining Air Leakage Rate by Fan Pressurization*. West Conshohocken, PA: ASTM International.

ASTM 2012a. ASTM D7663-12. *Standard Practice for Active Soil Gas Sampling in the Vadose Zone for Vapor Intrusion Evaluations* West Conshohocken, PA: ASTM International.

ASTM 2012b. ASTM D3273-12, *Standard Test Method for Resistance to Growth of Mold on the Surface of Interior Coatings in an Environmental Chamber*. West Conshohocken, PA: ASTM International.

Bahnfleth, W., and J. Firrantello. 2017. Field measurement and modeling of UVC cooling coil irradiation for HVAC energy use reduction. ASHRAE Research Project 1738 Final Report.

Dean, E. 1981. *Energy Principles in Architectural Design*. Sacramento: California Energy Commission. (Out of print, but photocopy versions are available by request from Publications, California Energy Commission).

Despenic, M., S. Chraibi, T. Lashina, and A. Rosemann. 2017. Lighting preference profiles of users in an open office environment. *Building and Environment* 116:89–107. https://doi.org/10.1016/j.buildenv.2017.01.033.

Emmerich, S.J., and A.K. Persily, Analysis of U. S. Commercial Building Envelope Air Leakage Database to Support Sustainable Building Design. *International Journal of Ventilation* 12(4): 331-343.

Firrantello, J., W. Bahnfleth, R.D. Montgomery, and P. Kremer. 2016. Field Study of Energy Use-Related Effects of Ultraviolet Germicidal Irradiation of a Cooling Coil. *ASHRAE Transactions* 121:1.

Fisk, W.J., H. Destaillats, and M.A. Sidheswaran. 2011. Saving Energy and Improving IAQ through Application of Advanced Air Cleaning Technologies, *REVA Journal*, May 2011 27:29.

Field, C. 2008. Acoustic Design in Green Buildings. *ASHRAE Journal* 50(9): 60–70.

Garrison, L.E., J.M. Kunz, L.A. Cooley, M.R. Moore, C. Lucas, S. Schrag, J. Sarisky, and C.G. Whitney. 2016. Vital Signs: Deficiencies in Environmental Control Identified in Outbreaks of Legionnaires' Disease—North America, 2000–2014. *Morbidity and Mortality Weekly Report* 65(22).

GSA 2014. P-100, *Facilities Standards for the Public Buildings Service*. Washington, DC: U.S. General Services Administration.

HVCA 2002. TR/19, *Internal Cleanliness of Ventilation Systems. Guide to Good Practice*. London: Heating and Ventilating Constructors' Association

International Code Council (ICC), International Energy Conservation Code (IECC). 2015.
https://codes.iccsafe.org/public/document/code/545/9719040

Lau, J., X. Lin, and G. Yuill 2014. RP-1547—*CO_2-Based Demand Controlled Ventilation for Multiple Zone HVAC Systems*. Atlanta: ASHRAE.

Morrison G., R. Shaughnessy, and J. Siegel, 2014. *In-duct air cleaning devices: Ozone emission rates and test methodology*. Sacramento, CA: California Resources Board and the California Environmental Protection Agency.

NADCA. 2006. *ACR, The NADCA Standard for Assessment, Cleaning & Restoration of HVAC Systems*. Mt. Laurel, NJ: National Air Duct Cleaners Association.

NAIMA. 2007. *Cleaning Fibrous Glass Insulated Duct (Ductboard/Duct Board) Systems-Recommended Practices. Recommendations for Opening and Closing*

Insulated Ducts Before and After Cleaning. Alexandria, VA: North American Insulation Manufacturers Association.

NIST Climate Sustainability Tool. www.bfrl.nist.gov/IAQanalysis/software/CST-desc.htm

Persily, A. 2004. Building Ventilation and Pressurization as a Security Tool. *ASHRAE Journal* 46(9):18-24.

Stanke, D., B. Bradway, A. Hallstrom, and N. Bailey. 1998. *Managing Building Moisture*. Piscataway, NJ: Trane Inc.

USACE 2012. *U.S. Army Corps of Engineers Air Leakage Test Protocol for Building Envelopes*. Champaign, IL: U.S. Army Corps of Engineers Research and Development Center.

Wang, Y., C. Sekhar, W.P. Bahnfleth, K.W. Cheong, and J. Firrantello. 2016. Effectiveness of an Ultraviolet Germicidal Irradiation System in Enhancing Cooling Coil Energy Performance in a Hot and Humid Climate. *Energy and Buildings* 130: 321–29.

Online

Business and Institutional Furniture Manufacturers Association (BIFMA) www.bifma.org.

California Department of Public Health, Indoor Air Quality Program. https://ww2.arb.ca.gov/our-work/programs/indoor-air-quality.

Carpet and Rug Institute www.carpet-rug.org.

CDC, *What Clinicians Needs to Know About Legionaires' Disease* www.cdc.gov/legionella/downloads/fs-legionella-clinicians.pdf.

Collaborative for High Performance Schools www.chps.net/dev/Drupal/node.

The European Working Group for Legionella Infections (EWGLI) L http://ecdc.europa.eu/en/activities/surveillance/ELDSNet/Documents/EWGLI-Technical-Guidelines.pdf.

GreenGuard Environmental Institute www.greenguard.org/en/index.aspx.

PHIUS 2015. Passive Building Standard—North America. www.phius.org/phius-2015-new-passive-building-standard-summary

U.S. Environmental Protection Agency, Nonattainment Areas for Criteria Pollutants (Green Book) https://www.epa.gov/green-book

CHAPTER NINE

ENERGY CONVERSION AND DISTRIBUTION SYSTEMS

For there to be heating, cooling, lighting, and electric power throughout a building, the energy required by these functions is usually distributed from one or more central points.

This is usually accomplished through the flow of steam or a hydronic fluid, air, electrons (electricity), and sometimes a refrigerant. Air and hydronic flows in particular also serve the function of disposition of used air (exhaust) or liquid waste. Because the air and hydronic media being moved are often at different thermal levels (i.e., warm or cold), opportunities are offered to incorporate green design techniques. There are several practical techniques whereby energy from one flow stream can be transferred usefully to another.

Several GreenTips on energy recovery systems are included in this chapter.

ENERGY EXCHANGE

Cooling and Dehumidifying Media

Chilled Water (CHW). Circulating CHW is generally the least energy-efficient process for refrigerated air conditioning. However, in many cases it is the most cost effective, particularly for large buildings. The relatively low thermal efficiency results from it being a two-step process. The most common process uses a refrigerant circulated at a low temperature through an electrically powered recycling vapor compression system. The refrigerant then chills a secondary refrigerant (water) to the temperature needed for cooling or both cooling and dehumidifying the air supplied to the occupied zones. CHW can also be produced by a heat-driven absorption process that uses a mixture of water with lithium bromide or ammonia.

Some installations can benefit from energy reduction by using thermal storage using either CHW or ice, if further reduced in temperature and melted for later use.

The CHW is pumped through air-conditioning coils that may be in a central plant room, in several building zones, or in individual rooms. Further energy losses result from heat gained through insulation on the CHW piping.

Liquid Desiccant (LD). Circulating LD is a single-step process that uses heat (i.e., that from natural gas or from liquid petroleum gas [LP], solar, or waste heat sources) when available. LD generally does not need piping to be insulated.

Synthetic Refrigerant (SR). Circulating SR from an electrically powered recycling vapor-compression system can be directly used in air-conditioning coils that are in the same housing or plant room. Alternatively, the SR can be distributed to air-conditioning coils in several building zones or in individual rooms (rooms with variable refrigerant flow [VRF] systems).

Natural Refrigerant (NR). Ammonia, hydrocarbons, and CO_2 should also be considered.

Heating Media

Hot Water. Distributing hot water throughout the building is generally the most efficient heating energy distribution method. Hot-water heating systems are classified as low-, medium-, or high-temperature water:

- Low-temperature water systems operate at temperatures of 250°F (121°C) and below.
- Medium-temperature water systems operate at temperatures between 250°F and 350°F (121°C and 177°C).
- High-temperature water systems operate at temperatures above 350°F (177°C).

Steam. Because generating steam requires heating water above its boiling point and therefore more energy than sensibly heating water, steam is considered an inefficient way to transport heating energy for space and domestic water heating, because such high temperatures are not needed for these applications. Nonetheless steam has some advantages and disadvantages.

Advantages

- Steam flows to the terminal usage point without aid of external pumping.
- Steam systems are not greatly affected by the height of the distribution system, which has a significant impact on a water system.
- Steam distribution can readily accommodate changes in the system terminal equipment.
- Major steam distribution repair does not require piping draindown. The thermodynamics of steam use are effective and efficient.

Disadvantages

- Steam equipment, such as steam traps and condensate pumps, requires frequent maintenance and replacement.
- Steam piping systems often have dynamic pressure differentials that are not easily controlled by typical steam control appurtenances.
- Returning condensate by gravity is not always possible. Lifting the steam back to the source can be a challenge because of pressure differential dynamics, space constraints, and wear and tear on equipment.
- Venting of steam systems (through pressure relief valves, flash tanks, boiler feed tanks, condensate receivers, or safety relief valves,) must be done properly and must be brought to the outside of the building.
- High-pressure steam systems require a full-time boiler plant operator, which adds to operating costs.

Classification

Steam distribution is either one-pipe or two-pipe. One-pipe distribution is defined as where both the steam supply and the condensate travel through a single pipe connecting the steam source and the terminal heating units. Two-pipe distribution is defined as where the steam supply and the steam condensate travel through separate pipes. Two-pipe steam systems are further classified as gravity return or vacuum return.

Steam systems are also classified according to system operating pressure. Low pressure is defined to be 15 psig (103 kPa) or less, medium pressure is defined to be between 15 and 50 psig (103 and 345 kPa), and high pressure is defined to be over 50 psig (345 kPa).

Selection of steam pressure is based on the constraints of the process served. The level of system energy rises with the system pressure. Higher steam pressure may allow smaller supply distribution pipe sizes, but it also increases the temperature difference across the pipe insulation and may result in more heat loss. Higher steam pressure also dictates the use of pipes, valves, and equipment that can withstand the higher pressure. This translates to higher installation costs.

Piping

Supply and return piping must be installed to recognize the thermodynamics of steam and to allow unencumbered steam and condensate flow. Piping that does not slope correctly—that is, it is installed with unintended water traps or has leaks—will not function properly and will increase system energy use. Careful installation will result in efficient and effective operation. If steam or condensate leaks from the piping system, additional water must be added to make up for the losses. Makeup water is chemically treated and is an operating cost.

Condensate Return

The energy conservation and operational problems that arise when using steam often occur because the system design and/or installation does not properly address the issue of returning condensate to the boiler or cogeneration plant. The simplest and most efficient way to return condensate is via gravity. When this is not possible, the designer needs to clearly understand issues of lift, condensate rate, and pressure differential to ensure that operational problems (e.g., water hammer and reduction in capacity) do not occur.

Efficient Steam System Design and Operation Tips

- Preheat boiler plant makeup water with waste heat.
- Use ecofriendly chemical treatment.
- Recover flash steam, making sure to understand pressure differential of flash steam compared to system pressure serving equipment where flash steam recovery is used.
- Minimize use of pumped condensate return.
- Do not use steam pressure for lifting condensate return.
- Consider clean steam generators where steam serves humidification or sterilization equipment.
- Consider the use of blowdown heat recovery when economically feasible

Sources of Further Information

Manufacturers of equipment-control valves, steam traps, and other devices are valuable resources. There are multiple websites that contain system design information; perform a search on the Internet for steam piping design. Chapter 11, Steam Systems, of the *ASHRAE Handbook—HVAC Systems and Equipment* will also prove useful (ASHRAE 2016). See the "References and Resources" section at the end of this chapter for additional resources.

ENERGY DELIVERY METHODS

Distribution Paths (Ducts/Pipes/Wires)

Ducts, pipes, and wires are used to move media. Proper sizing is a balancing act between energy use and cost and between material use and first cost. In terms of space consumed in running these carriers, and in terms of the energy carried for the cross section of the carrier involved, wires (through electricity) have the capacity for carrying the most energy, followed by pipes (hydronics), then followed by ducts (air). Another consideration is that the different energy-carrying media have different characteristics and capabilities in terms of meeting the requirements of the spaces served. The above factors will have some influence on determining the type of HVAC system to be used.

Chilled-water systems distribute cold water to terminal cooling coils to provide dehumidification and cooling of conditioned air or cooling of a process. They can also serve cooling panels in occupied spaces. Chilled-water panels, which serve as a heat sink for heat radiated from occupants and other warm surfaces to the radiant panel, can be used to reduce the sensible load normally handled solely by mechanical air cooling. The percentage by which the load can be reduced depends upon the panel surface area and dew-point limitations, which are necessary to avoid any possibility of condensation.

Condenser water systems connect mechanical refrigeration equipment to outdoor heat dissipation devices (e.g., cooling towers or water- or air-cooled condensers). These, in turn, reduce the temperature of the condenser water by rejecting heat to the outdoor environment.

Media Movers (Fans/Pumps)

Basics: Power, Flow, and Pressure. If air conditioning (heating/cooling) could be produced exactly where it is needed throughout a building, overall system efficiency would increase, because there would be no additional energy used to move (distribute) conditioned water or air. For acoustic, aesthetic, logistic, and a variety of other reasons, this ideal is seldom realized. Therefore, fans and pumps are used to move energy in the form of water and air. Throughout this process, the goal is to minimize system energy consumption.

Minimization of a media mover's power at full-load and part-load conditions is the goal. Understanding how fan and pump power change with flow and pressure is imperative.

$$\text{Power} \sim \text{Flow} \times \text{Pressure}$$

As flow drops, so does the pressure differential through pipes, chillers, and coils.

Pressure drop through these devices varies approximately with $flow^{1.85}$. This, in turn, means that the power required to cause flow through these devices varies approximately as $flow^{2.85}$. In an ideal world, power changes with the cube of the flow. While the 3.0 exponent is not fully achieved in practical application, it is nevertheless clear that reducing flow can drastically reduce energy use. For example, reducing the flow by 20% will result in a 50% reduction in energy use.

However, there are some pressure drops in typical systems that do not change as flow decreases. These include the pressure differential set point that many system controls use, cooling tower static lift, and pressure drop across balancing devices. Therefore, for flow through these system components, power varies directly with flow.

To reduce pressure drop, the pipe or duct size should be maximized and valve and coil resistance minimized. (Duct and pipe sizes are discussed later in this chapter.) Coil sizes should be maximized within the space allowed to reduce pressure drop on both the water side and the air side. The ideal selection requires striking an economic balance between first cost and projected energy savings (i.e., operating cost).

Chilled-Water Pumps. Historically, a design chilled-water temperature differential ΔT across air-handling unit cooling coils of 10°F (5.5°C) was used, which resulted in a flow rate of 2.4 gpm/ton (2.6 L/min per kW). In recent years, the 60% increase in required minimum chiller efficiency from a 3.8 coefficient of performance (ASHRAE 1975) to a 6.3 coefficient of performance (0.56 kW/ton) for large (>600 ton) water-cooled centrifugal chillers operating at full load (ASHRAE 2016) has led to a reexamination of the assumptions used in designing hydronic media flow paths and in selecting movers (pumps) with an eye to reducing energy consumption.

Taylor et al. (1999) came to the following conclusion:

> *...the trend for most applications is that higher chilled-water* ΔT*s result in lower energy costs, and they will always result in the same or lower first costs.*

Simply stated, increase the temperature difference in the chilled-water system to reduce the chilled-water pump flow rate and increase chiller efficiency with warmer return water. This reduces installed cost and operating costs. The *CoolTools Chilled Water Plant Design and Performance Specification Guide* (Taylor et al. 1999) recommends starting with a chilled-water temperature difference of 12°F to 20°F (7°C to 11°C). It is important to understand that for the chiller plant to use a higher chilled-water ΔT, you must start at the load coils (i.e., the air-handling units).

Condenser Water Pumps. In the same manner, design for condenser water flow has traditionally been based on a 10°F (5.6°C) ΔT, which equates to 3 gpm/ton (3.2 L/min/kW). Today's chillers will give approximately a 9.4°F (5.2°C) ΔT with that flow rate. The CoolTools guide states, "Higher ΔTs will reduce first cost (because pipes, pumps, and cooling towers are smaller), but the net energy-cost impact may be higher or lower depending on the specific design of the chillers and tower" (Taylor et al. 1999).

The CoolTools document states in its summary that there are times you can "have your cake and eat it too." In most cases larger ΔTs and the associated lower flow rates will not only save installation cost but will usually save energy over the course of the year. This is especially true if a portion of the first cost savings is reinvested in more efficient chillers. (Taylor et al. 1999). The document further recommends a design method that starts with a condenser water temperature difference of 12°F to 18°F (7°C to 10°C).

Thus, reducing chilled- and condenser-water flow rates (conversely, increasing the ΔTs) can not only reduce operating cost but, more importantly, can free funds from being applied to the less-efficient infrastructure and allow them to be applied toward increasing overall efficiency elsewhere.

Variable-Flow Systems. The previous discussion suggests that variable-flow air and water systems are an excellent way to reduce system energy consumption. Variable flow (either air or water) is required by Standard 90.1 (ASHRAE 2016) for many applications, but it may be beneficial in even more applications than the standard requires. Today's most used technology for reducing flow is the variable-frequency drive.

HYDRONICS

Pumping heated water and chilled water is common system design practice in many buildings. Water is often diluted with an antifreeze fluid to avoid freezing in extremely cold conditions. There are many approaches in using these *hydronic* systems.

Piping, Flow Rates, and Pumping

Each of these systems uses two pipes—a supply and a return—to make up the piping circuit, and each uses one or more pumps to move the water through the circuit. Information on the design and characteristics of these district energy and various hydronic systems can be found in Chapters 12–15 of the 2016 *ASHRAE Handbook—HVAC Systems and Equipmen*t (ASHRAE 2016).

Cost-effective design depends on consideration of the system constraints. Piping must be sized to provide the required load capacities and arranged to provide necessary flow at full- and part-load conditions. Design is determined by several system characteristics and selections including

- Supply and return water temperatures
- Flow rates at individual heat transfer units
- System flow rate at design condition
- Piping distribution arrangement
- System water volume
- Equipment selections for pumps, boilers, chillers, and coils
- Temperature control strategy

Pumping energy can be a significant portion of the energy used in a building. In fact, pumping energy is roughly equal to the inverse fifth power of pipe diameter, so a small increase in pipe size has a dramatic effect on lower pumping energy. Traditionally, it was common to select heating water flow for coils based on a temperature difference of 20°F (11.1 K) between the supply and return. Flow rate in gal/min (gpm) was calculated by dividing the heating load in Btu/h by 10,000 (1 Btu/lb·°F × 8.33 lb/gal × 60 min/h × 20°F temperature difference) (L/s was calculated by dividing the kW heating load by 4.187 [specific heat capacity of water, kJ/kg·K] × 11.1 K). As long as the cost of energy was cheap, this method was widely used.

Flow rate can easily be reduced by one-half of the 20°F (11.1 K) value by using a 40°F (22.2 K) temperature difference. The impact on pump flow rate is significant. The temperature difference selected depends upon the ability of the system to function with lower return water temperatures. Certain types of boilers can function with the low return water temperatures, while others cannot. Care must be taken in selecting the boiler type, coupled with supply and return water design temperatures. In specific instances, a low return water temperature could damage the boiler due to the condensation of combustion gases.

Lower flow rates could allow smaller pipe sizes, and pipe size, along with flow, affects pumping energy. A goal should be established for the pump power to be selected. A small increase in some or all of the pipe distribution sizes could reduce the pump energy needed for the system. When this goal is established and attained in the finished design, the concept and energy usage will be achieved. A reasonable goal can be expressed using the water transport factor equation adjusted to reflect kilowatts (multiply horsepower by 0.746). Measurements of efficient designs indicate a performance of 0.026 kW/ton (0.007 kW/kW$_R$ [with kW$_R$ being refrigeration cooling capacity]) being served as a reasonable goal for 10°F (5.6 K) ΔT systems. Adjusting the flow rate and ΔP variables in this formula will quickly show the benefits of larger pipes or lower flow rates (greater ΔT).

For instance, calculate the power (kW) required with the modified water transport factor equation:

$$\text{pump hp} = \frac{(Q)(\Delta P)(sg)}{(3968)(\eta_P)(\eta_M)} \quad \text{(I-P)}$$

$$\text{pump kW} = \frac{(Q)(\Delta P)(sg)(9.81)}{(1000)(\eta_P)(\eta_M)} \quad \text{(SI)}$$

where

Q = flow rate, gpm (L/s)
ΔP = pump head, ft (m)
sg = specific gravity
η_P = pump efficiency
η_M = electric motor efficiency

Create a performance index by solving the equation for 1 ton of cooling. That produces an answer in kilowatts per ton (kilowatts per kilowatts of refrigeration). A typical condenser system might use 3 gpm/ton (0.054 L/s·kW$_R$) of cooling and a ΔP of 100 ft of water (300 kPa) with an 82% efficient pump and a 92% efficient motor. Inserting these variables into the above equation, the derived answer for 1 ton (3.5 kW_R) of cooling would be 0.075 kW or 0.075 kW/ton (0.021 kW/kW_R).

Compare this index to the performance of an efficient design that increases piping and fitting size, and therefore reduces ΔP to, say, 30 ft (90 kPa) total dynamic head (TDH), keeps the water flow (gpm or L/s) the same, and uses an 85% efficient pump with a 92% efficient motor. (Because the most efficient motors cost the same as less efficient motors, almost everyone is now using premium efficiency motors.) If we solve again for kilowatts, we get 0.022 kW/ton (0.006 kW/kW_R).

This shows us we can reduce pumping energy by 71% by lowering the TDH from 100 to 30 ft (300 to 90 kPa) and selecting a more efficient pump. One author of this guide has personally measured systems with the characteristics indicated above. If the average cost of electricity per kWh is $0.08, and we are pumping for a 1000 ton (3517 kW_R) chiller, the annual operating cost difference would be more than $37,000/yr.

If the system lasts 20 years, the improved system would save $740,000 in electricity costs. Of course, the obvious question is: "How much does it cost to increase pipe and fitting size and pump efficiency?" The answer will vary by project, but keep in mind that efficient pumps cost no more than inefficient ones. Larger pipes and fittings cost more than smaller pipes and fittings and, in one author's experience, the cost of increasing the pipe size one size (going from a 10 in. [250 mm] pipe to a 12 in. [300 mm] pipe) is recovered in electrical cost in less than 18 months.

Using the formula above and carefully measuring existing projects, we can establish performance goals for our designs such as the ones in Table 9-1.

The reader is cautioned that the percentage savings shown in Table 9-1 are individual savings from each component and should not be mistaken with the overall

Table 9-1: Example Performance Goals

Components	Typical kW/ton (kW/kW_R)	Efficient kW/ton (kW/kW_R)	Improvement, kW/ton (kW/kW_R)	% Savings
Chiller	0.56 (0.159)	0.42 (0.119)	0.14 (0.04)	25
Cooling tower	0.045 (0.0128)	0.012 (0.0034)	0.033 (0.0094)	73
Condenser water pump	0.0589 (0.0167)	0.022 (0.0063)	0.0369 (0.0104)	63
Chilled-water pump	0.0765 (0.0218)	0.026 (0.0074)	0.0505 (0.0144)	66
Total waterside system	0.7404 (0.210)	0.48 (0.136)	0.2064 (0.0739)	35

building energy performance, which is a more holistic number involving all components.

At times, reductions in pipe sizes to reduce first cost are suggested as value engineering. However, energy usage of the building may be greatly impacted: pump size and horsepower could well be increased. To be truly valid, value engineering should also include refiguring the life-cycle cost of owning and operating the building. These factors can also be applied to chilled-water systems, except that chilled-water systems have a smaller range of temperatures within which to work.

System Volume

In small buildings, water system volume may relate closely to boiler or water-chiller operation. When pipe distribution systems are short and of small water volume, both boilers and water chillers may experience detrimental operating effects. Manufacturers of water chillers state that system water volume should be a minimum of 3 to 10 gal/installed ton of cooling (0.054 to 0.179 L/s·kW$_R$). In a system smaller than this and under light cooling load conditions, thermal inertia coupled with the reaction time of chiller controls may cause the units to short-cycle or shut down on low-temperature safety control.

Similar detrimental effects may occur with small modular boilers in small systems. Under light load conditions, boilers may experience frequent short cycles of operation. An increase in system volume may eliminate this condition.

Antifreeze Additions

Fluid properties can greatly affect system performance. Antifreeze generally increases pressure drop and decreases heat transfer effectiveness. This leads to reduced system efficiency and perhaps increased cost because of the need for larger components. Therefore, the design professional should first examine the system to determine if antifreeze is an absolute necessity or whether water, with proper antifreeze safeguards, could be used.

If antifreeze is truly necessary, remember the following:

- Determine whether it is freeze or burst protection that is being sought. If an affected component (such as a chiller) does not need to be operated during freezing conditions, perhaps only burst protection is necessary. The amount of antifreeze needed can be greatly reduced, although slush may form in the pipes.
- Use only the minimum antifreeze necessary to provide protection; higher concentrations will simply reduce performance.
- Balance all environmental aspects of the antifreeze. Understand that while some antifreeze solutions are viewed as less toxic, they can significantly increase system installation and operating costs. Often, the greatest environmental cost of antifreeze is the increased energy consumption.
- Consult the manufacturer for more information on burst and freeze protection.

AIR

Using air as a means of energy distribution is almost universal in buildings, especially as a means of providing distributed cooling to spaces that need it. A key characteristic that makes air so widely used is its importance in maintaining good indoor air quality. Thus, air distribution systems are not only a means for energy distribution, they also serve the essential role of providing fresh or uncontaminated air to occupied spaces.

That said, it is important to understand that properly distributing air to heat and cool spaces in a building is less efficient from an energy usage perspective than using hydronic or steam distribution. Air distribution systems are often challenging to design because, for the energy carried per cross-sectional area, they take up the most space in a ceiling cavity and are frequent causes of space conflicts among disciplines (i.e., structural members, plumbing lines, heating/cooling pipes, light fixtures, etc.). Many approaches have been tried to better coordinate duct runs with other services, or even to integrate them in some cases.

Another challenging aspect of air system design is that there are temperature limitations on supply air (because air is the energy medium that directly contacts occupants). While care must always be taken in how air is introduced into an occupied space, it is especially critical the colder the supply temperature gets. Low-temperature air supply systems offer many advantages in terms of green design, but an especially critical design aspect is avoiding occupant discomfort at the supply air/occupant interface.

Most of the same principles that were discussed under hydronic energy distribution systems apply to air systems with respect to temperature differences, carrier size, and driver power and energy. Thus, there are plenty of opportunities for applying green design techniques to such systems and for seeking innovative solutions.

Energy Usage

The *Advanced Variable Air Volume (VAV) Systems Design Guide* (CEC 2005) recommendations for duct design include the following:

- Run ducts as straight as possible to reduce pressure drop, noise, and first cost.
- Use standard length straight ducts and minimize both the number of transitions and joints.
- Use round spiral duct wherever it can fit within space constraints.
- Use radius elbows rather than square elbows with turning vanes whenever space allows.
- Use either conical or 45° taps at VAV box connections to medium pressure duct mains.
- Specify sheet metal inlets to VAV boxes; do not use flex duct.
- Avoid consecutive fittings because they can dramatically increase pressure drop.

- For VAV system supply air duct mains, use a starting friction rate of 0.25 in. to 0.30 in. per 100 feet (2 to 2.5 Pa/m). Gradually reduce the friction rate at each major juncture or transition down to a minimum friction rate of 0.10 in. to 0.15 in. per 100 feet (0.8 to 1.2 Pa/m) at the end of the duct system.
- For return air shaft sizing, maximum velocities should be in the 800 to 1200 fpm (4 to 6 m/s) range through the free area at the top of the shaft (highest airflow rate).
- To avoid system effect, fans should discharge into duct sections that remain straight for as long as possible, up to 10 duct diameters from the fan discharge to allow flow to fully develop.
- Use duct liner only as much as required for adequate sound attenuation. Avoid the use of sound traps.

In addition, the following are additional ways to reduce energy usage when designing air distribution systems:

- Require SMACNA's Seal Class A for all duct systems.
- Use VAV concepts for constant-volume systems.
- Use static pressure reset logic for all VAV applications.
- Minimize pressure drops of air-handling unit components and duct-mounted accessories (e.g., sound attenuators, dampers, diffusers, filter, energy wheels, etc.)

Sources of Further Information

Manufacturers of equipment—fans, ductwork, air-handling units, dampers, and other air devices—are valuable sources. There are also multiple websites that contain system design information. Air systems are also discussed in Chapters 20 and 21 of the 2017 *ASHRAE Handbook—Fundamentals* (ASHRAE 2017), and in Chapters 19 through 28 of the 2016 *ASHRAE Handbook—HVAC Systems and Equipment* (ASHRAE 2016). Another source for information is SMACNA (www.smacna.org).

Refer also to the *Advanced Variable Air Volume (VAV) System Design Guide* published by the California Energy Commission (CEC 2007) for a detailed discussion of variable-flow air systems.

ENERGY CONVERSION/GENERATION EQUIPMENT

Heat Generators

Considerable improvements in the seasonal efficiency of conventional heating plant equipment (e.g., boilers and furnaces) have been made over the last several decades. Designers should verify claims of equipment manufacturers by reviewing documented data of this equipment to prove the efficiency ratings are accurate.

There are two general types of boilers: fire-tube and water-tube. Boilers are classified as high-pressure or low-pressure and steam boiler or hot-water boiler. Boilers that operate higher than 15 psig (103 kPa) are called high-pressure boilers.

A hot-water boiler, strictly speaking, is not a boiler. It is a fuel-fired hot-water heater. Because of its similarities in many ways to a steam boiler, the term *hot-water boiler* is used.

Among steam, hot-water, water tube, and fire tube boilers, there are conventional atmospheric boilers and condensing boilers. Condensing boilers are far more efficient than conventional atmospheric boilers. See GreenTip #9-17 for more information on condensing boilers.

Cooling Generators

Chilled-water plants are most often used in medium to large facilities. Their benefits include higher efficiency, reduced maintenance costs, and redundant capacity (in comparison to decentralized plants).

Because chilled-water temperature can be closely controlled, chilled-water plants have an advantage over direct-expansion systems, because they allow air temperatures to be closely controlled.

Generally, a chilled-water plant consists of the following elements:

- Chillers
- Chilled-water pumps
- Condenser-water pumps (for water-cooled systems)
- Cooling towers (for water-cooled systems) or air-cooled condensers
- Associated piping, connections, and valves

Often, chilled-water plant equipment will be in a single, central location, allowing system control, maintenance, and problem diagnostics to be performed efficiently. Chilled-water plants also allow redundancy to be easily designed into the system by designing for firm capacity. Firm capacity is calculated with the largest piece of equipment not operating. With firm capacity, one piece of equipment can be maintained or repaired and the system still has the ability to meet peak loads.

Chiller Types

Electric chillers used within chiller plants employ either a scroll-, reciprocating-, screw-, or centrifugal-type compressor (in order of increasing size). Models can be offered with either air-cooled or water-cooled condensers, with the exception that today's centrifugal compressors are water-cooled only.

Steam-driven turbines or absorption chillers, powered by steam, hot water, natural gas, or other hot gases, are used in many central plants to balance steam and electric demands in combined heat and power plants, and to offset high electric demand or consumption charges.

The various chiller types are used most often, though not exclusively, in the following situations.

Electric Motor-Driven Chillers:

- Low to moderate electric consumption and demand charges prevail
- Air-cooled heat dissipation is preferred
- Condenser heat recovery is desired

Steam-Driven and Absorption Chillers:

- Part of a combined heat and power plant
- Low fuel (e.g., natural gas) costs prevail
- High electric demand charges prevail
- Plentiful source of heat available (its main use usually for other functions)

Heat Pumps. A heat pump is another means of generating cooling as well as heating using the same piece of equipment. There is usually an array of them used for a project, and they are generally distributed throughout the building (i.e., they are not part of a central plant). Buildings that have consistent demands for both chilled and hot water, such as hospitals, are good candidates to use a larger heat pump as a central plant. (See the GreenTips at the end of this chapter on various heat pump system types.)

Thermal Energy Storage (TES). TES is a technique that has been encouraged by electricity pricing schedules where the off-peak rate is considerably lower than the on-peak rate. Cooling, in the form of chilled water or ice, is generated during off-peak hours and stored for use during on-peak hours. Although not refrigeration equipment per se, the technique can usually reduce the size of refrigeration equipment—or obviate the need for adding a chiller to an existing plant.

The characteristics, merits, and cost factors of TES for cooling, as well as numerous reference sources, are presented in GreenTip #9-12.

ENERGY CONVERSION/GENERATION PLANT DESIGN

When a designer puts together a chilled-water plant, there are many design parameters to optimize. They include fluid flow rates and temperatures, pumping options, plant configuration, and control methods. For each specific application, the design professional should understand the client's needs and desires and implement the chiller plant options that best satisfy them.

Between July 2011 and June 2012, *ASHRAE Journal* published a series of five articles describing a systematic approach to the design of chilled-water plants (Taylor 2011a, 2011b, 2011c, 2012a, and 2012b). This series of articles also summarized the ASHRAE Self-Directed Learning (SDL) course Fundamentals of Design and Control of Central Chilled-Water Plants and the research that was performed to support its development.

"Chilled-water distribution system selection" (*ASHRAE Journal*, July 2011). This article discussed distribution system options, such as primary-secondary and primary-only pumping, and provided a simple application matrix to assist in selecting the best system for the most common applications (Taylor 2011a).

"Condenser water distribution system selection" (*ASHRAE Journal*, September 2011). This article discussed piping arrangements for chiller-condensers and cooling towers, including the use of variable-speed condenser water pumps and water-side economizers (Taylor 2011b).

"Pipe sizing and optimizing ΔT" (*ASHRAE Journal*, December 2011). This article discussed how to size piping using life-cycle costs, then how to use pipe sizing to drive the selection of chilled-water and condenser water temperature differences (ΔTs) (Taylor 2011c).

"Chillers and cooling tower selection" (*ASHRAE Journal*, March 2012). This article addressed how to select chillers using performance bids and how to select cooling tower type, control devices, tower efficiency, and wet-bulb approach (Taylor 2012a).

"Optimized control sequences" (*ASHRAE Journal*, June 2012). This article included a discussion of how to optimally control chilled-water plants, focusing on all variable-speed plants (Taylor 2012b).

The intent of the SDL (and these articles) is to provide simple yet accurate advice to help designers and operators of chilled-water plants to optimize life-cycle costs without having to perform rigorous and expensive life-cycle cost analyses for every plant. In preparing the SDL, a significant amount of simulation, cost estimating, and life-cycle cost analysis was performed on the most common water-cooled plant configurations to determine how best to design and control them. The result is a set of improved design parameters and techniques that provide much higher-performing chilled-water plants than common rules-of-thumb and standard practices.

Plant Configuration (Multiple Chillers). The most prevalent chiller plant configuration is the primary-secondary (decoupled) system. This system allows the flow rate through each chiller to remain constant, yet accommodates a reduction in pumping energy, since the system water flow rate varies with the load.

Becoming more common are variable-primary-flow systems that also vary the flow through the chiller evaporators. New chiller controls allow this. Often these systems can be installed at a reduced cost when compared to primary-secondary systems since fewer pumps (and their attendant piping, connections, valves, fittings, and electrical draws) are required. These systems can also save energy in comparison to the primary-secondary configuration, due to more efficient chiller operation and reduced pumping energy.

The designer should always review the overall use of energy within a facility and employ systems (including heat recovery systems) that interact with one another so as to minimize the overall energy consumption of the entire chilled-water plant.

Cooling System Heat Sinks

The function of building HVAC systems is to exchange a significant amount of thermal energy between the building and the surrounding environment, which could be air, ground, or a water body. Historically, air has been the primary source of heat rejection from buildings because of its availability. A good example of that process is a cooling tower, which is covered in the next section. In the last decade, the ground has emerged as a good and green choice for heat rejection (refer to GreenTip #9-11, Ground-Source Heat Pumps).

Designers should pay attention when selecting unconventional heat rejection sources or sinks, as some may have indirect environmental impacts. For example, systems have been installed that use nearby deep water in a lake or ocean as a heat sink, which results in significant energy savings. This technique has been used in Scandinavian systems for approximately 20 years and more recently in colder regions in North America. However, this heat rejection technique, if not studied properly, may increase the water temperature in the source and may affect the aquatic wildlife. The possible net energy savings are impressive, but these must be balanced against potential adverse impacts on local aquatic environment.

Cooling Tower Systems

Cooling towers, which are generally the heat rejection equipment in a water-cooled system, are a very efficient method of heat rejection to air, especially in dry climates. Cooling towers remove heat by evaporation and can cool close to the ambient wet-bulb temperature (the difference between the leaving water and ambient wet-bulb being defined as the *approach temperature*). Unless the air is totally saturated (i.e., 100% relative humidity), the wet-bulb temperature is always lower than the dry-bulb. Thus, water cooling allows more efficient condenser operation than air cooling. In a typical cooling tower operation, about 1% of the recirculated water flow is evaporated. This evaporation will cool the remaining 99% of the water for reuse. In addition to evaporation, some recirculated water must be bled from the system to prevent soluble and semisoluble minerals from reaching too high a concentration. This bleed or blowdown is usually sent to a publicly owned treatment works (POTW). The HVAC designer should always consult with the plumbing engineer on the local regulations governing the discharge of cooling tower blowdown.

Refer to Chapter 12, Water Efficiency, for a more detailed discussion on the water treatment of cooling tower.

Types of Cooling Towers. With respect to drawing air through the tower, there are two types of cooling towers:

- **Natural Draft.** Uses buoyancy via a tall chimney. Warm, moist air naturally rises because of the density differential compared to the dry, cooler outdoor air. Warm, moist air is less dense than drier air at the same pressure. This moist air buoyancy produces an upwards current of air through the tower.
- **Mechanical Draft.** Uses power-driven fan motors to force or draw air through the tower. There are two types of mechanical draft towers:
  - *Induced Draft.* A mechanical draft tower with a fan at the discharge (at the top) that pulls air up through the tower. The fan induces hot, moist air out of the discharge. This produces low entering and high exiting air velocities, reducing the possibility of recirculation in which discharged air flows back into the air intake.
  - *Forced Draft.* A mechanical draft tower with a blower-type fan at the intake. The fan forces air into the tower, creating high entering and low exiting air velocities. The low exiting velocity is much more susceptible to recirculation. With the fan on the air intake, the fan is more susceptible to complications caused by freezing conditions. Another disadvantage is that a forced draft design typically requires more motor horsepower than an equivalent induced draft design. The benefit of the forced draft design is its ability to work with high static pressure. Such setups can be installed in confined spaces and even in some indoor situations.

With respect to water to air direction of flow, there are two types of cooling towers:

- **Crossflow.** Crossflow is a design in which the airflow is directed perpendicular to the water flow. Airflow enters one or more vertical faces of the cooling tower to meet the fill material. Water flows (perpendicular to the air) through the fill by gravity. The air continues through the fill and thus past the water flow into an open plenum volume. Lastly, a fan forces the air out into the atmosphere.
  Advantages of the crossflow design include the following:
  - Gravity water distribution allows smaller pumps and maintenance while in use.
  - Nonpressurized spray simplifies variable flow.
- **Counterflow.** In a counterflow design, the airflow is directly opposite to the water flow. Airflow first enters an open area beneath the fill media, and is then drawn up vertically. The water is sprayed through pressurized nozzles near the top of the tower, and then flows downward through the fill, opposite to the airflow.
  Advantages of the counterflow design include the following:
  - Spray water distribution makes the tower more freeze resistant.
  - Breakup of water in spray makes heat transfer more efficient.

Maintenance

An often overlooked method for minimizing environmental impact is maintenance. Cooling towers operate outdoors under changing conditions. Wind damage to inlet air louvers, excessive airborne contamination, clogging of water distribution nozzles, and mechanical problems can best be prevented and quickly corrected with periodic inspections and maintenance.

Variable-Speed Fans

The designer is encouraged to include variable-frequency drives (VFDs) on the cooling-tower fans. With the decreasing cost of these drives, their life-cycle cost is likely to be favorable. VFDs on fans can save energy by reducing the fan speed to better match the load, reduce the fan on/off cycling (thus reducing inrush currents), and can accomplish other tasks such as fan rotation reversal (beneficial in cold climates).

DISTRICT ENERGY SYSTEMS

District energy (DE) systems involve the provision of thermal energy (heating and/or cooling) from one or more central energy plants to multiple buildings or facilities via a network of interconnecting thermal piping. Generally, district heating systems deliver heat as steam or hot water, while district cooling systems deliver cooling as chilled water or chilled secondary coolant (such as an aqueous glycol or an aqueous sodium nitrite solution) or even as a refrigerant.

DE systems often deliver multiple significant positive impacts to the local building environments that they serve. These typical impacts include the areas outlined in the following sections.

Energy Consumption

Heating and/or cooling buildings using DE systems can affect overall energy consumption in various ways, from modest increases or decreases to very dramatic decreases in fuel and energy consumption. The energy consumed within the boundaries of DE-served buildings will, of course, be dramatically reduced compared to a baseline building with its own dedicated boilers and chillers. This energy reduction within the buildings will be offset by the energy consumed in the central DE plants and in distributing the thermal energy from the central plants to the customer buildings. If the central DE plant uses similar technology (e.g., gas boilers and electric chillers) as otherwise used in the individual buildings, there may be little or no net reduction in energy use. However, the larger (and generally more efficient) DE plant equipment more than offsets extra energy consumed in the distribution of the thermal energy to the buildings for at least a slight net reduction in overall energy consumption. Reductions in overall fuel and energy consumption are achieved through the ability of DE plants to more readily and more economically use alternative technologies than is the case for individual buildings. These technologies include, but are not limited to, dual-fuel boilers; alternative fuel boilers (including renewable fuels such as low-energy landfill gas, municipal solid waste, wood waste, etc.); high-efficiency boilers; high-efficiency chillers; alternative energy-efficient refrigerants (e.g., ammonia); nonelectric chiller plants (e.g., absorption chillers, engine-driven chillers, or turbine-driven chillers); hybrid chiller plants (with various combinations of electric and nonelectric chillers); energy-efficient series or series-parallel chiller configurations for high ΔT systems; thermal energy storage (TES); cogeneration of combined heating and power (CHP); trigeneration or combined cooling, heating, and power (CCHP); and the use of natural renewable thermal energy (such as geothermal heat for district heating systems and cold deep water sources [e.g., lakes or oceans] for district cooling systems). In addition, DE plants are better able to meet the changing loads of the system.

Central energy plants associated with DE systems (compared to the alternative of dispersed, multiple, smaller boilers and chiller plants within individual buildings), generally have higher levels of operational efficiency and reliability. This is because larger DE plants can more easily justify sophisticated design, automated optimized control systems, more attentive maintenance programs, and more highly trained and focused operations and maintenance personnel. Central energy plants are also better able to match the system load with central equipment versus part load, especially at part load throughout a campus.

Emissions

As is the case with energy consumption, DE serves to eliminate many emissions from the local building environment, such as boiler exhausts and chiller plant heat rejection. Some emissions are of course relocated to the site of the central DE plant. However, just as DE plants tend to have higher levels of energy efficiency, they tend to have lower levels of emissions vs. those associated with individually heated and cooled buildings. And for DE systems utilizing one or more of the alternative technologies (as cited above), the overall emissions can be significantly reduced in terms of air pollutants (i.e., SO_X, NO_X, and precipitates) and greenhouse gases (i.e., CO_2, NO_X, and some refrigerants).

Noise and Vibration

Without boilers and/or chillers in the buildings, the occupants of DE-served buildings experience a local building environment that is free from the potential noise and vibrations associated with such equipment.

Chemical Supplies and Blowdown

DE systems can also eliminate or greatly mitigate the presence of potentially harmful fuel and chemicals to be handled within the occupied buildings. The storage, handling, and disposal of fuel, boiler water treatment chemicals, refrigerants, condenser water treatment chemicals, and chilled-water treatment chemicals can all be removed to the location of the central DE plants. Thus, the potential for related chemical spills, disruptions, and associated hazards are avoided within the occupied building environments.

Space Usage and Aesthetics

DE systems provide improved space usage for the occupants of the individual buildings. Also, there is no longer a need for local boiler exhaust stacks and/or for local chiller plant heat rejection via, for example, roof-mounted cooling towers. In addition to the improved aesthetics of having buildings without such stacks and towers, multiple and sometimes unsightly exhaust plumes from stacks and towers are also removed from the local building environment.

Other Factors for Consideration

DE systems do require additional infrastructure (e.g., a piping distribution network from the central plant to the various buildings within the network). A full-fledged, life-cycle assessment of the net benefits from a DE system should take items including construction of the additional piping network and additional space requirements of distributed equipment into consideration.

Where DE Systems are Used

Routinely used on university campuses, DE systems are also often used for other institutional applications (including schools, hospital and medical facilities, airports, military installations, and other federal, state, and local government facilities), for privately owned multibuilding commercial/industrial facilities, and for thermal utilities serving urban business districts. DE systems serve as few as two buildings or as many as many hundreds of buildings. The ideal times to consider utilizing DE for serving the heating and/or cooling needs of a building are either during master planning and new construction or during expansion or renovation of buildings or their HVAC systems.

DISTRIBUTED ELECTRICITY GENERATION

One opportunity for energy conservation in buildings is the use of on-site generation systems to provide both distributed electric power and thermal energy (otherwise wasted heat from the generation process), which can be used to meet the thermal loads of the building.

Distributed generation (DG) provides electricity directly to the building's electrical systems to offset loads that would otherwise have to be met by the utility grid (see Figure 9-1). The waste heat from that generation process goes through a heat recovery mechanism where it may be provided as heat to meet loads for conven-

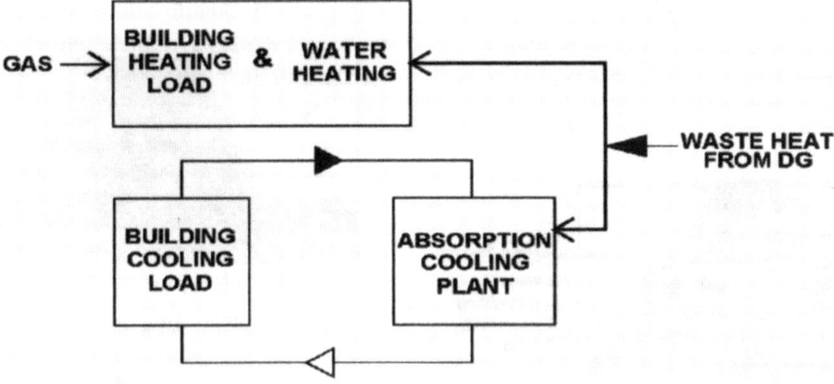

Image courtesy of Malcolm Lews, CTG Energetics, Inc.

Figure 9-1 Thermal uses of waste heat.

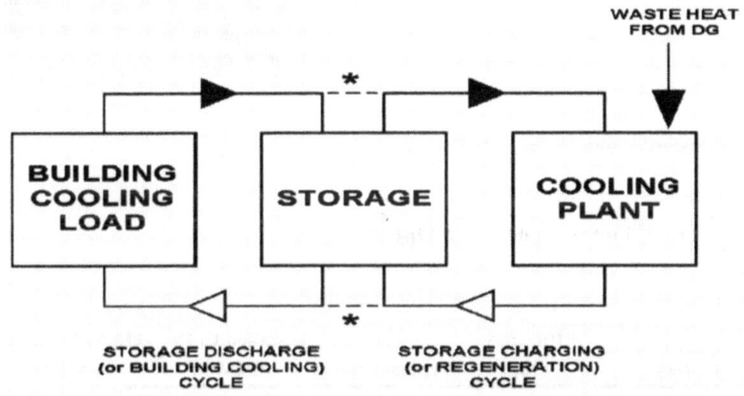

Image courtesy of Malcolm Lews, CTG Energetics, Inc.

Figure 9-2 Thermal energy storage and waste heat usage.

tional heating (such as space heating, reheat, and domestic water heating) or for specialized processes. Alternatively, that heat energy, if at a sufficiently high temperature, can be used to power an absorption chiller to produce chilled water to meet either space- or process-cooling loads. This is shown in Figure 9-2.

Any timing differences between the generation of the waste heat from the DG system and the thermal needs of the building can be handled by a chilled-water TES system. This concept may also allow downsizing the absorption cooling system (so it does not need to be sized for the peak cooling load).

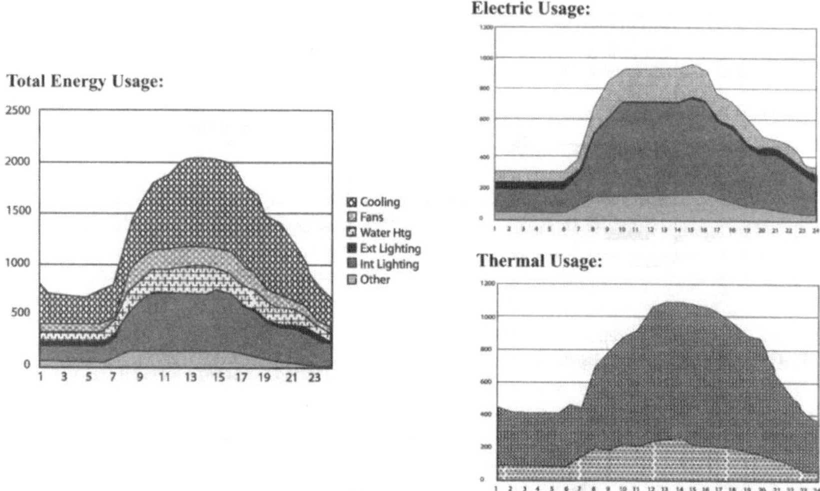

Image courtesy of Malcolm Lews, CTG Energetics, Inc.

Figure 9-3 Relationship between electric and thermal energy.

The overall usable energy value from the fuel input to the generation process is only about 30% (or less) if there is no waste heat recovery, but it can be more than 70% if most or all the waste heat is able to be used. This increased system efficiency can have a radical impact on the economics of the energy systems, because almost two-and-one-half times the useful value is obtained from the fossil fuel purchased. System sizing is generally done by evaluating the relative electrical and thermal loads over the course of the typical operating cycle and then selecting the system capacity to meet the lesser of the thermal or electrical loads. See Figure 9-3 showing comparative thermal and electrical energy for typical office buildings.

If the DG system is sized for the greater of the two, then there will be a net waste of energy produced, because it is seldom economical to sell electricity back to the grid. A key design issue that arises here is whether or not the system is being designed to improve efficiency, as discussed above, or as a baseload on-site generation system for purposes of improving the reliability of the electric and/or thermal energy supply. Either of these is a legitimate design criterion, although the goal, from a green design standpoint, almost always focuses on the energy-efficient strategy.

DG Technologies

Technologies that can be used for this type of generation include engine-driven generators, microturbine-driven generators, or fuel cells. Typically, each uses natural gas as the input fuel. There are advantages and disadvantages associated with these technologies, which are summarized briefly in Table 9-2.

Engine-Driven Generator (EDG). This technology has been around the longest of the three and is in many ways the least-expensive option. It produces a relatively high temperature of waste heat that can be more effectively used by the heat recovery systems. Disadvantages, however, include air pollution, acoustical impacts, and

Table 9-2: Comparison of Power Generation Options

Generation Option	Efficiency, %	Typical Size	Installed Costs, $/kW	Operations and Maintenance Costs, $/kWh
Engine driven natural gas cogeneration with heat recovery	70%–80%	100kW–5MW	$3300–$1600	$0.007–$0.02
Turbine-driven natural gas cogeneration with heat recovery	70%–80%	1MW–10MW	$5000–$3000	$0.003–$0.01
Fuel cell natural gas cogeneration with heat recovery	56%–80%	200kW–1MW	$8000–$5000	$0.070–$0.090
Wind turbine	N/A	5kW–1MW	$8500–$2500	$0.001–$0.007
Photovoltaic	8%–20%	1kW–2MW	$4800–$3500	$0.003–$0.005

Note the following:

Fuel cells vary greatly in cost by type of technology (e.g., phosphoric acid fuel cell [PAFC] at low end, solid oxide fuel cell [SOFC] at high end), size of application, and gas cleanup requirements (e.g., biogas treatment).

Installed and O&M cost information for PV and wind turbine: www.nrel.gov/analysis/tech-lcoe-re-cost-est.html.

Levelized cost of energy across multiple power options: http://en.openei.org/apps/TCDB/.

Cogeneration information (Crosby 2004): www.ashrae.org/File%20Library/docLib/Public/200412182611_326.pdf.

noise/vibration from the engines. EDG sets are available in sizes ranging from approximately 40 kW to several thousand kilowatts.

Microturbine Generator. As of this writing, microturbine generators are somewhat more expensive in first cost than EDGs, but they have less air pollution and less severe acoustic and vibration impacts than EDGs. They also have longer operating lives and a projected lower cost.

Fuel Cells. Fuel cells are the most advanced form of power generation in terms of being a clean and green technology. They generate virtually no air pollution, have minimal acoustic and vibration impacts, and are considered the wave of the future. At this point, however, the cost of fuel cell equipment is high, so it is the least attractive of these options from an economic standpoint. It is anticipated that this will change as continued development of the technology evolves over the next several years.

CCHP Systems

Combined heating and power is also referred to as *cogeneration*, and when combined with cooling they are called *trigeneration*. The larger, industrial-scale versions of such integrated energy systems have been in use through most of the 20th century. With the current focus shifting to distributed power generation for the reasons cited above, a new scale of CCHP has emerged, involving more integrated design of the components. These CCHP systems typically consist of one of the DG technologies described above (e.g., reciprocating engines or microturbines [possibly fuel cells, in the future] fired by natural gas, a heat exchanger to recover heat from the exhaust stream and/or jacket water, and, if there is a cooling demand, an absorption chiller). An intermediate medium such as hot water can be used to transfer heat between the exhaust stream of the prime-mover and the chiller, but lately a lot of development has gone into a dedicated heat recovery unit that also forms the generator of the absorption chiller (Rosfjord et al. 2004). Such integrated energy systems are admittedly in their infancy in the United States, but there is a significant drive by organizations including the U.S. Department of Energy (DOE) toward commercializing these systems as packaged systems.

The traditional application of CHP has been to directly use the waste heat for space heating and/or domestic hot-water heating. Larger central CCHP facilities use the waste heat to produce space or process cooling. The former is put together of off-the-shelf equipment, while the latter follows an integrated design approach with a better opportunity at optimization.

To illustrate the benefits of the DG-based CCHP systems, the following elementary view of the energy efficiencies, both at the component and system level, is offered. A typical reciprocating gas-fired engine has a thermal efficiency of about 35%; 65% of the energy in the fuel is ordinarily not used and wasted through the stack. The exhaust temperatures leaving the engine (>500°F [>260°C]) are typically high enough to drive, at the very least, a single-effect absorption chiller with a coefficient of performance (COP) in the vicinity of 0.7. Factoring in the actual amount of available heat in the exhaust stream (temperature/enthalpy and flow rate), this translates to overall fuel utilization rates/efficiencies that are as high as 80%. The ratio of electrical to thermal load carrying capability shifts a bit when the prime mover is a microturbine or a fuel cell, with typical generation efficiencies being 25% (based on higher heating value) and 40%, respectively. However, the overall fuel utilization rate is relatively unchanged. The waste heat can also be directly used for desiccant dehumidification (regeneration).

Thus, the integrated energy systems have achieved the following:

- Brought the power generation closer to the point of application/load (through distributed generation), eliminating transmission and distribution losses, etc.
- Removed or reduced the normal electric and primary fuel consumption by independent pieces of equipment providing cooling, heating, and/or dehumidification (e.g., separate electric chiller and boiler), thereby substantially improving overall fuel utilization rates, inclusive of the power generation process
- Removed or reduced emissions of CO_2 and other combustion by-products associated with the operation of the cooling, heating, and/or dehumidification equipment

Challenges include the matching of electrical and thermal loads, given the diversity of energy usage patterns in buildings. The consultant/contractor can determine if the heating/cooling components of the CCHP system will play a primary or complementary role (the latter involving other, more conventional equipment to fill in the missing thermal load). Numerous studies were done in this regard, including one by Ryan (2003). Further resources are given in GreenTip #9-7.

REFERENCES AND RESOURCES

Published

ASHRAE. 1998. *Fundamentals of Water System Design*. Atlanta: ASHRAE.

ASHRAE. 2016. *ASHRAE Handbook—HVAC Systems and Equipment*. Atlanta: ASHRAE.

ASHRAE. 2017. *ASHRAE Handbook—Fundamentals*. Atlanta: ASHRAE.

CEC. 2007. *Advanced Variable Air Volume (VAV) System Design Guide*, Public Document. Sacramento, CA: California Energy Commission.

Cho, Y.I., S. Lee, and W. Kim. 2003. Physical water treatment for the mitigation of mineral fouling in cooling-tower water applications. *ASHRAE Transactions* 109(1):346–57.

Crosby, D.A. 2004. Cogeneration: Engineering natural gas-driven systems. *ASHRAE Journal*, February.

EDR. 2009. *Chilled Water Plant Design Guide*. Energy Design Resources. https://energydesignresources.com/resources/publications/design-guidelines/design-guidelines-cooltools-chilled-water-plant.aspx.

Patnaik, V. 2004. Experimental verification of an absorption chiller for BCHP applications. *ASHRAE Transactions* 110(1):503–507.

Rosfjord, T., T. Wagner, and B. Knight. 2004. UTC microturbine CHP product development and launch. Presented at the 2004 DOE/CETC Annual Workshop on Microturbine Applications, January 20–22, Marina del Rey, CA.

Taylor, S., P. Dupont, M. Hydeman, B. Jones, and T. Hartman. 1999. *The CoolTools Chilled Water Plant Design and Performance Specification Guide*. San Francisco, CA: PG&E Pacific Energy Center.

Taylor, S. 2011a. Optimizing Design and Control of Chilled-Water Plants: Chilled-water distribution system selection. *ASHRAE Journal*, July.

Taylor, S. 2011b. Optimizing Design and Control of Chilled-Water Plants: Condenser water distribution system selection. *ASHRAE Journal*, September.

Taylor, S. 2011c. Optimizing Design and Control of Chilled-Water Plants: Pipe sizing and optimizing ΔT. *ASHRAE Journal*, December.

Taylor, S. 2012a. Optimizing Design and Control of Chilled-Water Plants: Chilled-water distribution system selection. *ASHRAE Journal*, March.

Taylor, S. 2012b. Optimizing Design and Control of Chilled-Water Plants: Optimized control sequences. *ASHRAE Journal*, June.

Torcellini, P.A., M. Long, and R. Judkoff. 2004. Consumptive water use for U.S. power production. *ASHRAE Transactions* 110(1):96–100.

Trane. 2000. *Multiple-Chiller-System Design and Control Manual*. Lacrosse, WI: Trane Company.

Online

U.S. Environmental Protection Agency, Combined Heat and Power (CHP) Partnerships www.epa.gov/chp/chp-partnership-partners.

U.S. Environmental Protection Agency, Green Chemistry www.epa.gov/greenchemistry/.

ASHRAE GreenTip #9-1

Variable-Flow/Variable-Speed Pumping System

GENERAL DESCRIPTION

In most hydronic systems, variable flow with variable-speed pumping can be a significant source of energy savings. Variable flow is produced in chilled- and hot-water systems by using two-way control valves and in condenser-water cooling systems by using automatic two-position isolation valves interlocked with the chiller machinery's compressors. In most cases, variable flow alone can provide energy savings at a reduced first cost, because two-way control valves cost significantly less to purchase and install than three-way valves. In condenser-water systems, even though two-way control valves may be an added first cost, they are still typically cost-effective, even for small (1 to 2 ton [3.5 to 7 kW_R]) heat pump and air-conditioning units. (Standard 90.1 [ASHRAE 2016] requires isolation valves on water-loop heat pumps and some amount of variable flow on all hot-water and chilled-water systems.)

Variable-speed pumping can dramatically increase energy savings, particularly when it is combined with demand-based pressure reset controls. Variable-speed pumps are typically controlled to maintain the system pressure required to keep the most hydraulically remote valve completely open at design conditions. The key to getting the most savings is placement of the differential pressure transducer as close to that remote load as possible. If the system serves multiple hydraulic loops, multiple transducers can be placed at the end of each loop, with a high-signal selector used to transmit the signal to the pumps. With direct digital control (DDC) systems, the pressure signal can be reset by demand and controlled to keep at least one valve at or near 100% open. If valve position is not available from the control system, a trim-and-respond algorithm can be used.

Even with constant-speed pumping, variable-flow designs save some energy, because the fixed-speed pumps ride up on their impeller curves, using less energy at reduced flows. For hot-water systems, this is often the best life-cycle cost alternative, as the added pump heat will provide some beneficial value. For chilled-water systems, it is typically cost effective to control pumps with variable-speed drives. It is very important to rightsize the pump and motor before applying a variable-speed drive, as a means of keeping drive cost down and performance up.

WHEN/WHERE IT'S APPLICABLE

Variable-flow design is applicable to chilled-water, hot-water, and condenser-water loops that serve water-cooled, air-conditioning, and heat-pump units. The limitations on each of these loop types are as follows:

- Chillers require a minimum flow through the evaporators. (chiller manufacturers can specify flow ranges if requested.) Flow minimums on the evaporator side can be achieved via hydronic distribution system design using either a primary/secondary arrangement or primary-only variable flow with a bypass line and valve for minimum flow.
- Some boilers require minimum flows to protect the tubes. These vary greatly by boiler type. Flexible bent water tube and straight water tube boilers can take huge ranges of turndown (close to zero flow). Fire tube and copper tube boilers, on the other hand, require a constant-flow primary pump.

Variable-speed drives on pumps can be used on any variable-flow system. As described above, they should be controlled to maintain a minimum system pressure. That system pressure can be reset by valve demand on hot-water and chilled-water systems that have DDC control of the hydronic valves.

PRO

- Both variable-flow and variable-speed control save significant energy.
- Variable-speed drives on pumps provide a "soft" start, extending equipment life.
- Variable-speed drives and two-way valves are self-balancing.

- Application of demand-based pressure reset significantly reduces pump energy and decreases the occurrence of system overpressurization, causing valves near the pumps to lift.
- Variable-speed systems are quieter than constant-speed systems.

CON

- Variable-speed drives add cost to the system and may not be cost-effective on hot-water systems.
- Demand-based supply pressure reset can only be achieved with DDC of the heating/cooling valves.
- Variable flow on condenser-water systems with open towers requires supplementary measures be taken to keep the fill wet on the cooling towers. Cooling towers with rotating spray heads or wands can accept a wide variation in flow rates without causing dry spots in fill. Fitting the cooling tower with variable-speed fans can take advantage of lower flow rates (there is more free area) to reduce fan energy while providing the same temperature of condenser water.

KEY ELEMENTS OF COST

The following provides a possible breakdown of the various cost elements that might differentiate a variable-flow/variable-speed system from a conventional one and an indication of whether the net cost for the hybrid option is likely to be lower (L), higher (H), or the same (S). This assessment is only a perception of what might be likely, but obviously it may not be correct in all situations. There is no substitute for a detailed cost analysis as part of the design process. The listings below may also provide some assistance in identifying the cost elements involved.

First Cost

- Hydronic system terminal valves: two-way versus three-way (applicable to hot-water and chilled-water systems) L
- Bypass line with two-way valve or alternative means (if minimum chiller flow is required) H

- Hydronic system isolation valves: two-position vs. none (applicable to condenser-water systems) H
- Cooling tower wet-fill modifications (condenser-water systems) H
- Variable-speed drives and associated controls H
- DDC system (may need to allow demand-based reset) or pressure transducers H
- Design fees H

Recurring Cost

- Pumping energy L
- Testing and balancing of hydronic system L
- Maintenance H
- Commissioning H

SOURCES OF FURTHER INFORMATION

ASHRAE. 2016. ANSI/ASHRAE/IES Standard 90.1-2016, *Energy Standard for Buildings Except Low-Rise Residential Buildings*. Atlanta: ASHRAE.

CEC. 2002. *Part II: Measure Analysis and Life-Cycle Cost—Pulse-power water treatment systems for cooling towers*. Sacramento, CA: Energy Efficiency & Customer Research & Development, Sacramento Municipal Utility District.

HPAC. 2004. Innovative grocery store seeks LEED certification. *HPAC Engineering* 27:31.

Torcellini, P.A., N. Long, and R. Judkoff. 2004. Consumptive water use for U.S. power production. *ASHRAE Transactions* 110(1):96-100.

Trane. 2005, *California Building Energy Standards*, P400-02-012. Sacramento, CA: California Energy Commission.

Taylor, S.T. 2002. Primary-only vs. primary-secondary variable-flow chilled-water systems. *ASHRAE Journal* 44(2):25-29.

Taylor, S.T., and J. Stein. 2002. Balancing variable flow hydronic systems. *ASHRAE Journal* 44(10):17-24.

ASHRAE GreenTip #9-2

Variable-Refrigerant-Flow (VRF) Systems

GENERAL DESCRIPTION

VRF systems have been used in Asia and Europe for almost 25 years. The main advantage of a VRF system is its ability to respond to fluctuations in space load conditions. By comparison, conventional direct expansion (DX) systems offer limited or no modulation in response to changes in the space load conditions. The problem worsens when conventional DX units are oversized, or during part-load operation (because the compressors cycle frequently). A simple VRF system is composed of an outdoor condensing unit and several indoor evaporators. The systems are interconnected by refrigerant piping with integrated oil and refrigerant management controls, which allow each individual thermostat to modulate its corresponding electronic expansion valve to maintain its space temperature set point.

There are two basic types of VRF multisplit systems: heat pump and heat recovery (Figure 9-5). Both heat pump and heat recovery VRF systems are available in air-to-air and water-source (water-to-refrigerant) configurations. Heat pumps can operate in heating or cooling mode. A heat recovery system, by managing the refrigerant through a gas flow device, can simultaneously heat and cool, with some indoor fan-coil units in heating and some in cooling, depending on the requirements of each building zone. The majority of VRF systems are equipped with variable-speed compressors. Often called *variable-frequency drives* or *inverter compressors*, this component responds to indoor temperature changes, varying the speed to operate only at the levels necessary to maintain a constant and comfortable indoor environment. Heat recovery systems increase VRF efficiency because, when operating in simultaneous heating and cooling, energy from one zone can be transferred to meet the needs of another.

WHEN/WHERE IT'S APPLICABLE

VRF systems offer controls that match the space heating/cooling loads to that of the indoor coil over a range of operation. Variable-

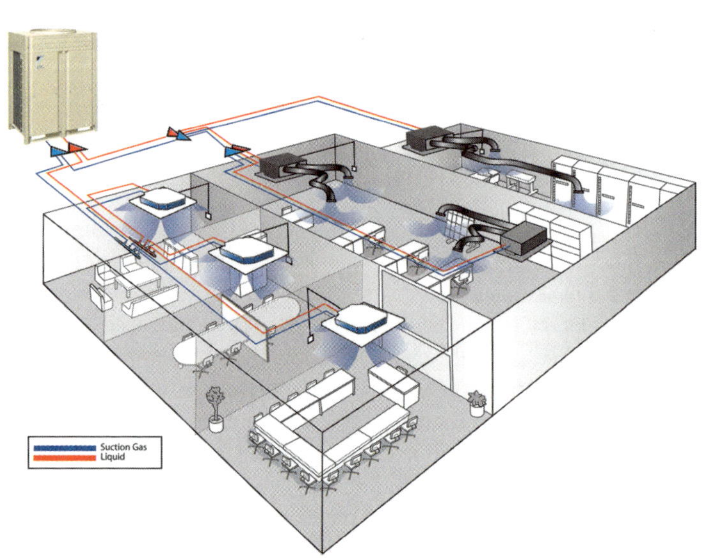

Image courtesy Daikin AC

Figure 9-5 Heat pump system in cooling mode.

speed compressors and fans in the outdoor units modulate their speed, saving energy at part-load conditions.

VRF systems are best suited to buildings with diverse zoning especially buildings requiring individual control, such as office buildings, residential, schools, or hotels.

PRO

- A single condensing unit can serve multiple indoor units.
- VRF systems are generally modular and can easily be modified, which makes it easy to adapt the system to expansion and/or reconfiguration.
- VRF indoor unit capacities are generally smaller, allowing more individual zones and individual zone controls. Systems can be designed with individual space zoning and controls.
- Variable-speed compressors enable capacity modulation, which translates to higher part load efficiencies in VRF systems.

CON

- Indoor units in VRF systems generally do not have high latent capacities and are not suitable for applications requiring a high percentage of outdoor air.
- The external static pressure available for ducted indoor sections is limited. For ducted indoor sections, the permissible ductwork lengths and fittings must be kept to a minimum.
- There is the potential for refrigerant leaks in the building.

KEY ELEMENTS OF COST

The following provides a possible breakdown of the various cost elements that might differentiate a VRF system from a conventional one and an indication of whether the net cost for the hybrid option is likely to be lower (L), higher (H), or the same (S). This assessment is only a perception of what might be likely, but it obviously may not be correct in all situations. There is no substitute for a detailed cost analysis as part of the design process. The listings below may also provide some assistance in identifying the cost elements involved.

First Cost

- Conventional heat pumps/DX systems H
- Ground-source heat pumps L
- Refrigerant piping L
- Installation costs H
- Controls S
- Design fees H

Recurring Cost

- Overall energy cost L
- Maintenance L

SOURCES OF FURTHER INFORMATION

Published

Afify, R. 2008. Designing VRF systems. *ASHRAE Journal* 52-55.

ASHRAE. 1975. ASHRAE Standard 90-75, *Energy Conservation in New Building Design*. Atlanta: ASHRAE.

ASHRAE. 2016a. *ASHRAE Handbook—HVAC Systems and Equipment*. Atlanta: ASHRAE.

ASHRAE. 2016b. ANSI/ASHRAE/IES Standard 90.1-2016, *Energy Standard for Buildings Except Low-Rise Residential Buildings*. Atlanta: ASHRAE.

ASHRAE. 2017. *ASHRAE Handbook—Fundamentals*. Atlanta: ASHRAE.

Aynur, T., Y. Hwang, and R. Radermacher. 2008a. Experimental evaluation of the ventilation effect on the performance of a VRV system in cooling mode—Part I: Experimental evaluation. *HVAC&R Research* (now *Science and Technology for the Built Environment*) 14(4):615–30.

Aynur, T., Y. Hwang, and R. Radermacher. 2008b. Simulation evaluation of the ventilation effect on the performance of a VRV system in cooling mode—Part II: Simulation evaluation. *HVAC&R Research* 14(5):783-95.

CEC. 2005. *Advanced Variable Air Volume System Design Guide*. Sacramento, CA: California Energy Commission.

Cendón, S.F. 2009. New and Cool: Variable Refrigerant Flow Systems. *AIArchitect*.

Goetzler, W. 2007. Variable refrigerant flow systems. *ASHRAE Journal* 49(9):24–31.

NADCA. *General Specifications for the Cleaning of Commercial Heating, Ventilating and Air Conditioning Systems*. Washington, DC: National Air Duct Cleaners Association.

Pearson, F. 2003. ICR 06 Plenary Washington. *ASHRAE Journal* 45(10).

SMACNA. 2005. *HVAC Duct Construction Standards*, 3d ed. Chapter 1. Chantilly, VA: Sheet Metal and Air Conditioning Contractors National Association, Inc.

Taylor, S., P. Dupont, M. Hydeman, B. Jones, and T. Hartman. 1999. *The CoolTools Chilled Water Plant Design and Perfor-*

mance Specification Guide. San Francisco, CA: PG&E Pacific Energy Center.

Taylor, S. 2011a. Optimizing Design and Control of Chilled-Water Plants: Chilled-water distribution system selection. *ASHRAE Journal*, July.

Taylor, S. 2011b. Optimizing Design and Control of Chilled-Water Plants: Condenser water distribution system selection. *ASHRAE Journal*, September.

Taylor, S. 2011c. Optimizing Design and Control of Chilled-Water Plants: Pipe sizing and optimizing ΔT. *ASHRAE Journal* 53, December.

Taylor, S. 2012a. Optimizing Design and Control of Chilled-Water Plants: Chilled-water distribution system selection. *ASHRAE Journal*, March.

Taylor, S. 2012b. Optimizing Design and Control of Chilled-Water Plants: Optimized control sequences. *ASHRAE Journal*, June.

Online

Armstrong Intelligent Systems Solutions www.armstrong-intl.com.

Spriax Sarco Design of Fluid Systems, Steam Utilization. www.spiraxsarco.com/global/us/Training/Documents/Design_of_Fluid_Systems_Steam_Utilization.pdf.

ASHRAE GreenTip #9-3

Displacement Ventilation

GENERAL DESCRIPTION

With a ceiling supply and return air system, the ventilation effectiveness may be compromised if sufficient mixing does not take place. While there are no data suggesting that cold air supplied at the ceiling will do so, it is possible that a fraction of the supply air may bypass directly to the return inlet ("short circuit") without mixing at the occupied level when heating from the ceiling. For example, when heating with a typical overhead system with supply temperatures exceeding 15°F (8.3°C) above room temperature, ventilation effectiveness will approach 80% or less. In compliance with Table 6.2 of ASHRAE Standard 62.1-2016, zone air distribution effectiveness is only 0.8, so ventilation rates must be multiplied by 1/0.8 or 1.25. While proper system design and diffuser selection can alleviate this problem, another potential solution is displacement ventilation.

In displacement ventilation, conditioned air with a temperature slightly lower than the desired room temperature is supplied horizontally at low velocities at or near the floor. Returns are located at or near the ceiling. The supply air is spread over the floor and then rises by convection as it picks up the load in the room. Displacement ventilation does not depend on mixing. Instead, the stale, polluted air is literally displaced and forced up and out of the return or exhaust grille. Ventilation effectiveness may actually exceed 100%, and Table 6.2 of ASHRAE Standard 62.1-2016 indicates a zone air distribution effectiveness of 1.2 must be used.

Displacement ventilation is a fairly common practice in Europe, but its acceptance in North America has been slow, primarily because of the conventional placement of ductwork at the ceiling level and North America's more extreme climatic conditions.

WHEN/WHERE IT'S APPLICABLE

Displacement ventilation is typically used in industrial plants and data centers, but it can be applied in almost any application where a conventional overhead forced-air distribution system could be used and the load permits.

Because the range of supply air temperatures and discharge velocities is limited to avoid discomfort to occupants, displacement ventilation has a limited ability to handle high heating or cooling loads if the space served is occupied. Some designs use chilled ceilings or heated floors to overcome this limitation. When chilled ceilings are used, it is critical that building relative humidity be controlled to avoid condensation. Another means of increasing cooling capacity is to recirculate some of the room air.

Some associate displacement ventilation solely with underfloor air distribution and the perceived higher costs associated with it. In fact, most underfloor pressurized plenum, air distribution systems do not produce true displacement ventilation but, rather, well-mixed airflow in the lower section of the conditioned space. It can, however, be a viable alternative when considering systems for modern office environments where data cabling and flexibility concerns may merit a raised floor.

PRO

- Displacement ventilation offers the potential for improved thermal comfort and IAQ because of increased ventilation effectiveness.
- There is reduced energy use because of extended economizer availability associated with higher supply temperatures.

CON

- It may add complexity to the supply air ducting.
- It is more difficult to address high heating or cooling loads.
- There are perceived higher costs.

KEY ELEMENTS OF COST

The following provides a possible breakdown of the various cost elements that might differentiate a system using displacement ventilation from one that does not and an indication of whether the net cost is likely to be lower (L), higher (H), or the same (S). This assessment is only a perception of what might be likely, but it obviously may not be correct in all situations. There is no substitute for a detailed cost analysis as part of the design process. The listings below may also provide some assistance in identifying the cost elements involved.

First Cost

- Controls S
- Equipment S
- Distribution ductwork S/H
- Design fees S

Recurring Cost

- Energy cost L
- Maintenance of system S
- Training of building operators S/H
- Orientation of building occupants S/H
- Commissioning cost S

SOURCES OF FURTHER INFORMATION

Advanced Buildings Technology and Practice www.advancedbuildings.org.

ASHRAE. 2000. Interpretation IC-62-1999-30 (August) of *ANSI/ASHRAE Standard 62.1-1999, Ventilation for Acceptable Indoor Air Quality*. Atlanta: ASHRAE.

ASHRAE. 2017. *ASHRAE Handbook—Fundamentals*. Atlanta: ASHRAE.

ASHRAE. 2016. ASHRAE Standard 62.1-2016, *Ventilation for Acceptable Indoor Air Quality*. Atlanta: ASHRAE.

Bauman, F., and T. Webster. 2001. Outlook for underfloor air distribution. *ASHRAE Journal* 43(6). Atlanta: ASHRAE.

Public Technology Inc., U.S. Department of Energy and U.S. Green Building Council. 1996. *Sustainable Building Technical Manual-Green Building Design, Construction and Operations*.

ASHRAE GreenTip #9-4

Dedicated Outdoor Air Systems (DOAS)

GENERAL DESCRIPTION

ASHRAE Standard 62.1 describes in detail the ventilation required to provide a healthy indoor environment as it pertains to IAQ. Traditionally, designers have attempted to address both thermal comfort and IAQ with a single mixed-air system. But ventilation becomes less efficient when the mixed-air system serves multiple spaces with differing ventilation needs. If the percentage of outdoor air is simply based on the critical space's need, then all other spaces are overventilated. In turn, providing a separate dedicated outdoor air system (DOAS) may be the only reliable way to meet the true intent of ASHRAE Standard 62.1.

A DOAS uses a separate air handler to condition outdoor ventilation air before delivering it directly to occupied spaces. The air delivered to the space from the DOAS should not adversely affect thermal comfort (i.e., too cold, too warm, too humid); therefore, many designers call for systems that deliver neutral air. However, there is a strong argument for supplying cool dry air and decoupling the latent conditioning as well as the IAQ components from the thermal comfort (sensible only) system.

The only absolute in a DOAS is that the ventilation air must be delivered directly to the space from a separate system. Control strategy, energy recovery, and leaving air conditions are all variables that can be defined by the designer.

WHEN/WHERE IT'S APPLICABLE

While a DOAS can be applied in any design, it is most beneficial in a facility with multiple spaces with differing ventilation needs. A DOAS can be combined with any thermal comfort conditioning system, including, but not limited to, all-air systems, fan-coil units, and hydronic radiant cooling. Note, however, that a design incorporating a separate 100% outdoor air unit delivering air to the mixed-air intakes of other units is not a DOAS as defined here. While this type of system may have benefits, such as using less energy or providing more accurate humidity control, it still suffers from the multiple space dilemma described above.

PRO

- A DOAS ensures compliance with ASHRAE 62.1 for proper multiple space ventilation and adequate IAQ.
- It reduces a building's energy use when compared to mixed-air systems that require overventilation of some spaces in order to ensure adequate ventilation.
- It allows the designer to decouple the latent load from the sensible load, hence providing more accurate space humidity control.
- It allows easy airflow measurement and balance and keeps ventilation loads off main HVAC units.

CON

- Depending on overall design (thermal comfort and IAQ), it may add additional first cost associated with providing parallel systems.
- Depending on overall design, it may require additional materials with their associated embodied energy costs.
- Depending on overall design, there may be more systems to maintain.
- With two airstreams, proper mixing may not occur when distributed to the occupied space.
- The total airflow of two airstreams may exceed airflow of a single system.

KEY ELEMENTS OF COST

The following provides a possible breakdown of the various cost elements that might differentiate a building with a DOAS from one with another system and an indication of whether the net cost is likely to be lower (L), higher (H), or the same (S). This assessment is only a perception of what might be likely, but it obviously may not be correct in all situations. There is no substitute for a detailed cost analysis as part of the design process. The listings below may also provide some assistance in identifying the cost elements involved.

First Cost

• Controls	H
• Equipment	S/H
• Distribution ductwork	S/H
• Design fees	S/H

Recurring Cost

• Energy cost	S/L
• Maintenance of system	S/H
• Training of building operators	S/H
• Orientation of building occupants	S
• Commissioning cost	S/H

SOURCES OF FURTHER INFORMATION

ASHRAE. 2016. ASHRAE Standard 62.1-2016, *Ventilation for Acceptable Indoor Air Quality*. Atlanta: ASHRAE.

ASHRAE. 2017. *ASHRAE Design Guide for Dedicated Outdoor Air Systems*. Atlanta: ASHRAE.

Coad, W.J. 1999. Conditioning ventilation air for improved performance and air quality. *Heating/Piping/Air Conditioning*, September.

Morris, W. The ABCs of DOAS. *ASHRAE Journal* 45(5).

Mumma, S.A. 2001. Designing dedicated outdoor air systems. *ASHRAE Journal* 43(5).

ASHRAE GreenTip #9-5

Ventilation Demand Control Using CO_2

GENERAL DESCRIPTION

A significant component of indoor environmental quality (IEQ) is the indoor air quality (IAQ). ASHRAE Standard 62.1 describes in detail the ventilation required to provide a healthy environment. However, providing ventilation based strictly on the peak occupancy using the ventilation rate procedure (Section 6.1) will result in overventilation during some periods. Any positive impact on IAQ brought on by overventilation could be outweighed by the costs associated with the energy required to condition the ventilation air. Demand-controlled ventilation can be done based on a number of different methods to determine the room occupancy, but by far the most common method is through the use of CO_2 measurement.

CO_2 can be used to measure or control the per-person ventilation rate in a given space and, in turn, allow the designer to introduce a ventilation demand control strategy. Simply put, the amount of CO_2 present in the air is an indicator of the number of people in the space relative to the ventilation being provided. In turn, this level can help determine if an adjustment in the amount of ventilation air that is being provided. CO_2-based ventilation control does not affect the peak design ventilation capacity required to serve the space as defined in the ventilation rate procedure, but it does allow the ventilation system to modulate in sync with the building's occupancy.

The key components of a CO_2 demand-based ventilation system are CO_2 sensors and a means by which to control the outdoor air intake (i.e., a damper with a modulating actuator). There are many types of sensors, and the technology is evolving while, at the same time, costs have dropped over the past decade or so. Sensors can be wall-mounted or mounted in the return duct, but it is recommended that the sensor be installed within the occupied space whenever possible.

WHEN/WHERE IT'S APPLICABLE

CO_2 demand control is best suited for buildings with a variable occupancy. The savings will be greatest in spaces that have a wide variance, such as gymnasiums, large meeting rooms, and auditoriums. For buildings with a constant occupancy rate, such as an office building or school, a simple nighttime setback scenario may be more appropriate for ventilation demand control, but CO_2 monitoring may still be used for verification that high IAQ is achieved.

PRO

- CO_2 demand control reduces a building's energy use as it relates to providing ventilation above that needed for adequate IAQ.
- It assists in maintaining adequate ventilation levels regardless of occupancy.

CON

- There is an added first cost associated with the sensors and additional controls.
- There are additional materials and their associated embodied energy costs.
- Evolving sensor technology may not be developed to full maturity.

KEY ELEMENTS OF COST

The following provides a possible breakdown of the various cost elements that might differentiate a building using a CO_2 ventilation demand control strategy from one that does not and an indication of whether the net cost is likely to be lower (L), higher (H), or the same (S). This assessment is only a perception of what might be likely, but it obviously may not be correct in all situations. There is no substitute for a detailed cost analysis as part of the design process. The listings below may also provide some assistance in identifying the cost elements involved.

First Cost

- Controls H
- Design fees S/H

Recurring Cost

- Energy cost L
- Maintenance of system S/H
- Training of building operators S/H
- Orientation of building occupants S/H
- Commissioning cost S/H

SOURCES OF FURTHER INFORMATION

Advanced Buildings Technologies and Practices www.advancedbuildings.org.

ASHRAE. 2016. ANSI/ASHRAE Standard 62.1-2016, *Ventilation for Acceptable Indoor Air Quality*. Atlanta: ASHRAE.

ASTM. 1998. *ASTM D 6245-1998: Standard Guide for Using Indoor Carbon Dioxide Concentrations to Evaluate Indoor Air Quality and Ventilation*. Philadelphia: ASTM International.

Lawrence, T.M. 2004. Demand-controlled ventilation and sustainability. *ASHRAE Journal* 46(12):117–21.

Schell, M., and D. Int-Hout. 2001. Demand control ventilation using CO_2. *ASHRAE Journal* 43(2):18–24.

ASHRAE GreenTip #9-6

Hybrid Ventilation

GENERAL DESCRIPTION

A hybrid ventilation system allows the controlled introduction of outdoor air ventilation into a building by both mechanical and passive means; it is sometimes called *mixed-mode ventilation*. It has built-in strategies to allow the mechanical and passive portions to work in conjunction with one another so as to not cause additional ventilation loads compared to what would occur using mechanical ventilation alone. It thus differs from a purely passive ventilation system, consisting of operable windows alone, which has no automatic way of controlling the amount of outdoor air load.

Two variants of hybrid ventilation are the changeover (or complementary) type and the concurrent (or zoned) type. With the former, spaces are ventilated either mechanically or passively, but not both simultaneously. With the latter variant, both methods provide ventilation simultaneously, though usually to zones discrete from one another.

Control of hybrid ventilation is obviously an important feature. With the changeover variant, controls could switch between mechanical and passive ventilation seasonally, diurnally, or based on a measured parameter. In the case of the concurrent variant, appropriate controls are needed to prevent "fighting" between the two ventilation methods.

WHEN/WHERE IT'S APPLICABLE

A hybrid ventilation system may be applicable in the following circumstances:

- When the owner and design team are willing to explore employing a nonconventional building ventilation technique that has the promise of reducing ongoing operating costs as well as providing a healthier, stimulating environment.
- When it is determined that the building occupants would accept the concept of using the outdoor environment to determine (at least, in part) the indoor environment, which may mean greater

variation in conditions than with a strictly controlled environment.

- When the design team has the expertise and willingness—and the charge from the owner—to spend the extra effort to create the integrated design needed to make such a technique work successfully.
- Where extreme outside conditions or a specialized type of building use do not preclude the likelihood of the successful application of such a technique.

Buildings with atria are particularly good candidates.

PRO

- Hybrid ventilation is an innovative and potentially energy-efficient way to provide outdoor air ventilation to buildings and, in some conditions, to cool them, thus reducing energy otherwise required from conventional sources (power plant).
- Corollary to the above, it could lead to a lower building life-cycle cost as the operational costs are lower.
- It would generally be expected to create a healthier environment for building occupants.
- It offers a greater sense of occupant satisfaction because of the increased ability to exercise some control over the ventilation provided.
- There is more flexibility in the means of providing ventilation; the passive variant can act as backup to the mechanical system and vice versa.
- It could extend the life of the equipment involved in providing mechanical ventilation because it would be expected to run less.

CON

- Failure to integrate the mechanical aspects of a hybrid ventilation system with the architectural design could result in a poorly functioning system. Some architectural design aspects could be constrained in providing a hybrid ventilation system, such as building orientation, depth of occupied zones, or grouping of spaces.

- Additional first costs could be incurred because two systems are being provided where only a single one would be provided otherwise, and controls for the passive system could be a major portion of the added cost.
- If automatic operable window openers are used, these could result in security breaches if appropriate safeguards and overrides are not provided.
- If integral building openings are used in lieu of, or in addition to, operable windows, pathways for the entrance of outside pollutants and noise or of unwanted insects, birds, and small animals would exist. If filters are used to prevent this, they could become clogged or could be an additional maintenance item to keep clear.
- Building operators may have to have special training to understand and learn how best to operate the system. Future turnovers in building ownership or operating personnel could negatively affect how successfully the system performs.
- Occupants would probably need at least some orientation so that they would understand and be tolerant of the differences in conditions that may prevail with such a system. Future occupants may not have the benefit of such orientation.
- Special attention would need to be given to certain safety issues, such as fire and smoke propagation in case of a fire.
- Although computer programs (such as computational fluid dynamics) exist to simulate, predict, and understand airflow within the building from passive ventilation systems, it would be difficult to predict conditions under all possible circumstances.
- Few codes and standards in the United States recognize and address the requirements for hybrid ventilation systems. This would likely result in local code enforcement authorities having increased discretion over what is acceptable.

KEY ELEMENTS OF COST

The following provides a possible breakdown of the various cost elements that might differentiate a hybrid ventilation system from a conventional one and an indication of whether the net cost for the hybrid option is likely to be lower (L), higher (H), or the same (S). This assessment is only a perception of what might be likely, but it obviously may not be correct in all situations. There is no substitute for a detailed cost analysis as part of the design process. The listings below may also provide some assistance in identifying the cost elements involved.

First Cost

• Mechanical ventilation system elements	S
• Architectural design features	H/L
• Operable window operators	H
• Integral opening operators/dampers	H
• Filters for additional openings	H
• Controls for passive system/coordination with mechanical system	H
• Design fees	H

Recurring Cost

• Energy for mechanical portion of system	L
• Maintenance of above	L
• Energy used by controls, mechanical operators	H
• Maintenance of passive system	H
• Training of building operators	H
• Orientation of building occupants	H
• Commissioning cost	H
• Occupant productivity	H

SOURCE OF FURTHER INFORMATION

Kosik, W.J. 2001. Design strategies for hybrid ventilation. *ASHRAE Journal* 43(10).

ASHRAE GreenTip #9-7

Combined Heating and Power (CHP) Systems

GENERAL DESCRIPTION

Other abbreviations that have been used to describe CHP systems are CCHP (includes *cooling*) and BCHP (building cooling, heating, and power). The goal, regardless of the abbreviation, is to improve system efficiencies or source fuel utilization by availing of the low-grade heat that is a by-product of the power generation process for heating and/or cooling duty. Fuel utilization efficiencies as high as 80% were reported (Adamson 2002). The resulting savings in operating costs, relative to a conventional system, are then viewed against the first cost, and simple payback periods of less than four years have been anticipated (LeMar 2002). This is particularly important, from a marketing perspective, for both the distributed-generation and the thermal equipment provider. This is because, by themselves, a microturbine manufacturer and an absorption chiller manufacturer, for example, would find it difficult to compete with a utility and an electric chiller manufacturer, respectively, as the provider of low-cost power and cooling. Last, but by no means least, the higher fossil fuel utilization rates result in reduced emissions of CO_2, the greenhouse gas with a more than 55% contribution to global warming (Houghton et al. 1990).

Gas engines, microturbines, and fuel cells have been at the center of CHP activity as the need for reliable power and/or grid independence has recently become evident. These devices are also being promoted to reduce the need for additional central-station peaking power plants. As would be expected, however, they come at a first cost premium, which can range from $1000 to $4000/kW (Ellis and Gunes 2002). At the same time, operating (thermal) efficiencies have remained in the vicinity of those of the large, centralized power plants, even after the transmission and distribution losses are taken into account. This is particularly true of engines and microturbines (which have a 25% to 35% thermal efficiency), while fuel cells promise higher efficiencies (of ~50%), albeit at the higher cost premiums ($3000 to $4000/kW).

On the thermal side, standard gas-to-liquid or gas-to-gas heat exchanger equipment can be used for the heating component of the CHP system. This transfers the heat from the exhaust gases to the process/hydronic fluid or air, respectively. For the cooling component, the size ranges of distributed power generators offer a unique advantage, in terms of flexibility, in the selection of the chiller equipment. These can be smaller-end (relative to commercial), water-lithium bromide single- or double-effect absorption chillers or larger-end (relative to residential) single-effect or generator-absorber heat exchange ammonia-water absorption chillers (Erickson and Rane 1994). Such chillers have a typical coefficient of performance of 0.7, and, as a rule of thumb, for thermal-to-electrical load matching, for every 4 kW of power generated, 1 ton of cooling may be achieved (Patnaik 2004). Figure 9-6 shows typical operating

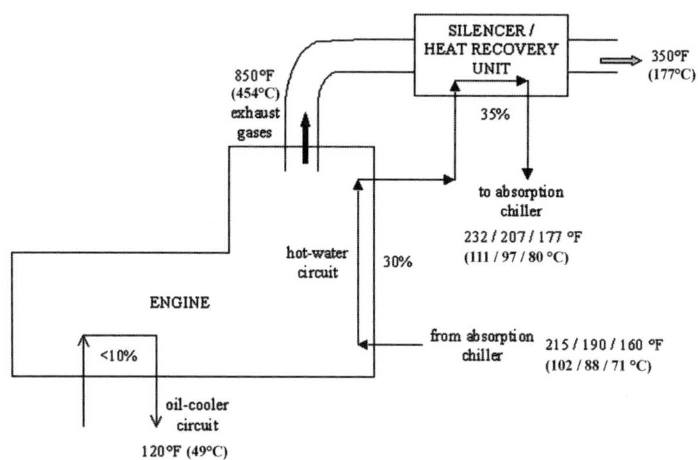

Image courtesy of Vikas Patnaik.

Figure 9-6 Schematic of CHP system consisting of a gas-fired reciprocating engine showing typical operating temperatures (Patnaik 2004).

conditions that an absorption chiller would see with a reciprocating engine.

WHEN/WHERE IT'S APPLICABLE

CHP is particularly suited for applications involving distributed power generation. Buildings requiring their own power generation, either due to a stringent power reliability and/or quality requirement or remoteness of location, must also satisfy various thermal loads (i.e., space heating or cooling, water heating, dehumidification). A conventional fossil-fuel-fired boiler and/or electric chiller can be displaced, to some extent if not entirely, by a heat recovery device (e.g., standard heat exchanger) and/or an absorption chiller driven by the waste heat from the power generator. Since the source of heating and/or cooling is waste heat that would ordinarily have been rejected to the surroundings, the operating cost of meeting the thermal demand of the building is significantly mitigated if not eliminated.

Economic analyses suggest that CHP systems are ideally suited for base-loaded distributed power generation and steady thermal (heating and/or cooling) loading. This is also the desirable mode of operation for the absorption chiller. Peak loading is then met by utility power. Alternatively, if utility power is used for base-loading and the DG meets the peaking demand, the thermal availability may be intermittent and require frequent cycling of the primary thermal equipment (i.e., the boiler and/or chiller).

PRO

- One of the primary advantages of CHP systems is the reduction in centralized (utility) peak-load generating capacity. This is especially true because one of the biggest contributors to summer peak loads is the demand for air conditioning. If some of this air-conditioning demand can be met by chillers fired by essentially free energy (waste heat), there is a double benefit.
- Additionally, DG-based CHP systems enable the following:
  - They bring the power generation closer to the point of application/load (i.e., distributed generation), eliminating transmission and distribution losses, etc.

- They remove or reduce the normal electric and primary fuel consumption by independent pieces of equipment providing cooling, heating, and/or dehumidification (e.g., a separate electric chiller and boiler), thereby substantially improving overall fuel utilization rates, inclusive of the power generation process.
- They remove or reduce emissions of CO_2 and other combustion by-products associated with the operation of the cooling, heating, and/or dehumidification equipment.

CON

- If the CHP system is to replace a conventional boiler/electric chiller system, the ratio of electrical to thermal load of the building must closely match the relative performance (i.e., efficiencies) of the respective equipment.
- Start-up times for absorption chillers are relatively longer than those for vapor-compression chillers, particularly when coupled to microturbines, which themselves have large time constants. Such systems would require robust and sophisticated controls that take these transients into account.

KEY ELEMENTS OF COST

The following provides a possible breakdown of the various cost elements that might differentiate a CHP system from a conventional one and an indication of whether the net cost for the alternative system is likely to be lower (L), higher (H), or the same (S). This assessment is only a perception of what might be likely, but it obviously may not be correct in all situations. There is no substitute for a detailed cost analysis as part of the design process. The listings below may also provide some assistance in identifying the cost elements involved.

First Cost

- Distributed power generator — H
- Heat recovery device/heat exchanger — L
- Absorption chiller — H
- Integrating control system — H

Recurring Cost

- Distributed power generator (engine/microturbine/fuel cell) — S/H/L
- Heat recovery device/heat exchanger — None
- Absorption chiller — None
- Integrating control system — H

SOURCES OF FURTHER INFORMATION

Technical sessions in recent meetings have been devoted to CHP, generally sponsored by technical committees on cogeneration systems (TC 1.10) and absorption/sorption heat pumps and refrigeration systems (TC 8.3). Presentations from these should be available on the ASHRAE technology portal (https://technologyportal.ashrae.org/).

REFERENCES

Adamson, R. 2002. Mariah Heat Plus Power™ Packaged CHP, applications and economics. Second Annual DOE/CETC/CANDRA Workshop on Microturbine Applications, January 17–18, University of Maryland, College Park, MD.

Ellis, M.W., and M.B. Gunes. 2002. Status of fuel cell systems for combined heat and power applications in buildings. *ASHRAE Transactions* 108(1):1032–44.

Erickson, D.C., and M. Rane. 1994. Advanced absorption cycle: Vapor exchange GAX. ASME International Absorption Heat Pump Conference, January 19–21, New Orleans, LA.

Houghton, J.T., G.J. Jenkins, and J.J. Ephraums, eds. 1990. *Climate Change: The IPCC Scientific Assessment.* Intergovernmental Panel on Climate Change. New York: Cambridge UP.

LeMar, P. 2002. Integrated energy systems (IES) for buildings: A market assessment. Final report by Resource Dynamics Corporation for Oak Ridge National Laboratory, Contract No. DE-AC05-00OR22725.

Patnaik, V. 2004. Experimental verification of an absorption chiller for BCHP applications. *ASHRAE Transactions* 110(1):503–507.

ASHRAE GreenTip #9-8

Low-NO_x Burners

GENERAL DESCRIPTION

Low-NO_x burners are natural gas burners with improved energy efficiency and lower emissions of NO_x. When fossil fuels are burned, nitric oxide and nitrogen dioxide are produced. These pollutants initiate reactions that result in the production of ozone and acid rain. The NO_x come from two sources: high-temperature combustion (thermal NO_x) and nitrogen bound to the fuel (fuel NO_x). For clean-burning fuels such as natural gas, fuel NO_x generation is insignificant.

In most cases, NO_x levels are reduced by lowering flame temperature. This can be accomplished by modifying the burner to create a larger (and therefore lower temperature) flame, injecting water or steam into the flame, recirculating flue gases, or limiting the excess air in the combustion process. In many cases a combination of these approaches is used. In general, reducing the flame temperature will reduce the overall efficiency of the boiler. However, recirculating flue gases and controlling the air-fuel mixture can improve boiler efficiency so that a combination of techniques may improve total boiler efficiency.

Natural-gas-fired burners with lowered NO_x emissions are available for commercial and residential heating applications. One commercial/residential boiler has a burner with inserts above the individual burners; this design reduces NO_x emissions by 30%. The boiler also has a wet base heat exchanger to capture more of the burner heat and reduce heat loss to flooring.

NO_x production is of special concern in industrial high-temperature processes, because thermal NO_x production increases with temperature. Processes include metal processing, glass manufacturing, pulp and paper mills, and cement kilns. Although natural gas is the cleanest-burning fossil fuel, it can produce emissions as high as 100 ppm or more.

A burner developed by the Massachusetts Institute of Technology and the Gas Research Institute combines staged-introduction combustion air, flue gas recirculation, and integral reburning to control NO_x emissions. These improvements in burner design result in a low-temperature, fuel-rich primary zone, followed by a low-temperature, lean secondary zone; these low temperatures result in lower NO_x formation.

In addition, any NO_x emission present in the recirculated flue gas is reburned, further reducing emissions. A jet pump recirculates a large volume of flue gas to the burner; this reduces NO_X emissions and improves heat transfer.

The low-NO_X burner used for commercial and residential space heating is larger in size than conventional burners, although it is designed for ease of installation.

WHEN/WHERE IT'S APPLICABLE

Low NO_X burners are best applied in regions where air quality is affected by high ground-level ozone and where required by law.

PRO

- Lowers NO_X and CO emissions, where that is an issue.
- Increases energy efficiency.

CON

- High cost.
- Higher maintenance.

KEY ELEMENTS OF COST

The following provides a possible breakdown of the various cost elements that might differentiate a low-NO_X system from a conventional one and an indication of whether the net cost for the alternative system is likely to be lower (L), higher (H), or the same (S). This is only a perception of what might be likely, but it may not be correct in all situations. There is no substitute for a detailed cost analysis as part of the design process. The listings below may provide some assistance in identifying the cost elements involved.

First Cost

- Conventional burner L
- Low NO_X burner H

Recurring Cost

- Maintenance H
- Possible avoidance of pollution fines L

SOURCE OF FURTHER INFORMATION

American Gas Association
www.aga.org.

ASHRAE GreenTip #9-9

Combustion Air Preheating

GENERAL DESCRIPTION

For fuel-fired heating equipment, one of the most potent ways to improve efficiency and productivity is to preheat the combustion air going to the burners. The source of this heat energy is the exhaust gas stream, which leaves the process at elevated temperatures. A heat exchanger, placed in the exhaust stack or ductwork, can extract a large portion of the thermal energy in the flue gases and transfer it to the incoming combustion air.

With natural gas, it is estimated that for each 50°F (10°C) the combustion air is preheated, overall boiler efficiency increases by approximately 1%. This provides a high leverage boiler plant efficiency measure, because increasing boiler efficiency also decreases boiler fuel usage. And, since combustion airflow decreases along with fuel flow, there is a reduction in fan-power usage as well.

There are two types of air preheaters: recuperators and regenerators. Recuperators are gas-to-gas heat exchangers placed on the furnace stack. Internal tubes or plates transfer heat from the outgoing exhaust gas to the incoming combustion air while keeping the two streams from mixing. Regenerators include two or more separate heat storage sections. Flue gases and combustion air take turns flowing through each regenerator, alternatively heating the storage medium and then withdrawing heat from it. For uninterrupted operation, at least two regenerators and their associated burners are required: one regenerator is needed to fire the furnace while the other is recharging.

WHEN/WHERE IT'S APPLICABLE

While theoretically any boiler can use combustion preheating, flue temperature is customarily used as a rough indication of when it will be cost-effective. However, boilers or processes with low flue temperatures but a high exhaust gas flow may still be good candidates and must be evaluated on a case-by-case basis. Financial justification is based on energy saved rather than on temperature differential. Some processes produce dirty or corrosive exhaust gases that can plug or attack a heat exchanger, so material selection is critical.

PRO

- Lowers energy costs.
- Increasing thermal efficiency lowers CO_2 emissions.

CON

- Additional material and equipment costs.
- Corrosion and condensation can add to maintenance costs.
- Low specific heat of air results in relatively low U-factors and less economical heat exchangers.
- Increasing combustion temperature also increases NO_x emissions.

KEY ELEMENTS OF COST

The following provides a possible breakdown of the various cost elements that might differentiate a building with a combustion preheat system from one without and an indication of whether the net cost is likely to be lower (L), higher (H), or the same (S). This assessment is only a perception of what might be likely, but it obviously may not be correct in all situations. There is no substitute for a detailed cost analysis as part of the design process. The listings below may also provide some assistance in identifying the cost elements involved.

First Cost

- Equipment costs H
- Controls S
- Design fees H

Recurring Cost

- Overall energy cost L
- Maintenance of system H
- Training of building operators H

SOURCES OF FURTHER INFORMATION

DOE. 2002. *Energy Tip Sheet #1*, May. Washington, DC: U.S. Department of Energy, Office of Industrial Technologies, Energy Efficiency, and Renewable Energy.

Fiorino, D.P. 2000. Six conservation and efficiency measures reducing steam costs. *ASHRAE Journal* 42(2):31–39.

ASHRAE GreenTip #9-10

Combination Space/Water Heaters

GENERAL DESCRIPTION

Combination space and water heating systems consist of a storage water heater, a heat delivery system (e.g., a fan coil or hydronic baseboards), and associated pumps and controls. Typically gas-fired, they provide both space and domestic water heating. The water heater is installed and operated as a conventional water heater. When there is a demand for domestic hot water, cold city water enters the bottom of the tank, and hot water from the top of the tank is delivered to the load. When there is a demand for space heating, a pump circulates water from the top of the tank through fan coils or hydronic baseboards.

The storage tank is maintained at the desired temperature for domestic hot water (e.g., 140°F [60°C]). Because this temperature is cooler than conventional hydronic systems, the space heating delivery system needs to be slightly larger than typical. Alternatively, the storage tank can be operated at a higher water temperature; this requires tempering valves to prevent scalding at the taps.

The water heater can be either a conventional storage-type water heater (either naturally venting or power vented) or a recuperative (condensing) gas boiler. Conventional water heaters have an efficiency of approximately 60%. By adding the space heating load, the energy factor increases because of longer runtimes and reduced standby losses on a percentage basis. Recuperative boilers can have efficiencies approaching 90%.

WHEN/WHERE IT'S APPLICABLE

These units are best suited to buildings that have similar space and water heating loads including dormitories, apartments, and condos. They are suited to all climate types.

PRO

- Reduces floor space requirements.
- Lowers capital cost.
- Improves energy efficiency.
- Increases tank life.

CON

- Only available in small sizes.
- All space heating piping has to be designed for potable water.
- No ferrous metals or lead-based solder can be used.
- All components must be able to withstand prevailing city water pressures.
- Some jurisdictions require a double-wall heat exchanger for such a scheme to be acceptable.

KEY ELEMENTS OF COST

The following provides a possible breakdown of the various cost elements that might differentiate a combination space and water heating system from a conventional one and an indication of whether the net cost for the hybrid option is likely to be lower (L), higher (H), or the same (S). This assessment is only a perception of what might be likely, but it obviously may not be correct in all situations. There is no substitute for a detailed cost analysis as part of the design process. The listings below may also provide some assistance in identifying the cost elements involved.

First Cost

- Conventional heating equipment L
- Combination space/domestic water heater H
- Sanitizing/inspecting space heating system H
- Piping and components able to withstand higher pressures H
- Floor space used L

Recurring Cost

- Heating energy L
- Maintenance L

SOURCES OF FURTHER INFORMATION

Sustainable Sources
www.greenbuilder.com.

UG. 2010. Wise Energy Guide. Chatham, ON: Union Gas. www.uniongas.com/-/media/residential/energyconservation/education/wiseEnergyGuide/WEG_Booklet_Web_Final.pdf

U.S. Department of Energy, Energy Efficiency and Renewable Energy
www.eere.energy.gov/.

ASHRAE GreenTip #9-11

Ground-Source Heat Pumps (GSHPs)

GENERAL DESCRIPTION

A GSHP extracts solar heat stored in the upper layers of the earth; the heat is then delivered to a building. Conversely, in the summer season, the heat pump rejects heat removed from the building into the ground rather than into the atmosphere or a body of water.

GSHPs can reduce the energy required for space heating, cooling, and service water heating in commercial/institutional buildings by as much as 50%. GSHPs replace the need for a boiler in winter by utilizing heat stored in the ground; this heat is upgraded by a vapor-compressor refrigeration cycle. In summer, heat from a building is rejected to the ground. This eliminates the need for a cooling tower or heat rejection device and also lowers operating costs, because the ground is cooler than the outdoor air. (See Figure 9-7 for an example of a GSHP system.)

There are numerous types of GSHP loop systems. Each has its advantages and disadvantages. Visit the Geoexchange Geothermal Heat Pump Consortium website (www.geoexchange.org) for a more detailed description of the loop options.

Water-to-air heat pumps are typically installed throughout a building with ductwork serving only the immediate zone; a two-pipe water distribution system conveys water to and from the ground-source heat exchanger. The heat exchanger field consists of a grid of vertical boreholes with plastic U-tube heat exchangers connected in parallel.

Simultaneous heating and cooling can occur throughout the building, as individual heat pumps, controlled by zone thermostats, can operate in heating or cooling mode as required.

Unlike conventional boiler/cooling, tower-type, water-loop heat pumps, the heat pumps used in GSHP applications are generally designed to operate at lower inlet water temperature. GSHPs are also more efficient than conventional heat pumps, with higher COPs and energy efficiency ratios. Because there are lower water temperatures in the two-pipe loop, piping needs to be insulated to prevent sweating. In addition, a larger circulation pump is needed because the units are slightly larger in the perimeter zones, requiring larger flows.

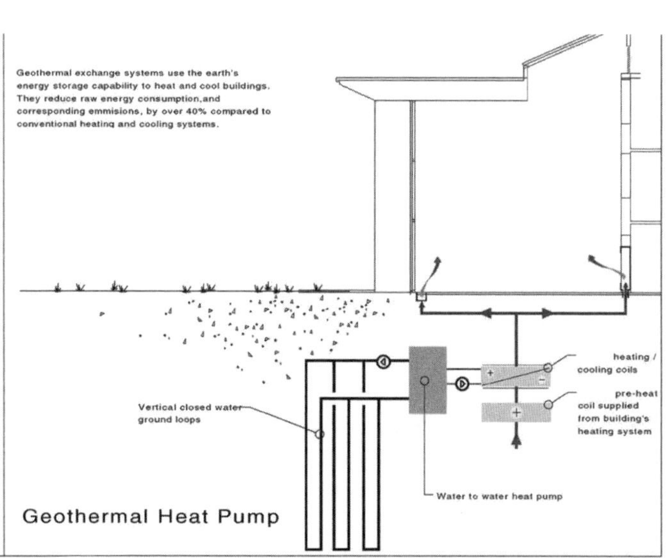

Image courtesy of Buro Happold.

Figure 9-7 Schematic example of GSHP closed-loop system.

GSHPs reduce energy use and, hence, atmospheric emissions. Conventional boilers and their associated emissions are eliminated, since no supplementary form of energy is usually required. Typically, single-packaged heat pump units have no field refrigerant connections and have significantly lower refrigerant leakage compared to central chiller systems.

GSHP units have life spans of 20 years or more. The two-pipe water-loop system typically used allows for unit placement changes to accommodate new tenants or changes in building use. The plastic piping used in the heat exchanger should last as long as the building itself.

When the system is disassembled, attention must be given to the removal and recycling of the hydrofluorocarbon (HFC) refrigerants used in the heat pumps and the antifreeze solution typically used in the ground heat exchanger.

WHEN/WHERE IT'S APPLICABLE

The most economical application of GSHPs is in buildings that require significant space/water heating and cooling over extended hours of operation. Examples are retirement communities, multifamily complexes, large office buildings, retail shopping malls, and schools. Building types not well-suited to the technology are buildings where space and water heating loads are relatively small or where hours of use are limited.

GSHPs are generally not suited for all climates, especially ones that are mostly very hot or very cold. In such climates where the total numbers of hours of cooling and heating per year are not close, the ground will not have a healthy charging/discharging cycle and its temperature will be affected over the years, resulting in system performance degradation.

PRO

- Requires less mechanical room space.
- Requires less outdoor equipment.
- Does not require roof penetrations, maintenance decks, or architectural blends.
- Relatively little operational noise.
- Reduces operation and maintenance costs.
- Requires simple controls only.
- Requires less space in ceilings.
- Loop piping, carrying low-temperature water, does not have to be insulated.
- Installation costs are lower than for many central HVAC systems.

CON

- Requires surface area for heat exchanger field.
- Higher initial cost.
- Requires additional site coordination and supervision.

KEY ELEMENTS OF COST

The following provides a possible breakdown of the various cost elements that might differentiate a GSHP system from a conventional one and an indication of whether the net cost for it is likely to be lower (L), higher (H), or the same (S). This assessment is only a perception of what might be likely, but it obviously may not be correct in all situations. There is no substitute for a detailed cost analysis as part of the design process. The listings may also provide some assistance in identifying the cost elements involved.

First Cost

- Conventional heating/cooling generators L
- Heat pumps H
- Outside piping system H
- Heat exchanger field H
- Operator training H
- Design fees H

Recurring Cost

- Energy cost (fossil fuel for conventional) L
- Energy cost (electricity for heat pumps) H
- Maintenance L

SOURCES OF FURTHER INFORMATION

ASHRAE. 1995. *Commercial/Institutional Ground-Source Heat Pump Engineering Manual*. Atlanta: ASHRAE.

Earth Energy Society of Canada www.earthenergy.ca.

Kavanaugh, S.P., and K. Rafferty. 1997. *Ground-Source Heat Pumps: Design of Geothermal Systems for Commercial and Institutional Buildings*. Atlanta: ASHRAE.

Natural Resources Canada, RETScreen (software for renewable energy analysis) www.retscreen.net.

ASHRAE GreenTip #9-12

Thermal Energy Storage (TES) for Cooling

GENERAL DESCRIPTION

There are several suitable media for storage of cooling energy, including the following:

- Chilled water
- Ice
- Calcium chloride solutions (brine)
- Glycol solutions
- Concentrated desiccant solutions

Active thermal storage systems use a building's cooling equipment to remove heat, usually at night, from an energy storage medium for later use as a source of cooling. The most common energy storage media are ice and chilled water. These systems decouple the production of cooling from the demand for cooling, (i.e., the plant output does not have to match the instantaneous building cooling load). This decoupling increases flexibility in design and operations, thereby providing an opportunity for a more efficient air-conditioning system than with an alternative that does not include storage. Before applying active thermal storage, however, the design cooling load should be minimized.

Although many operating strategies are possible, the basic principle of a TES system is to reduce peak building cooling loads by shifting a portion of peak cooling production to times when the building cooling load is lower. Energy is typically charged, stored, and discharged on a daily or weekly cycle. The net result is an opportunity to run a chiller plant at peak efficiency during the majority of its operating period. A system without storage, on the other hand, has to follow the building cooling load, and the majority of its operation is at part-load conditions. Part-load operation of chiller plants comes at the expense of efficiency.

Several buildings have demonstrated site energy reductions with the application of TES, as discussed in both the "Pro" and "Sources of Further Information" sections that follow.

In addition to the potential for site energy reduction, operation of TES systems can reduce energy resource consumption. This reduction is due to a shift toward using energy during periods of low aggregate electric utility demand. As a result, transmission and distribution losses are lower and power plant generating efficiencies can be higher because the load is served by base-load plants. Thermal storage can also have beneficial effects on CHP systems by flattening thermal and electric load profiles.

The ASHRAE *Design Guide for Cool Thermal Storage* (Dorgan and Elleson 1993) covers cool storage application issues and design parameters in some detail.

WHEN/WHERE IT'S APPLICABLE

TES systems tend to perform well in situations where there is variability in loads. Successful applications of TES systems have included commercial office buildings, schools, worship facilities, convention centers, hotels, health care facilities, industrial processes, and turbine inlet air cooling.

PRO

- Because TES allows downsizing the refrigeration system, the resulting cost savings (which may include avoiding having to add such equipment on an existing project) may substantially or entirely cover the added incremental cost of the storage system proper (also see the first con in the next section). However, if the first cost is more than another design option, there are still life-cycle cost benefits due to a significant reduction in utility costs.
- The addition of a TES system allows the size of refrigerating equipment to be reduced, since it will have to meet an average cooling load rather than the peak cooling load. Reduced refrigeration equipment size means less on-site refrigerant usage and a lower probability of environmental impacts due to direct effects.
- Because TES allows operation of the refrigeration system at or near peak efficiency during all operating hours, the annual

energy usage may be lower than systems without storage that must operate at lower part-load ratios to meet instantaneous loads. In addition, since off-peak hours are usually at night when lower ambient temperatures prevail, lower condensing temperatures required for heat rejection would tend to increase refrigeration efficiency. A number of carefully documented examples of energy savings can be found in the literature, including Bahnfleth and Joyce (1994); Fiorino (1994); and Goss et al. (1996).

- Because TES systems shift the consumption of site energy from on-peak to off-peak periods, the total energy resources required to deliver cooling to the facility will be lower (Reindl et al. 1995; Gansler et al. 2001). In addition, in some electric grids, the last generation plants to be used to meet peak loads may be the most polluting per kW of energy produced (Gupta 2002). In such cases, emissions would be further reduced by the use of TES.
- TES enables the practical incorporation of other high-efficiency technologies such as cold-air distribution systems and nighttime heat recovery.
- TES can be effective at preventing or delaying the need to construct additional power generation and transmission equipment.
- Liquid desiccant can be circulated in plastic pipes and does not need insulation.

CON

- Compared to a conventional system, the thermal storage element proper (i.e., the water tank or ice tank) and any associated pumping, piping accessories, and controls add to the incremental capital cost. If the system's refrigeration equipment can be reduced in size sufficiently (see the first pro listed in the previous section), this burden may be mitigated substantially or balanced out.
- The need to generate cooling at evaporator temperatures lower than conventional ones tends to decrease refrigeration efficiency. This reduction may be overcome, however, by factors that increase efficiency (see the third pro).
- Successful TES systems require additional efforts in the design phase of a project.

- TES systems will require increased site space usage. The impact of site space usage can be mitigated by considering ice storage technologies.
- Because a thermal storage system departs from the norm of system operation, continued training of facility operations staff is required, as are procedures for propagating system knowledge through a succession of facilities personnel.
- Ice requires special control of the melt rate to prevent uneven melting and to maximize performance.
- Calcium chloride brine needs management to prevent corrosion. Glycol needs management to prevent corrosion and toxicity.
- Liquid desiccant needs small resistant heating to be above 77°F (25°C) to prevent crystallization (similar to compressor sumps to prevent condensation).

KEY ELEMENTS OF COST

The following provides a possible breakdown of the various cost elements that might differentiate a TES system above from a conventional one and an indication of whether the net cost for the alternative option is likely to be lower (L), higher (H), or the same (S). This assessment is only a perception of what might be likely, but it obviously may not be correct in all situations. There is no substitute for a detailed cost analysis as part of the design process. The listings below may also provide some assistance in identifying the cost elements involved.

First Cost

- Storage element (e.g., chilled water, ice, glycol, and brine tanks) (Desiccant cost is higher than chilled water and brine but similar to glycol) H
- Additional pumping/piping re-storage element H
- Chiller/heat rejection system L
- Controls H
- Electrical (regarding chiller/heat rejection system) S/L
- Design fees H
- Operator training H
- Commissioning S/H
- Site space H

Recurring Cost

• Electric energy	L
• Gas supply with low electrical demand	H
• Operator training (ongoing)	H
• Maintenance training	L

SOURCES OF FURTHER INFORMATION

Bahnfleth, W.P., and W.S. Joyce. 1994. Energy use in a district cooling system with stratified chilled-water storage. *ASHRAE Transactions* 100(1):1767–78.

CEC. 1996. Source energy and environmental impacts of thermal energy storage. Completed by Tabors, Caramanis & Assoc. for the California Energy Commission.

Dorgan, C., and J.S. Elleson. 1993. *Cool Storage Design Guide*. Atlanta: ASHRAE.

Duffy, G. 1992. Thermal storage shifts to saving energy. *Eng. Sys.* July/August: 32–38.

Elleson, J.S. 1996. *Successful Cool Storage Projects: From Planning to Operation*. Atlanta: ASHRAE.

Fiorino, D.P. 1994. Energy conservation with stratified chilled-water storage. *ASHRAE Transactions* 100(1):1754–66.

Galuska, E.J. 1994. Thermal storage system reduces costs of manufacturing facility (Technology Award case study). *ASHRAE Journal*, March.

Gansler, R.A., D.T. Reindl, T.B. Jekel. 2001. Simulation of source energy utilization and emissions for HVAC systems. *ASHRAE Transactions* 107(1):39–51.

Goss, J.O., L. Hyman, and J. Corbett. 1996. Integrated heating, cooling and thermal energy storage with heat pump provides economic and environmental solutions at California State University, Fullerton. *Proceedings of the EPRI International Conference on Sustainable Thermal Energy Storage, Minneapolis, MN,* pp. 163–67.

Lawson, S.H. 1988. Computer facility keeps cool with ice storage. *HPAC* 60(88):35–44.

Mathaudhu, S.S. 1999. Energy conservation showcase. *ASHRAE Journal* 41(4):44–46.

O'Neal, E.J. 1996. Thermal storage system achieves operating and first-cost savings (Technology Award case study). *ASHRAE Journal* 38(4).

Reindl, D.T., D.E. Knebel, and R.A. Gansler. 1995. Characterizing the marginal basis source energy and emissions associated with comfort cooling systems. *ASHRAE Transactions* 101(1):1353–63.

ASHRAE GreenTip #9-13

Double-Effect Absorption Chillers

GENERAL DESCRIPTION

Chilled-water systems that use fuel types other than electricity can help offset high electric prices, whether those high prices are caused by consumption or demand charges. Absorption chillers use thermal energy (rather than electricity) to produce chilled water. A double-effect absorption chiller using high-pressure steam (115 psig [793 kPa]) has a COP of approximately 1.20. Some double-effect absorption chillers use medium-pressure steam (60 psig [414 kPa]) or 350°F to 370°F (177°C to 188°C) hot water, but with lower efficiency or higher cost. ASHRAE Standard 90.1-2016 requires a minimum COP ≥ 1.0 for a double-effect absorption chiller when operating at full load.

Double-effect absorption chillers are available from several manufacturers. Most are limited to chilled-water temperatures of 40°F (4.3°C) or above, since water is the refrigerant. The interior of the chiller experiences corrosive conditions; therefore, the manufacturer's material selection is directly related to the chiller life. The more robust the materials, the longer the life.

WHEN/WHERE IT'S APPLICABLE

Double-effect absorption chillers can be used in the following applications:

- When natural gas prices (used to produce steam) are significantly lower than electric prices.
- When the design team and building owner wish to have fuel flexibility to hedge against changes in future utility prices.
- When there is steam available from an on-site process; an example is steam from a turbine.
- When a steam plant is available but lightly loaded during the cooling season. Many hospitals have large steam plants that run at extremely low loads and low efficiency during the cooling season. By installing an absorption chiller, the steam plant efficiency can be increased significantly during the cooling season.
- At sites that have limited electric power available.

- In locations where district steam is available at a reasonable price (e.g., New York City).

PRO

- Reduces electric charges.
- Allows fuel flexibility, since natural gas, No. 2 fuel oil, propane, or waste steam may be used to supply thermal energy for the absorption chiller.
- Uses water as the refrigerant, making it environmentally friendly.
- Allows system expansion even at sites with limited electric power.
- When the system is designed and controlled properly, it allows versatile use of various power sources.

CON

- Cost of an absorption chiller will be roughly double that of an electric chiller of the same capacity, as opposed to 25% more for a single-effect absorption machine.
- Size of an absorption chiller is larger than an electric chiller of the same capacity.
- Although absorption chiller efficiency has increased in the past decade, the amount of heat rejected is significantly higher than that of an electric chiller of similar capacity. This requires larger cooling towers, condenser pipes, and cooling tower pumps.
- Few plant operators are familiar with absorption technology.

KEY ELEMENTS OF COST

The following provides a possible breakdown of the various cost elements that might differentiate an absorption chiller system from a conventional one and an indication of whether the net cost for the hybrid option is likely to be lower (L), higher (H), or the same (S). This assessment is only a perception of what might be likely, but it obviously may not be correct in all situations. There is no substitute for a detailed cost analysis as part of the design process. The listings below may also provide some assistance in identifying the cost elements involved.

First Cost

- Absorption chiller H
- Cooling tower and associated equipment H
- Electricity feed L
- Design fees H
- System controls H

Recurring Cost

- Electric costs L
- Chiller maintenance S
- Training of building operators H

SOURCES OF FURTHER INFORMATION

ASHRAE. 2014. *ASHRAE Handbook—Refrigeration*. Chapter 18, Absorption Equipment. Atlanta: ASHRAE.

ASHRAE. 2016a. *ASHRAE Handbook—HVAC Systems and Equipment*. Chapter 50, Room Air Conditioners and Packaged Terminal Air Conditioners. Atlanta: ASHRAE.

Trane Co. 1999. *Trane Applications Engineering Manual, Absorption Chiller System Design*, SYS-AM-13. Lacrosse, WI: Trane Company.

ASHRAE GreenTip #9-14

Gas Engine-Driven Chillers

GENERAL DESCRIPTION

Chilled-water systems that use fuel types other than electricity can help offset high electric prices, regardless of whether those high prices are caused by consumption or demand charges. Gas engines can be used in conjunction with electric chillers to produce chilled water. Depending on chiller efficiency, a gas engine-driven chiller may have a cooling COP of 1.6 to 2.3.

Some gas engines are directly coupled to a chiller's shaft. Another option is to use a gas engine and switchgear. In such cases, the chiller may be operated either by using electricity from the engine or from the electric utility.

WHEN/WHERE IT'S APPLICABLE

A gas engine is applicable in the following circumstances:

- When natural gas prices are significantly lower than electric prices.
- When the design team and building owner wish to have fuel flexibility to hedge against changes in future utility prices.
- At sites that have limited electric power available.

PRO

- Reduces electric charges.
- Allows fuel flexibility if installed as a hybrid system (i.e., part gas engine and part electric chiller, so the plant may use either gas engine or electricity from utility).
- Allows system expansion even at sites with limited electric power.
- When the system is designed and controlled properly, these chillers allow for use of various fuel sources.
- May be used in conjunction with an emergency generator if switchgear is provided.

CON

- Added cost of gas engine.
- Additional space required for engine.
- Due to the amount of heat rejected being significantly higher than for similar capacity electric chiller, larger cooling towers, condenser pipes, and cooling tower pumps may be required.

- Site emissions are increased.
- Noise from engine may need to be attenuated, both inside and outside.
- Significant engine maintenance costs.

KEY ELEMENTS OF COST

The following provides a possible breakdown of the various cost elements that might differentiate a gas engine-driven chiller from a conventional one and an indication of whether the net cost is likely to be lower (L), higher (H), or the same (S). This assessment is only a perception of what might be likely, but it obviously may not be correct in all situations. There is no substitute for a detailed cost analysis as part of the design process. The listings below may also provide some assistance in identifying the cost elements involved.

First Cost

- Gas engine H
- Cooling tower and associated equipment H
- Electricity feed L
- Site emissions H
- Site acoustics H
- Design fees H
- System controls H

Recurring Cost

- Electric costs L
- Engine maintenance H
- Training of building operators H
- Emissions costs H

SOURCE OF FURTHER INFORMATION

NBI. 1998. *Gas Engine-Driven Chillers Guideline*. Fair Oaks, CA: New Buildings Institute and Southern California Gas Company.

ASHRAE GreenTip #9-15

Desiccant Cooling and Dehumidification

GENERAL DESCRIPTION

There are two basic types of open-cycle desiccant process: solid and liquid desiccant. Each of these processes has several forms, and these should be investigated to determine the most appropriate for the particular application. Both of these systems need to have the air being conditioned come in good contact with the desiccant, during which moisture is absorbed from the air and the temperatures of both the air and desiccant are coincidently raised.

The moisture absorption process is caused by desiccant having a lower surface vapor pressure than the air. As the temperature of the desiccant rises, its vapor pressure rises, and its useful absorption capability lessens. Some systems, particularly liquid types, have cooling of air and desiccant coincident with dehumidification. This can allow the need for less space and equipment.

The dehumidified air then has to be cooled by other means. Two supply air arrangements are available. One method uses a mixture of recycled return air and dehumidified outdoor ventilation air as supply air to the building. Moisture and contaminants such as volatile organic compounds can be absorbed by the desiccant and recycled; particles of solid or liquid desiccant may also be carried over into the ducts and to building occupants.

The other arrangement combines energy recovery from building exhaust air that is typically much cooler and less humid than outdoor ventilation air. By dehumidifying the exhaust air to a sufficiently low humidity ratio (i.e., moisture content), it can be used to indirectly cool outdoor air for supply to the building that has not contacted desiccant. Using the recovered energy, this arrangement can be used to process the total ventilation air requirement, even up to 100% from outside.

So that desiccant can be reused, it has to be redried by a heating process generally called *reactivation* or *regeneration* (for solid types) or *reconcentration* (for liquid types). The redrying can be either direct (by contact with heated outdoor air with the desiccant) or indirect. Indirect may be preferable, particularly in high humid climates, because of the higher temperature needed to maximize the vapor difference and drying potential. The energy storage benefit for liquid desiccant was discussed in GreenTip #9-12.

Rotary desiccant dehumidifiers use solid desiccants such as silica gel to attract water vapor from the moist air. Humid air, generally referred to as *process air*, is dehumidified in one part of the desiccant bed while a different part of the bed is dried for reuse by a second airstream known as *reactivation air*. The desiccant rotates slowly between these two airstreams, so that dry, high-capacity desiccant leaving the reactivation air is available to remove moisture from the moist process air.

Process air that passes through the bed more slowly is dried more deeply, so for air requiring a lower dew point, a larger unit (and slower velocity) is required. The reactivation air inlet temperature changes the outlet moisture content of the process air. In turn, if the designer needs dry air, it is generally more economical to use high reactivation temperatures. On the other hand, if the leaving humidity need not be especially low, inexpensive, low-grade heat sources (e.g., waste heat or rejected cogeneration heat) can be used.

The process air outlet temperature is higher than the inlet temperature primarily because the heat of sorption of the moisture removed is converted to sensible heat. The outlet temperature rises roughly in proportion to the amount of moisture that is removed. In most comfort applications, provisions must be made to remove excess sensible heat from the process air following reactivation. Cooling is accomplished with cooling coils, and the source of this cooling affects the operating economics of the system.

WHEN/WHERE IT'S APPLICABLE

In general, applications that require a dew point at or below 40°F (4.3°C) may be candidates for active desiccant dehumidification. Examples of such candidates include facilities handling hygroscopic materials; film drying; the manufacturing of candy, chocolate, or chewing gum; the manufacturing of drugs and chemicals; the manufacturing of plastic materials; packaging of moisture-sensitive products; and the manufacturing of electronics. Supermarkets often use desiccant dehumidification to avoid condensation on refrigerated casework. And when there is a need for a lower dew point and a convenient source of low-grade heat for reactivation is available, rotary desiccant dehumidifiers can be especially economical.

PRO

- Desiccant equipment tends to be very durable.
- Often this is the most economical means to dehumidify below a 40°F (4.3°C) dew point.
- It eliminates condensate in the airstream, in turn, limiting the opportunity for mold growth.

CON

- Desiccant usually must be replaced, replenished, or reconditioned every 5 to 10 years.
- In comfort applications, simultaneous heating and cooling may be required.
- The process is not especially intuitive, and the controls are relatively complicated.

KEY ELEMENTS OF COST

The following provides a possible breakdown of the various cost elements that might differentiate a building with a rotary desiccant dehumidification system from one without and an indication of whether the net cost is likely to be lower (L), higher (H), or the same (S).

First Cost

- Equipment costs H
- Regeneration (heat source and supply) H
- Ductwork S
- Controls H
- Design fees S

Recurring Cost

- Overall energy cost S/H
- Maintenance of system H
- Training of building operators H
- Filters H

SOURCE OF FURTHER INFORMATION

ASHRAE. 2016. *ASHRAE Handbook—HVAC Systems and Equipment*. Atlanta: ASHRAE.

ASHRAE GreenTip #9-16

Indirect Evaporative Cooling

GENERAL DESCRIPTION

Evaporative cooling of supply air can be used to reduce the amount of energy consumed by mechanical cooling equipment. Two general types of evaporative cooling—direct and indirect—are available. The effectiveness of either of these methods is directly dependent on the extent that dry-bulb temperature exceeds wet-bulb temperature in the supply airstream.

Direct evaporative cooling introduces water directly into the supply airstream, usually with a spray or wetted media. As the water absorbs heat from the air, it evaporates. While this process lowers the dry-bulb temperature of the supply airstream, it also increases the air moisture content.

Two forms of indirect evaporative cooling (IEC) are described below.

1. **Coil/cooling Tower IEC.** This type uses an additional water-side coil to lower supply air temperature. The added coil is placed ahead of the conventional cooling coil in the supply airstream and is piped to a cooling tower where the evaporative process occurs. Because evaporation occurs elsewhere, this method of precooling does not add moisture to the supply air, but it is somewhat less effective than direct evaporative cooling. A conventional cooling coil provides any additional cooling required.
2. **Plate Heat Exchanger IEC.** This is composed of sets of parallel plates arranged into two sets of passages separated from each other. In a typical arrangement, exhaust air from a building is passed through one set of passages, during which it is wetted by water sprays. A stream of outdoor air is coincidently passed through the other set of passages and is cooled by heat transfer through the plates by the wetted exhaust air before being supplied to the building. Alternatively, the exhaust air may be replaced by a second stream of outdoor air. The wetted air is reduced in dry-bulb temperature to be close to its wet-bulb temperature. The stream of dry air is cooled to be close to the dry-bulb temperature

of the wetted exhaust air. In some applications, the cooled stream of outdoor air is passed through the coil of a direct expansion refrigeration unit, where it is further cooled before being supplied to the building. This system is an efficient way for an all outdoor air supply system. The plates in the heat exchanger can be formed from various metals and polymers. Consideration needs to be given to preventing the plate material from corroding.

WHEN/WHERE IT'S APPLICABLE

This may be used in climates with low wet-bulb temperatures where significant amounts of cooling are available. In such climates, the size of the conventional cooling system can also be reduced.

In more humid climates, indirect evaporative cooling can be applied during nonpeak seasons. It is especially applicable for loads that operate 24 h/day for many days of the year.

PRO

- Indirect evaporative cooling can reduce the size of the conventional cooling system.
- It reduces cooling costs during periods of low wet-bulb temperature.
- It does not add moisture to the supply airstream (in contrast, direct evaporative cooling does add moisture).
- It may be designed into equipment such as self-contained units.
- There is no cooling tower or condenser piping in the plate heat exchanger IEC that is described.

CON

- Air-side pressure drop (typically 0.2 to 0.4 in. of water [50 to 100 Pa]) increases due to an additional coil in the airstream.
- To make water cooler in the coil/cooling tower IEC, the cooling tower fans operate for longer periods of time and consume more energy.
- For the coil/cooling tower IEC, condenser piping and controls must be accounted for during the design process.

KEY ELEMENTS OF COST

The following provides a possible breakdown of the various cost elements that might differentiate an IEC system from a conventional one and an indication of whether the net cost for the hybrid option is likely to be lower (L), higher (H), or the same (S). This assessment is only a perception of what might be likely, but it obviously may not be correct in all situations. There is no substitute for a detailed cost analysis as part of the design process. The listings below may also provide some assistance in identifying the cost elements involved.

First Cost

- Indirect cooling coil H
- Decreased conventional cooling system capacity L
- Condenser piping, valves, and control H

Recurring Cost

- Cooling system operating cost L
- Supply fan operating cost H
- Tower fan operating cost H
- Maintenance of indirect coil S

SOURCES OF FURTHER INFORMATION

ASHRAE. 2015. *ASHRAE Handbook—HVAC Applications*. Chapter 52, Evaporative Cooling. Atlanta: ASHRAE.

ASHRAE. 2016. *ASHRAE Handbook—HVAC Systems and Equipment*. Chapter 41, Evaporative Air-Cooling Equipment. Atlanta: ASHRAE.

ASHRAE GreenTip #9-17

Condensing Boilers

GENERAL DESCRIPTION

Condensing boilers are heat-producing equipment that recover heat from their wasted flue gases, cooling them to a point where water condensation occurs.

The flue of any boiler usually contains carbon dioxide, nitrous oxides, and water vapor with a dew-point temperature of about 138°F–140°F (58.8°C–60°C). Any return water temperature to the boiler less than this temperature will result in water vapor condensation (and hence corrosion). Conventional boilers avoid this problem by limiting return water temperatures and keeping the flue temperature hot enough. (Conventional boilers have flue temperatures in the range of 250°F–350°F [121°C–176°C]). This results in about 10%–15% of the heat produced by conventional boilers being lost through the flue, making their annual fuel utilization efficiency (AFUE) in the range of 70%–80%.

Condensing boilers, on the other hand, are designed to handle the corrosive nature of condensing water vapor. They usually include a second heat exchanger or an oversized single heat exchanger that will extract heat from the flue, thus cooling it below the dew-point temperature. Water vapor in the flue will condense and mix with the carbon dioxide and nitrous oxides to form carbonic and nitric acid, respectively. Flue temperatures of condensing boilers are generally in the 120°F–140°F (48.8°C–60°C) range. Return water temperatures to a condensing boiler are theoretically not bound by any minimum limit. However, typical values range from 130°F down to 80°F (54.4°C down to 26.6°C). The lower the return water temperature, the more condensation will occur and the more efficient the condensing boiler will be. Typical AFUEs of condensing boilers are in the 90%–98% range.

WHEN/WHERE IT'S APPLICABLE

Condensing boilers are widely used in Europe where certain countries, like the United Kingdom, have mandated efficiency levels that can only be met by them. The U.S. market, although late in adoption, is currently showing a tendency to outrun Europe in the efficiency race.

Although condensing boilers can theoretically be used on any new system or to retrofit any existing system, they are generally ideal and cost effective for low-temperature applications such as radiant floor heating, swimming pool heating, and water source heat pumping.

PRO

- Very high efficiency
- Lower CO_2 and NO_x emissions due to lower combustion temperatures and the dilution of some of these gases with the condensing water
- Can accommodate a wider range of supply temperatures without the need for mixing valves
- No warm-up cycle required since the boiler can accommodate lower return water temperatures
- Can be coupled with other low-temperature equipment such as a heat recovery chiller
- Federal tax rebate (depending on year) for residential application. State and local incentives for residential and light commercial applications in many other jurisdictions

CON

- Higher capital cost
- A drain discharge is required which is generally corrosive
- Stack material must be able to withstand the corrosive nature of the flue
- In very cold climates, care should be exercised in running the flue stack outdoors to avoid freezing
- In retrofit applications, high-temperature coils may need to be upsized
- Lower equipment life

KEY ELEMENTS OF COST

The following provides a possible breakdown of the various cost elements that might differentiate a condensing boiler system from a conventional one and an indication of whether the net cost for an option is likely to be lower (L), higher (H), or the same (S). This assessment is only a perception of what might be likely, but it obviously may not be correct in all situations. There is no substitute for a detailed cost analysis as part of the design process. The listings below may also provide some assistance in identifying the cost elements involved.

First Cost

- Boiler H
- Flue Stack S/H
- Additional plumbing requirements H

Recurring Cost

- Energy consumption L
- Testing and balancing (TAB) S
- Maintenance S/H
- Commissioning S

SOURCE OF FURTHER INFORMATION

ASHRAE. 2016. *ASHRAE Handbook—HVAC Systems and Equipment*. Chapter 32, Boilers. Atlanta: ASHRAE.

CHAPTER TEN

Energy Sources

RENEWABLE AND NONRENEWABLE ENERGY SOURCES

There are often discussions about using renewable energy sources as a way to power the world, but little is done to actually implement it. This chapter focuses on ways to use renewable energy to offset nonrenewable energy sources. Various definitions of a renewable energy source (RES) exist. The *International Green Construction Code* (IgCC) defines this as "Energy derived from solar radiation, wind, waves, tides, biogas, biomass, or geothermal energy," while ANSI/ASHRAE/USGBC/IES Standard 189.1 emphasizes on-site renewable energy systems defined as "photovoltaic, solar thermal, geothermal energy, and wind energy systems...located on the building project." This is in contrast to the common nonrenewable energy sources such as coal, oil, natural gas, and nuclear.

The use of renewable energy is separated between using the energy source on site versus paying for the renewable resource. Today, many utility companies offer the ability to purchase renewable energy mixes from their generation portfolio. These utilities generate their own renewable energy with either large-scale wind or solar facilities. Note that quite often, large-scale hydroelectric plants are not considered a renewable resource because of the size of the environmental impact of such facilities. *Green energy*, as it is sometimes called, can also be purchased by third-party resellers of energy. The concept is that renewable energy is put into the utility grid and you, as the end user, can purchase that power somewhere else on the grid. The bottom line is that you can, in most areas, purchase green power, and you are not limited by the default utility offering.

According to the U.S. Department of Energy's (DOE) Energy Information Administration (EIA), renewable energy consumption exceeded 9 quadrillion Btu in 2016 and accounted for over 60% of the new capacity additions. In the European Union (EU), renewable energy contributed 10.3% of the total energy consumption in 2014 (www.energy.eu/#renewable).

To further strengthen the exploitation of RES, the EU officially adopted a 20-20-20 Renewable Energy Directive on December 17, 2008, setting the following ambitious targets for 2020: cutting greenhouse gas emissions by at least 20% of 1990 levels; cutting energy consumption by 20% through improved energy efficiency; and boosting the use of RES to 20% of total energy production (currently at about 8.5%). Under the terms of the directive, for the first time each member state has a legally binding renewable energy target for 2020 and by June 2010 each state was to have drawn up a National Action Plan detailing plans to meet their 2020 targets (EU 2008).

The designer has little say over how energy at the source will be provided. Electricity may be generated from renewable or nonrenewable sources, and each source may have broad implications for national or industry interests, the environment, and/or economics. However, there is little a designer can do about it—at least in choosing between conventional nonrenewable energy sources. This subject is addressed in more detail in the Chapter 34 of the *ASHRAE Handbook—Fundamentals* (ASHRAE 2017), and the designer is referred to that source for more specific data.

In many cases, renewable energy can be considered free. The issue is that this free energy source usually needs some capital equipment to concentrate the diffuse nature of renewable energy into a useful form for the building. One way to illustrate this characteristic is to think of a gallon jug of, say, fuel oil compared to an array of hydronic solar collectors. The fuel oil could provide hot water, on demand, for hours in a simple water heater in a corner of a boiler room, but the fuel oil will consume fossil fuels and produce emissions. The equivalent job done by solar collectors would require an array of collectors on a roof, plus a tank, piping, and some controls (a simple thermosiphon-type solar collector can operate without controls) but would produce no emissions and would operate with free solar energy.

While consideration of renewables is a highly touted element of green design, the design team should be well aware of the key characteristics of whichever renewable is being considered and develop creative strategies accordingly. In the next sections, we focus on two main renewable energy sources (other than hydropower) in the world today.

WATER-ENERGY NEXUS

The term *water-energy nexus* is used in reference to the interdependent and inseparable nature of two of our most precious resources, water and energy. From large-scale utilities to the built environment, water usage requires energy to extract and deliver for consumption (e.g., hydraulic fracturing and biofuels), while the generation of electricity and associated energy resources used for that purpose (such as fossil and nuclear fuels) demand significant amounts of water for production. With approximately 8% of global energy generation used for pumping, treatment and transportation of water resources—and approximately 15% of the world's total water usage used for energy production, each resource will continue to face rising demands and constraints as a consequence of economic and population growth and climate change.

Though both water and energy are critical resources today, each has different characteristics. The earth has a fixed amount of water in various forms (such as seas, glaciers, underground aquifers, precipitation, vapor, and that embodied in living organisms). No water in any form leaves or enters the planet. What the earth has is what it must make the best use of.

With a fixed amount of water on earth, the task becomes ensuring that the forms and quality of water available (especially potable water) are sufficient to serve the needs of the population. While water is not consumed in the sense of being "used up," it can become contaminated, requiring energy and other inputs to clean, or transformed into a form not usable at this location (such as through evaporation). Another key aspect of water is its uneven distribution across the planet, with some areas having plenty and others where water is scarce.

Energy, on the other hand, does leave and impinge on the earth through solar gain and radiation from the earth to space, and has for millennia. Much of the energy that has impacted the earth has been stored over those millennia in the form of fossil fuels and nuclear fuel as well as raised water bodies and waterfalls. We rely to a large degree on these stored energy forms, but energy continues to affect the earth, and humankind has developed means to capture and use a portion of that energy on a timely, albeit small-scale basis (such as solar and wind energy systems).

With that increasing energy demand as well as naturally occurring water constraints such as droughts, heat waves, or human-induced shortages, we can expect the demands on our water resources to increase. In addition to increasing demand, changing temperatures, shifting precipitation patterns, increasing variability, and more extreme weather will yield significant uncertainty regarding water availability.

Water and energy, in their various classifications, have generally been viewed in individual "silos," which have limited the adoption of integrated solutions. To properly address the challenges and opportunities posed by the water-energy nexus, emphasis on policy incentives and sustainable engineering solutions promoting optimal usage efficiency of each resource, as well as advanced technologies promoting both water and energy conservation, will be required.

SOLAR

Solar energy is the primary energy source that fuels the growth of the Earth's natural capital and drives wind and ocean currents that also can provide alternative energy sources. Since the beginning of time, solar energy has been successfully harnessed for human use. Early civilizations and some modern ones used solar energy for many purposes: food and clothes drying, heating water for baths, heating adobe and stone dwellings, etc. Solar energy is free and available to anyone who wishes to use it.

Solar thermal heating for domestic hot-water and space heating has grown considerably over the years and is well established in several countries. Global installed capacity is estimated at about 127 × 10^{10} Btu/h (374 gigawatts of thermal energy [GW_{th}]) for glazed flat-plate (286 × 10^9 Btu/h [84 GW_{th}] or and evacuated-tube collectors (900 × 10^9 Btu/h [264 GW_{th}]) at 84 × 10^9 Btu/h (24.5 GW_{th}) for unglazed plastic collectors, and 4.1 × 10^9 Btu/h (1.2 GW_{th}) for glazed and unglazed air collectors (Weiss et al. 2008). In North America (i.e., United States and Canada), swimming pool heating is dominant with an installed capacity of about 66.9 × 10^9 Btu/h (19.6 GW_{th}) of unglazed plastic collectors. In other countries, flat-plate and evacuated-tube collectors are used to generate hot water for sanitary use and space heating. Solar energy is also used for applications such as industrial or agricultural processes.

For many, a key impediment to increased solar use is economics. The cost of some solar technologies is perceived to be high compared to the fossil-based energy source it is offsetting. For example, while the simple payback of solar PV systems tends to be rather long (although much improvement has occurred over the past couple of decades), the recent increase in the cost of energy and advances in solar energy justify a fresh look at the applications and the engineering behind those applications. In addition, public policy in many areas encourages solar and other renewable energy applications through tax incentives and encouraging or requiring repurchase of excess electrical energy generated by the utility provider.

The applicability and, consequently, the economics and public policy incentives available with different solar energy system types and applications depend greatly on the location, which determines the technical factors (such as solar resources available), as well as the nontechnical (such as the country's political situation).

Solar Thermal Applications

Solar energy thermal applications range from low-temperature applications (swimming pool heating or domestic hot water) up to medium- or high-temperature applications (space heating, absorption cooling or steam production for electrical generation). Figure 10-1 shows examples of typical installations.

The most common solar energy thermal application is for domestic hot water (DHW) production. Solar hot water can be used as the sole source of domestic hot water or to preheat incoming supply water. This can be as simple as having an uninsulated tank in a hot attic in a southern climate to having a batch water heater on the roof of a building. The same solar collectors can be used to deliver thermal energy for space heating. A typical installation for the combined production of DHW and space heating (i.e., solar combi systems) includes the solar collectors, the heat storage tank, and a boiler used as an auxiliary heater. Combi systems require a larger collector area than a DHW system to meet the higher loads. It is possible to use a heat storage tank and a DHW storage vessel, but it may also be suitable to combine them in a single storage tank (with a high vertical stratification) to meet the different operating temperatures for space heating and DHW. To assess and compare

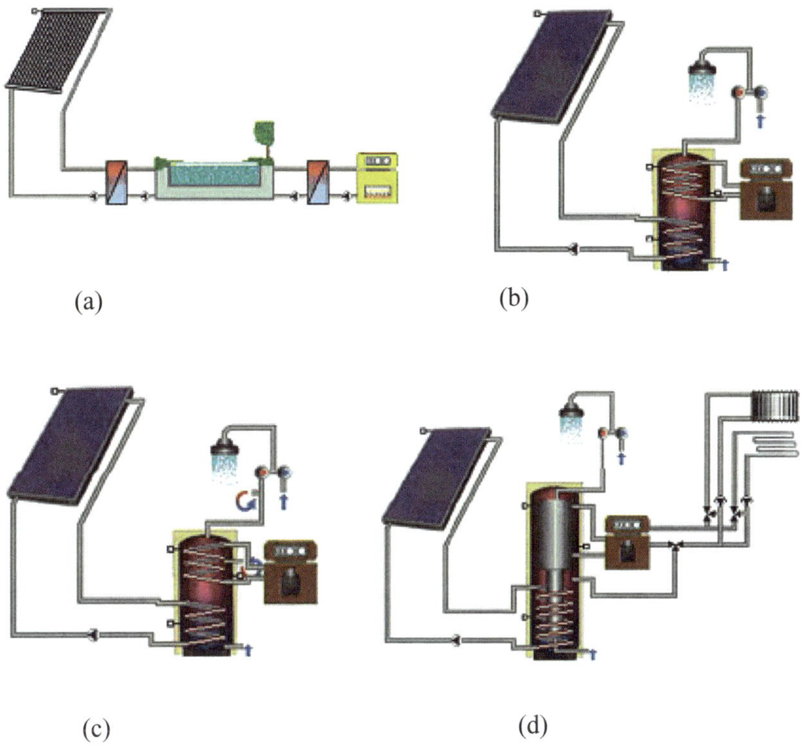

Figure 10-1 Examples of typical solar installations: (a) swimming pool, (b) domestic hot water, (c) domestic hot water for dwellings, and (d) solar combi systems for domestic hot water and space heating for houses. (Images generated using T-SOL 4.5 Expert; Valentin Energie software, www.valentin.de.)

performances of different designs for solar combi systems, the International Energy Agency (IEA) launched Task 26 to address issues in this area (http://task26.iea-shc.org/). Standardized classification and evaluation processes and design tools were developed for these systems, along with proposals for the international standardization of combi system test procedures. The primary standard used to define the assessment of solar thermal collector performance is ISO 9806 (ISO 2013). Criteria for the design and installation of solar thermal systems are also defined in ICC 900/SRCC 300-2015 (ICC 2015).

Besides preheating water, solar energy can be used to preheat incoming air before it is introduced to a building using systems integrated with the overall building design or collectors. This is a simple technological approach that can be economically viable and is well suited for northern climates or areas with high building heating load. The DOE has published a bulletin on these transpired solar collectors (DOE 2006).

The main drawback of solar combi systems has been the fact that during summer the available high solar radiation and the heat produced from the solar collectors cannot be fully used, thus making the system financially less attractive and limiting its use to the low DHW summer demand. In addition, there are some technical problems related to stagnation (when there is low thermal demand and hence low flow through the collector). This could potentially mean that the medium in the solar collector loop vaporizes during a period of low flow rate and high solar radiation availability. This generally should not occur with low-temperature collectors such as unglazed ones, but may be an issue for higher-temperature collectors. The Solar Rating and Certification Corporation Standard 100 specifies that all collectors be designed to withstand stagnation conditions without "significant degradation" (SRCC 2014). That standard defines stagnation as "the solar collector temperature at which the energy gain is balanced by the heat loss." Because high building cooling loads generally coincide with high solar radiation, the readily available solar heat from the existing solar collectors can be exploited by a heat-driven cooling machine, thus extending the use of the solar field throughout the year (solar hot water [SHW] and space heating in winter and SHW and cooling in summer). Combining solar heating and cooling is usually referred to as a solar combi-plus system that can increase the total solar fraction (i.e., the percentage of energy for a use that the solar system can provide, usually expressed on an annual basis).

Europe has the most sophisticated market for different solar thermal applications, including systems for hot-water production, plants for space heating of single- and multifamily houses and hotels, large-scale plants for district heating, and a growing number of systems for air-conditioning, cooling, and industrial applications.

The capital cost of solar thermal systems generally increases with higher working fluid temperatures. The higher the delivery temperature, the lower the efficiency, and the more solar collector area is generally required to deliver the same net energy. This is due to parasitic thermal losses that are inherent in solar collector design. Different solar collector types provide advantages and disadvantages, depending on the application, and there are significant cost differences among each solar collector type. Some common collector types are discussed here. (See Figure 10-2 for examples of hardware.)

Flat-Plate Solar Collectors. These are best suited for processes requiring low-temperature working fluids (80°F to 160°F [27°C to 71°C]) and can deliver 80°F (27°C) fluid temperatures, even during overcast conditions. The term *flat-plate collector* gen-

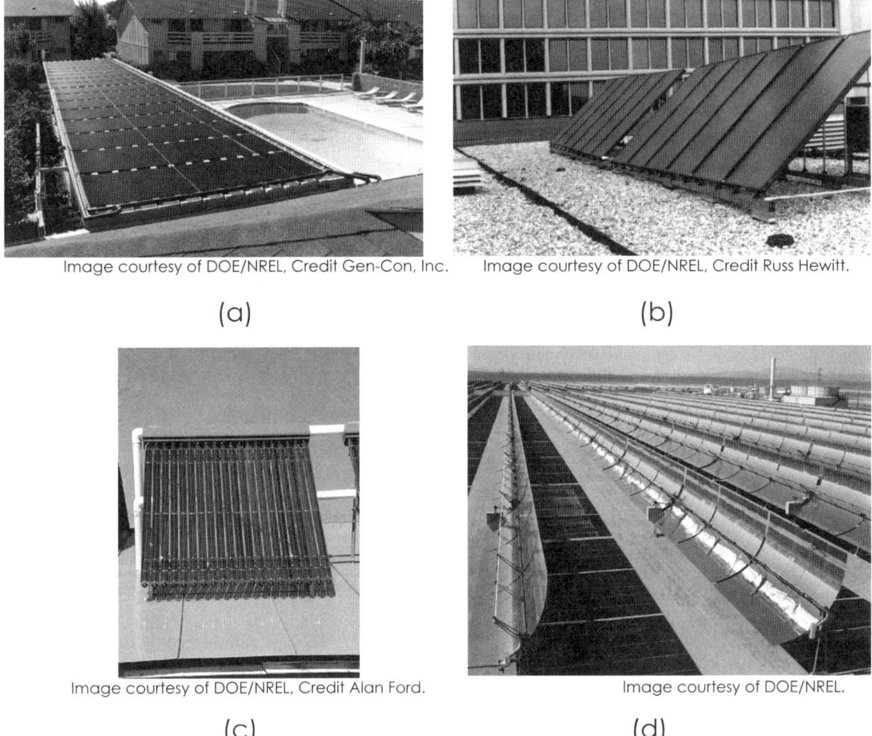

Image courtesy of DOE/NREL, Credit Gen-Con, Inc.

(a)

Image courtesy of DOE/NREL, Credit Russ Hewitt.

(b)

Image courtesy of DOE/NREL, Credit Alan Ford.

(c)

Image courtesy of DOE/NREL.

(d)

Figure 10-2 Examples of solar hardware: (a) unglazed plastic collector, (b) glazed flat-plate collector, (c) evacuated-tube collector, and (d) concentrating collectors.

erally refers to a hydronic coil-covered absorber housed in an insulated box with a single- or double-glass cover that allows solar energy to heat the absorber. The glazing used is a special high-transmissivity version. Heat is removed by a fluid running through the hydronic coils. Its design makes it more susceptible to parasitic losses than an evacuated tube collector, but more efficient in solar energy capture, because flat-plate collectors convert both direct and indirect solar radiation into thermal energy. This makes flat-plate collectors the preferred choice for domestic hot-water and other low-temperature heating applications. Coupling a water-source heat pump with low fluid temperatures with solar collectors provides heating efficiencies that are higher than ground-source heat pump (GSHP) applications and standard natural gas furnaces.

Typical flat-plate solar collector performance curves are illustrated in Figure 10-3. The collector's efficiency η is expressed as a function of the solar collector's working

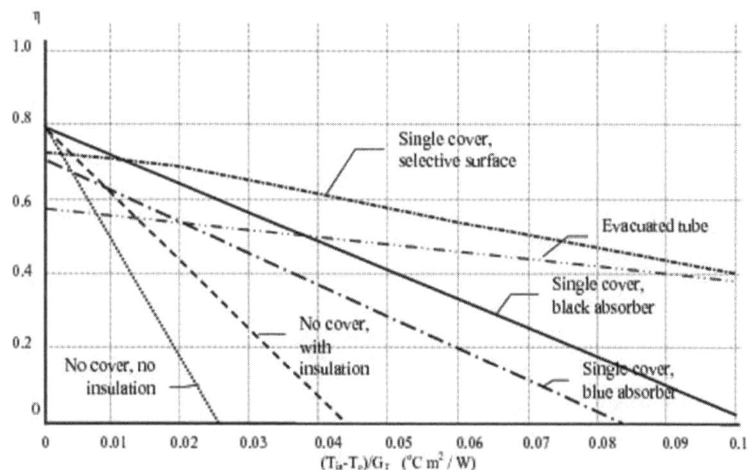

Image courtesy of C.A. Balaras.

Figure 10-3 Typical flat-plate solar collector performance curves.

fluid inlet temperature T_{in}, the ambient temperature T_o, and the total solar radiation incident on the collector's surface G_T.

Evacuated-Tube Collectors. This type of collector involves encasing the absorbing device inside of an evacuated tube. The evacuated tube minimizes parasitic losses even at elevated temperatures. Three different design types are generally used in evacuated tube collectors: U-tubes, flooded and heat pipe. The heat pipe type of collector does not circulate the working fluid through the evacuated tube, but rather relies on the heat pipe to transfer the heat gained to the transfer working fluid.

Another variation on the evacuated-tube type collector involves a series of small absorbers consisting of small-diameter (approximately 3/8 in. [10 mm]) copper tubing encased in a clear, cylindrical, evacuated glass tube. Because of the relatively small absorber area, significantly more collector area is required than with the flat-plate or concentrating collector.

Concentrating Collectors. This refers to the use of a parabolic trough reflector that focuses the solar radiation falling within the reflector area onto a centrally located absorber. Another concentrating collector type uses concentrating lenses (such as a Fresnel lens) to focus the solar radiation. Concentrating collectors are best suited for processes requiring high-temperature working fluids (300°F to 750°F [150°C to 400°C]) and do not operate under overcast conditions. This type of collector converts only direct solar radiation, which varies dramatically with sky clearness and air quality. Concentrating collectors often will use solar trackers that rotate on one or two axes to track the sun and collect the available direct solar radiation, although some designs do not use tracking.

Other Collector Types. Other types of solar thermal collectors include thermosiphon systems, which use natural convection to circulate the working fluid without a need for a pump. These systems are more common in areas where freezing temperatures are not a concern, and are simple and robust, making them a cost-effective option. Some solar thermal collectors are combined with photovoltaic (PV) panels. One option is where the electricity generated is directly connected to resistive heating elements in a water heater. Another option has a solar thermal collector integrated on the back of a PV module. This allows for an increased overall PV efficiency because of the lower temperatures, although a thermal sink may be needed if no thermal load condition exists to avoid stagnation conditions and resulting higher PV module temperatures.

Pros and cons of different solar collector types and other limitations of active solar systems can be found in a number of sources. Many of the key contributions came during the initial energy concerns of the 1970s, and a few are listed in the References and Resources section at the end of this chapter.

Solar Thermal Systems for Building Cooling

The main heat-driven cooling technologies that can be used for solar cooling include the following:

- Closed-cycle systems (e.g., LiBr/H_2O and H_2O/NH_3 absorption systems [see Chapter 18 of the *ASHRAE Handbook—Refrigeration* {ASHRAE 2014}] and adsorption cycle systems [see Chapter 32 of the *ASHRAE Handbook—Fundamentals* {ASHRAE 2017}]). They produce chilled water that can be used in combination with any air-conditioning equipment, such as air-handling units, fan-coil systems, and chilled ceilings.
- Open-cycle systems (e.g., desiccant systems [see Chapter 32 of the *ASHRAE Handbook—Fundamentals* {ASHRAE 2017}]). The term *open cycle* is used to indicate that the refrigerant is discarded from the system after it provides the cooling effect. New refrigerant is supplied in its place in an open-ended loop.

Solar-assisted cooling systems use solar thermal collectors connected to thermal-driven cooling devices. They consist of several main components (Figure 10-4): solar collectors, heat storage, heat distribution system, heat-driven cooling unit, optional cold storage, an air-conditioning system with appropriate cold distribution, and an auxiliary subsystem (which is integrated at different places in the overall system and is used as an auxiliary heater parallel to the collector or the collector/storage, as an auxiliary cooling device, or both). International research and development efforts, along with monitoring data from several installations are available

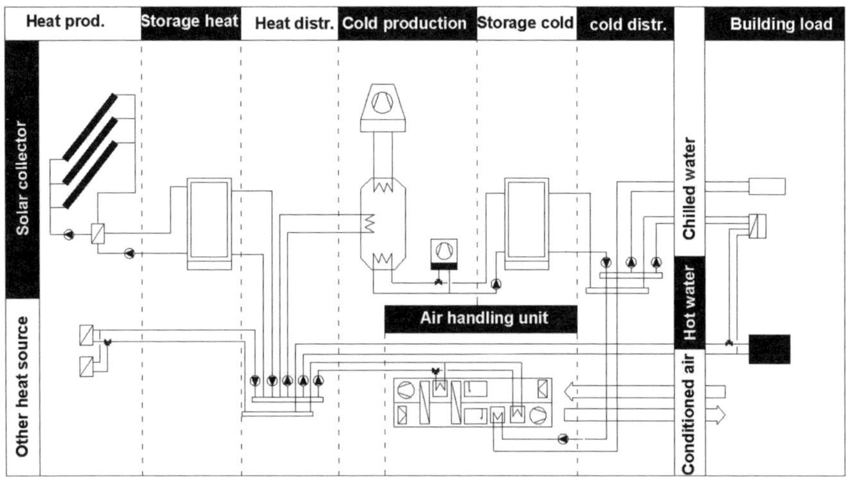

Image courtesy of Hans-Martin Henning.

Figure 10-4 Schematic description of a solar air-conditioning system showing the integration of different component options.

from Task 38 on solar air conditioning and refrigeration from the International Energy Agency (IEA).

The average specific solar collector area averages 136 ft^2/ton (3.6 m^2/kW refrigeration), ranging from 19 to 208 ft^2/ton (0.5 to 5.5 m^2/kW refrigeration), depending on the employed technology. Adsorption and absorption systems typically use more than 76 ft^2/ton (2 m^2/kW refrigeration) and usually less than 189 ft^2/ton (5 m^2/kW refrigeration). Overall, H_2O/NH_3 systems require larger specific collector areas than $LiBr/H_2O$ systems and, as a result, the installations are usually more expensive. The initial overall cost per installed cooling capacity in averages about \$13,400/ton or 4000 Euro/kW_R at 2017 exchange rates, excluding the cost for distribution networks among the system and the application and the delivery units.

Each technology has specific characteristics that match the building's HVAC design, loads, and local climatic conditions. A good design must first exploit all available solar radiation and then cover the remaining loads from conventional sources. Proper calculations for collector and storage size depend on the employed solar cooling technology. Hot-water storage may be integrated between the solar collectors and the heat-driven chiller to dampen the fluctuations in the return temperature of the hot water from the chiller. The storage size depends on the application; if cooling loads mainly occur during the day, then a smaller storage unit will be necessary than when the loads peak in the evening. The goal should be to avoid heating the hot-water storage by the backup heat source as much as possible. The storage's only function is to store excess heat of the solar system and to make it available when sufficient solar heat is not available. An overview of European efforts on solar combi and combi-plus systems, summarizing their main design, and operational and performance characteristics, is available in Balaras et al. (2010).

Evaluation of Solar Thermal Systems

The percentage of energy a solar system can provide is known as the solar fraction. The original method for evaluating solar thermal systems was developed by Sanford Klein at the University of Wisconsin and is known as the F-Chart method (Klein and Beckman 2001). This method provides an accurate assessment of the amount of energy a solar thermal system will provide. This modeling provides the designer the ability to vary system parameters (e.g., collector area, storage volume, operating temperature, and load) to optimize system design. The method was originally developed to aid engineers who did not have computer resources available to do the complex analyses required for solar thermal calculations. More recently, several commercially available software packages have become available that use closed-form energy balance modeling of the system.

The cost effectiveness of thermal solar systems is also dependent on having a constant load for the energy the solar system provides. Since space heating requirements are generally seasonal in most climates, it is advisable that energy from the solar system have more uses than space heating alone. Domestic water heating is usually a much steadier year-round load, though typically not very substantial. Solar cooling can extend use of solar collectors during the summer. The fact that peak cooling demand in summer is associated with high solar radiation availability offers an excellent opportunity to exploit solar energy with heat-driven cooling machines. The cost analysis of a proposed system would evaluate the energy provided with respect to the load variation expected (or known) versus the initial system cost, the cost for supplemental energy and overall solar availability.

Solar-Electric Systems

The direct conversion from sunlight to electricity is accomplished with PV systems. These systems continue to drop in price, and many states (as well as the federal government) offer tax incentives for the installation of these systems. The most common application is a grid-tied system where electricity is directly fed into the grid.

Other applications of PV may be attractive and provide additional value beyond the cost of offsetting utility power. Whenever power lines must be extended, PV should be investigated. It is cost-effective to use stand-alone PV applications for signs, remote lights, and blinking traffic lights. Many times, if an existing small grid line is available and additional power is needed, it is less expensive to add the PV system rather than increase the power line. This may be true of remote guard shakes, restroom facilities, or other outbuildings. PV should also be considered as part of the uninterruptible power system for a building, for example, as part of a battery backup system. Batteries can be centralized and the PV fed directly into this system. The cost can be less than other on-site generation and fuel storage.

Technical and legal factors have generally been worked out that have hindered the use of grid-tied solar PV systems in the past. In temperate climates, grid-tied PV systems provide a good method to reduce peak summer electrical demand, since peak solar gains generally correspond to peak air-conditioning demand. This was a major factor that led to the passing of the California Million Solar Roofs initiative in early 2006. The German 100,000 Roofs Program (HTDP), which began in early 1999, had a goal of 300 MW. It was successfully completed in 2003 with the parallel introduction of the German Renewable Energy Resources Act (EEG) that came into effect in 2000. The EEG and HTDP program secured commercially oriented PV investors (because they received a full payback of their investment).

According to the International Energy Agency, the cumulative installed capacity of PV systems around the world in 2015 has reached about 227 GWp (peak gigawatts).

The economic viability of PV systems is dependent on many factors. Improvements in technology and manufacturing techniques have lowered the cost of PV systems dramatically. For example, in the 10-year period from 2007–2016, the average PV system installed cost per peak watt dropped from about $9.50 to around $3.00 for residential systems and from around $7.75 to about $2.00 for commercial scale systems. (NREL 2015, Fu et. al. 2016).

The relative fraction of the total system cost due to the PV modules has dropped significantly as well, with the module plus inverter cost now representing only about one-third to one-half of the total installed cost (depending on if it is a smaller residential scale or larger utility scale system). The typical PV cell efficiencies (i.e., the ratio of electrical energy produced by a solar cell to the incident solar irradiance) currently range from less than 5% for the first generation of thin-film cells to more than 24% for the most advanced crystalline-silicon cells in laboratory conditions.

The integration of PV systems with building materials is a new development that may, in the long run, help make PV systems viable in most areas. One of the key ways to incorporate solar energy technologies is to incorporate them directly into the architecture of the building (see Figure 10-5). Trombe walls can replace spandrel panels and other glass facades; PV panels can become overhangs for the building or parking shading structures. Using these elements as part of the building doubles the value of the systems. An example of these technologies is the concept called *building-integrated PVs*. A picture of one concept is shown in Figures 10-5 and 10-6. These are being developed by the National Renewable Energy Laboratory (NREL), among others. PV collectors are addressed in GreenTip #10-3.

Image courtesy of DOE/NREL, Credit Ben Kroposki.

Image courtesy of DOE/NREL, Credit Lawrence Berkeley Lab.

Figure 10-5 Building-integrated PVs.

Image courtesy of DOE/NREL, Credit United Solar Ovonic.

Figure 10-6 Building-integrated PV roofing.

WIND

Using prevailing breezes and wind energy is one of the most promising alternative technologies today. Wind turbine design and power generation have become more reliable over the last 30 years. Consistency and velocity of available breezes are essential to successful application of wind for electrical generation and natural ventilation in buildings. Information on velocities, durations, and direction of winds in a project area is generally available from the National Oceanic and Atmospheric Administration (NOAA). That said, wind power systems installed on buildings or on actual new building project sites are still relatively rare; the energy availability is not that great and the negative factors discussed below are real and must be taken into consideration.

From a natural-ventilation perspective for passive cooling of buildings, other factors (e.g., temperature and relative humidity) must also be taken into consideration. Many areas can use outdoor air/natural breezes to provide passive cooling only during a limited period of the year, requiring designers wanting to use natural ventilation to carefully analyze both climatic conditions and cost to implement. The benefits of using natural ventilation must be weighed against the consistency of breezes, potential for higher relative humidity levels (which may affect indoor air quality), and the impact on occupant comfort and HVAC operation.

There are many excellent examples where natural ventilation provides the cooling needed throughout the year, but most are located in cool, arid climates. Even in moderate climates, such as that in Atlanta, GA, natural ventilation can provide passive cooling during some periods of the year, but in doing so, the designer must consider the impact on HVAC operation and building control. These elements may be very difficult to get to work effectively with operable windows. Additional programming and control points are required for successful operation. The potential of natural ventilation, its appropriate use, the design and dimensioning methodologies, the need for an integrated design approach, and how to overcome barriers are available in a handbook by Alvarez et al. (1998).

Incorporating one or more wind turbine generators at a building site is a more active approach. Initially, it should be recognized that there is a disadvantage of scale: one or two wind turbines alone are destined to be less cost-effective than a wind farm with hundreds. Wind turbines should first be evaluated by looking at the wind resource at the site. For most small-scale wind turbines, steady winds over 20 mph (9 m/s) are preferred to maximize the cost effectiveness. Other issues that add difficulty include noise, vibration, building geometry (if on the building), wind pattern interrupters, wildlife effect, periodic maintenance, safety, visual impact, and community acceptance; these must also be taken into account.

HYDRO

According to the International Hydropower Association (IHA), hydropower is "the world's largest source of renewable energy by potential capacity." In all of its forms, from conventional power generation to emerging technologies like ocean and wave energy, hydrokinetic, and tidal power, there is a role of all types and scales of hydropower. Pumped storage may be combined with other RES systems to meet power demand variations. The water may be pumped from a lower reservoir to an upper reservoir when demand for electricity is low (the power for pumping may be supplied from RES systems or the grid). The water is then released from the upper reservoir back to the lower reservoir to generate power using the hydrokinetic energy of the water flow through a generator (this is done during peak power demand or low generating output from RES systems).

The ability to use hydro is limited on an individual building scale. On rare occasions, older building renovations can take advantage of existing river dams and generate hydroelectric on site. This is especially true in the northeastern United

States, where the primary power supply in the 1800s was hydro. In some cases, small streams can be diverted and small turbines installed—this is called *low-impact hydro*, as a minimal dam is needed and water is returned to the same stream.

In some cases, pressure reduction values can be replaced with small turbines. Although this is not really free energy from the big picture, as the water was pressurized before being distributed to the site, it is better to recover some of the energy rather than essentially waste it during pressure reduction.

BIOMASS

Biomass is the conversion of plant and animal matter into useful energy. Several ideas should be investigated to be considered as an energy source in a building. In the European Union, "bioenergy contributes only to 3.7% of the total primary energy supply; however, it plays a considerable role in several European countries, such as Finland and Sweden where its contribution amounts to respectively 20% and 16% of the gross inland consumption" (European Biomass Industry Association).

Wood chip conversion to heat (and possibly then electricity) can replace boiler systems in buildings. Automatic feed systems can now take wood chips and make hot water as conveniently as oil and natural gas systems can. In many locations, wood chips are free for the cost of the transportation, saving further cost. Pellet burners are a similar concept, except for the additional cost of the wood pellets. Burners are also available for corn cobs. In some cases, manufacturers will lease equipment for the energy savings, resulting in no additional cost to the end user.

The production and distribution of bio-oils and biodiesel fuels derived from bio-based materials is growing in momentum. Heating systems are available to burn used vegetable oils. These systems are useful near the point of production of used vegetable oil—usually the food service industry. Grocery stores and restaurants may benefit from such technologies. Most diesel-fueled machines can run on bio-diesel without any (or minor) modifications. In addition, hauling and waste costs for the oil are eliminated, increasing the attractiveness of the technology.

REFERENCES AND RESOURCES

Published

Alvarez, S., E. Dascalaki, G. Guarracino, E. Maldonado, S. Sciuto, and L. Vandaele. 1998. *Natural Ventilation of Buildings—A Design Handbook*. F. Allard, ed. London: James & James Science Publishers Ltd.

ASHRAE. 2014. *ASHRAE Handbook—Refrigeration*. Atlanta: ASHRAE.

ASHRAE. 2017. *ASHRAE Handbook—Fundamentals*. Atlanta: ASHRAE.

Balaras, C.A., E. Dascalaki, P. Tsekouras, and A. Aidonis. 2010. High solar combi systems in Europe. *ASHRAE Transactions* 116(1):408–15.

Balaras, C.A., H.-M. Henning, E. Wiemken, G. Grossman, E. Podesser, and C.A. Infante Ferreira. 2006. Solar cooling—An overview of European applications and design guidelines. *ASHRAE Journal* 48(6):14–22.

Beckman, W., S.A. Klein, and J.A. Duffie. 1977. *Solar Heating Design*. New York: John Wiley & Sons, Inc.

DOE. 2006. Solar buildings. Transpired air collectors ventilation preheating. DOE/GO-102001-1288. Washington, DC: U.S. Department of Energy.

EU. 2008. Climate change: Commission welcomes final adoption of Europe's climate and energy package. Europa press release IP/08/1998, European Union, Brussels.

Feldman, D., G. Barbose, R. Margolis, M. Bolinger, D. Chung, R. Fu, J. Steel, C. Davidson, N. Dargouth and R. Wiser. 2015. Photovoltaic System Pricing Trends: Historical, Recent and Near-Term Projections. NREL/PR-6A20-64898. www.nrel.gov/docs/fy15osti/64898.pdf.

Fu, R., D. Chung, T. Lowder, D. Feldman, K. Ardani, and R. Margolis. 2016. U.S. Solar Photovoltaic System Cost Benchmark: Q1 2016. NREL Technical Report NREL/TP-6A20-66532. http://www.nrel.gov/docs/fy16osti/66532.pdf.

Henning. H-M., ed. 2004. *Solar Assisted Air-Conditioning in Buildings—A Handbook For Planners*. Wien: Springer-Verlag.

ICC. 2015. ICC 900/SRCC 300-2015: Solar Thermal System Standard. Washington, DC: International Code Council.

ISO. 2013. ISO 9806:2013 Solar energy—Solar thermal collectors—Test methods. Geneva: International Standards Organization.

Kreith, F., and J. Kreider. 1989. *Principles of Solar Engineering*, 2d ed. Washington, DC: Hemisphere Pub.

NREL. 2010. NREL highlights utility green power leaders. NREL press release NR-1910. Golden, CO: National Renewable Energy Laboratory.

Shiklomanov, I.A. 1993 World fresh water resources. In *Water in Crisis*. P.H. Gleick, ed. New York: Oxford University Press.

SRCC. 2014. SRCC Standard 100-2014-07: Minimum Standards for Solar Thermal Collectors. Cocoa, FL: Solar Rating and Certification Corporation.

Online

American Solar Energy Society
www.ases.org.

American Wind Energy Association
www.awea.org.

Centre for Analysis and Dissemination of Demonstrated Energy Technologies
www.caddet-re.org.

EIA. Table 10.8, Photovoltaic cell and module shipments by type, trade, and prices, 1982–2008. U.S. Energy Information Administration. www.eia.doe.gov/aer/txt/stb1008.xls.

EPA. 2010. *Guide to Purchasing Green Power*. Washington, DC: U.S. Environmental Protection Agency. www.epa.gov/greenpower/guide-purchasing-green-power.
EPIA. 2010. 2014 Global Market Outlook for Photovoltaics until 2014. European Photovoltaic Industry Association.
www.greenunivers.com/wp-content/uploads/2010/05/Global_Market_Outlook_for_Photovoltaics_until_20141.pdf.
ESTIF. 2010 Solar thermal markets in Europe —Trends and market statistics 2009. www.estif.org/fileadmin/estif/content/market_data/downloads/2009%20solar_thermal_markets.pdf.
Europe's Energy Portal
www.energy.eu/#renewable.
European Biomass Association
www.aebiom.org.
European Biomass Industry Association
www.eubia.org/.
European Photovoltaic Industry Association
www.epia.org.
European Renewable Energy Centres Agency
www.eurec.be.
European Solar Thermal Industry
http://solarheateurope.eu/welcome-to-solar-heat-europe/.
European Wind Energy Association
https://windeurope.org/.
EUROSOLAR—The European Association for Renewable Energies e. V.
www.eurosolar.org.
Global Wind Energy Council
www.gwec.net.
Green Energy In Europe
www.ecofys.com/.
Green Venture
www.greenventure.ca.
GWEC. 2010. Global wind 2009 report. Global Wind Energy Council, Brussels, Belgium. www.gwec.net/index.php?id=167.
International Energy Agency, Bioenergy
www.ieabioenergy.com.
International Energy Agency, Photovoltaic Power Systems Programme
www.iea-pvps.org.
International Energy Agency, Solar Heating And Cooling Programme
www.task38.iea-shc.org.
International Energy Agency, Task 38
www.ieabioenergy-task38.org/.

International Hydropower Association
www.hydropower.org.

International Hydropower Association Fact Sheet on the Hydropower Sustainability Assessment Protocol.
www.internationalrivers.org/campaigns/the-hydropower-sustainability-assessment-protocol.

International Solar Energy Society
www.ises.org.

Klein, S., and W. Beckman. 2001. *F-Chart Software.* The F-Chart Software Company., Madison, WI.
www.fchart.com.

National Hydropower Association
www.hydro.org.

National Oceanic and Atmospheric Administration
www.noaa.gov.

National Renewable Energy Laboratory, Solar Energy Research Facility
www.nrel.gov/pv/facilities_serf.html.

North Carolina Solar Center, DSIRE
www.dsireusa.org/.

Oak Ridge National Laboratory, Bioenergy Feedstock Information Network
web.ornl.gov/sci/ees/cbes/.

Solar Energy Industries Association
www.seia.org.

U.S. Department of Energy, Energy Efficiency and Renewable Energy, Bioenergy Technologies Office web page
www1.eere.energy.gov/biomass/for_consumers.html.

Weiss, W., I. Bergmann, and G. Faninger. 2008. *Solar Heat Worldwide*. Solar Heat & Cooling Programme, International Energy Agency, Paris, France.
https://www.iea-shc.org/data/sites/1/publications/Solar_Heat_Worldwide-2008.pdf.

Weiss, W., and F. Mauthner. 2010. Solar heat worldwide—Markets and contribution to the energy supply 2008.
https://www.iea-shc.org/data/sites/1/publications/Solar_Heat_Worldwide-2010.pdf.

World Meteorological Organization, World Radiation Data Centre
www1.eere.energy.gov/bioenergy/.

The World's Water
www.worldwater.org.

ASHRAE GreenTip #10-1

Passive Solar Thermal Energy Systems

GENERAL DESCRIPTION

Passive solar thermal energy systems use solar energy (mainly for space heating) via little or no use of conventional energy or mechanisms other than the building design and orientation. All above-grade buildings are passive solar. Making buildings collect, store, and use solar energy wisely then becomes the challenge for building designers. A building that intentionally optimizes passive solar heating can visually be a solar building, but many reasonably sized features that enhance energy collection and storage can be integrated into the design without dominating the overall architecture.

To be successful, a well-designed passive solar building needs (1) an appropriate thermal load (e.g., space heating); (2) aperture (e.g., clear, glazed windows); (3) thermal storage to minimize overheating and to use heat at night; (4) control, either manual or automatic, to address overheating; and (5) night insulation of the aperture so that there is not a net heat loss.

HIGH-PERFORMANCE STRATEGIES

The following strategies are general in nature and are presented as guidelines to help maximize the performance of a passive thermal solar design:

- Think about conservation first. Minimizing the heating load will reduce conventional and renewable heating systems' sizes and yields the best economics. Insulate, including the foundation, and seal the building well. Use quality exterior windows and doors.
- In the northern hemisphere, the aperture must face due south for optimal performance. If this is not possible, make it within ±10 degrees of due south. In the southern hemisphere, this solar aperture looks north.
- Minimize use of east- and west-facing glazing. They admit solar energy at nonoptimal times, at low angles that minimize storage, and are difficult to control by external shading. Also, reduce north glazing in the colder regions of the northern hemisphere, because this can lead to high heat loss rates.

- Use optimized and/or moveable external shading devices, such as overhangs, awnings, and side fins. Internal shading devices should not be relied upon for passive solar thermal control—they tend to cause overheating.
- Use high-mass direct gain designs. The solar collector (windows) and storage (floors and potentially walls) are part of the occupied space and typically have the highest solar savings fraction, which is the percent of heating load met by solar. Directly irradiated thermal masses are much more effective than indirect, thus floors or trombe walls are often best.
- Use vertical glazing. Horizontal or sloped windows and skylights are hard to control and insulate.
- Calculate the optimal thermal mass—it is often around 8 in. thick (about 200 mm)—high-density concrete for direct-gain floors over conditioned basements. The optimization should direct the designer to a concept that will capture and store the highest amount of solar energy without unnecessarily increasing cost or complexity. For all direct-gain surfaces, make sure they are of high-absorptivity (dark color) and are not covered by carpet, tile, much furniture, or other items that prevent or slow solar energy absorption. Be sure the thermal mass is highly insulated from the outdoor air or ground.
- Seeking a very high annual solar savings fraction f_s often leads to disappointment and poor economics, so keep expectations reasonable. Even a 15% annual fraction represents a substantial reduction in conventional heating energy use. A highly optimized passive solar thermal single-family house, in an appropriate climate, often only has about a 40% solar fraction. Combining passive with active solar, PVs, wind, and other renewable energy sources is often the most satisfactory way to achieve a very high annual solar savings fraction.
- An old solar saying, of unknown origin, is "the more passive a building, the more active the owner." Operating a passive solar building to optimize collection has thus been called *solar sailing* and requires time and experience. Be sure that passive solar and the building operator are good matches for each other.

KEY ELEMENTS OF COST

Passive solar energy systems must be engineered; otherwise, poor performance is likely.

Window sizes are typically larger than for conventional design. Some operable windows and/or vents, placed high and low, are needed for overheat periods. Proper solar control must also be foreseen.

Concrete floors and walls are commonly thicker in the storage portion of a passive solar building. A structural engineer's services are likely required.

Conventional backup systems are still needed. They will be used during cloudy and/or cold weather, so select high-efficiency equipment with low-cost fuel sources.

Night insulation must be used consistently or else there will be a net heat loss. Making the nightly installation and removal automatic is recommended, but costly.

SOURCES OF FURTHER INFORMATION

ASHRAE. 2015. *ASHRAE Handbook—HVAC Applications*. Chapter 35, Solar Energy Use. Atlanta: ASHRAE.

Balcomb, J.D. 1984. *Passive Solar Heating Analysis: A Design Manual*. Atlanta: ASHRAE.

Crosbie, M.J., ed. 1998. *The Passive Solar Design and Construction Handbook*, Steven Winter Associates. New York: John Wiley & Sons, Inc.

Online

Passive Solar Design.
http://passivesolar.sustainablesources.com/.

ASHRAE GreenTip #10-2

Active Solar Thermal Energy Systems

GENERAL DESCRIPTION

Both passive and active solar thermal energy systems rely on capture and use of solar heat. Active solar thermal energy systems differ from passive systems in the way that they use solar energy, as they use it primarily for space and water heating. This can allow for greatly enhanced collection, storage, and use of solar energy.

To be successful, a well-designed active solar system needs (1) an appropriate thermal load (e.g., potable water, space air, or pool heating); (2) collectors such as flat-plate solar panels; (3) thermal storage to use the heat at a later time; and (4) control, typically automatic, to optimize energy collection and storage and for freeze and overheat protection.

The working fluid or coolant that moves heat from the collectors to the storage device is typically water, a water/glycol solution, or air. The heat storage medium is often water but could be a rock-bed or a high-mass building for air-coolant systems. Freeze protection must be an important design consideration, except in for areas not subject to wintertime freezing conditions. The glycol solution can break down under repeated exposure to high temperatures in the panels, and thus there may be a need to drain and replace the circulating fluid.

The solar energy collectors are most often of fixed orientation and nonconcentrating, but they could be tracking and/or concentrating. Large surface areas of collectors, or mirrors and/or lens for concentration, are needed to gather heat and to achieve higher temperatures.

There are many different types of active solar thermal energy systems. For example, one type often used for potable water heating is flat-plate, pressurized water/glycol coolant, and two-tank storage. An internal double-wall heat exchanger is typically employed in one tank, known as the preheat tank, and the other tank, plumbed in series, is a conventional water heater. These preheat tanks are now widely available due to nonsolar use as indirect water heaters. One-tank systems typically have an electric-resistance heating element installed in the top of the special tank, but could also be designed for natural gas supplemental heat as well.

Some systems include a photovoltaic powered DC current circulation pump that circulate the working fluid through the collector when solar radiation is available. This also helps to avoid nighttime cooling of the fluid.

HIGH-PERFORMANCE STRATEGIES

The following strategies are general in nature and are presented as a guideline to helping maximize the performance of an active thermal solar design:

- Think about conservation first. Minimizing the heating load will reduce conventional and renewable heating systems' sizes, and yield the best economics.
- In the northern hemisphere, the solar collectors must face due south for optimal performance. If not possible, make them within ±10 degrees of due south. In the southern hemisphere, the solar collectors look north.
- When using solar collectors for collecting the most energy over the entire year (such as water heating) and mounted at a fixed angle, mount them at an angle according to the following recommendations:
    - For latitudes below 25°, use approximately the local latitude. water heating, use the local latitude plus 10°.
    - For latitudes above 50°, a more detailed analysis may be needed, but a rule of thumb might be the local latitude minus 10°.
    - For solar collectors used for space heating during the winter, use the local latitude plus 15°. This optimizes the solar heat gain during winter.
- Calculate the optimal thermal storage—about one day's heat storage (or less) often yields the best economics. Place the thermal storage device within the heated space, and be sure it is highly insulated, including under its base.

- Seeking a very high annual solar savings fraction f_S often leads to disappointment and poor economics, so keep expectations reasonable. Even a 25% annual fraction represents a substantial reduction in conventional energy use. A highly optimized, active solar thermal domestic water-heating system, in an appropriate climate, often has about a 60% solar fraction. Combining active solar with passive, PVs, wind, and other renewable energy sources is often the most satisfactory way to achieve a very high annual solar savings fraction.

KEY ELEMENTS OF COST

Active solar energy systems must be engineered; otherwise, poor performance is likely.

Well-designed, factory-assembled collectors are recommended, but are fairly expensive. Site-built collectors tend to have lower thermal performance and reliability.

Storage tanks must be of high quality and be durable. Water will eventually leak, so proper tank placement and floor drains are important. For rock storage, moisture control and air-entrance filtration are important for mold growth prevention.

Quality and appropriate pumps, fans, and controls can be somewhat expensive. Using surge protection for all the electrical components is recommended. Effective grounding, for lightning and shock mitigation, is normally required by building code. For liquid coolants in sealed loops, expansion tanks and pressure relief valves are needed. For domestic water heating, a temperature-limiting mixing valve is required for the final potable water to prevent scalding; even nonconcentrating systems can produce up to 180°F (82°C) or so water at times. All thermal components require insulation for safety and reduced heat loss.

Installation requires many trades: a contractor to build or install the major components, a plumber to do the piping and pumps (i.e., water systems), an HVAC contractor to install ducts (i.e., air systems) and/or space-heating heat exchangers, an electrician to provide power, and a controls specialist.

Conventional backup systems are still needed. They will be used during cloudy and/or cold weather, so select high-efficiency equipment with low-cost fuel sources.

SOURCES OF FURTHER INFORMATION

ASHRAE. 1988. *Active Solar Heating Systems Design Manual.* Atlanta: ASHRAE.

ASHRAE. 2015. *ASHRAE Handbook—HVAC Applications*. Chapter 35, Solar Energy Use. Atlanta: ASHRAE.

Burch, J., J. Salasovich, and T. Hillman. 2005. As Assessment of Unglazed Solar Domestic Water Heaters. ISES Solar World Congress, Orlando FL. NREL/CP-550-37759.

Howell, J., R. Bannerot, and G. Vliet. 1982. *Solar-Thermal Energy Systems: Analysis and Design.* New York: McGraw-Hill.

Online Resources

North American Board of Certified Energy Practitioners (NABEP) http://www.nabcep.org/.

Solar Energy Industries Association. Solar Heating and Cooling Roadmap Fact Sheet. www.seia.org/research-resources/solar-heating-and-cooling-roadmap-fact-sheet.

U.S. Department of Energy. Active Solar Heating https://energy.gov/energysaver/active-solar-heating.

ASHRAE GreenTip #10-3

Solar Energy System—PV

GENERAL DESCRIPTION

Light shining on a PV cell, which is a solid-state semiconductor device, liberates electrons, which are collected by a wire grid, to produce direct-current electricity.

The use of solar energy to produce electricity means that PV systems reduce greenhouse gas emissions, electricity costs, and resource consumption. Electrical consumption can be reduced. Because the peak generation of PV electricity coincides with peak air-conditioning loads (if the sun shines), peak electricity demands (from the grid) may be reduced, though it is unlikely without substantial storage capacity.

PV can also reduce electrical power installation costs where the need for trenching and independent metering can be avoided. The public appeal of using solar energy to produce electricity results in a positive marketing image for PV-powered buildings and, thus, can enhance occupancy rates in commercial buildings.

While conventional PV design has focused on the use of independent applications in which excess electricity is stored in batteries, grid-connected systems are becoming more common. In these cases, electricity generated in excess of immediate demand is sent to the electrical grid, and the PV-powered building receives a utility credit. Grid-connected systems are often integrated into building elements. Increasingly, PV cells are being incorporated into sunshades on buildings for a doubly effective reduction in cooling and electricity loads.

PV power is being applied in innovative ways. Typical economically viable commercial installations include the lighting of parking lots, pathways, signs, emergency telephones, and small outbuildings.

A typical PV module consists of 33 to 40 cells, which is the basic block used in commercial applications. Typical components of a module are aluminum, glass, polyvinyl fluoride (PVF) films, and rubber. The cell is usually silicon, with trace amounts of boron and phosphorus.

Because PV systems are made from a few relatively simple components and materials, the maintenance costs of PV systems are low. Manufacturers now provide 20-year warranties for PV cells.

PV systems are adaptable and can easily be removed and reinstalled in other applications. Systems can also be enlarged for greater capacity through the addition of more PV modules.

WHEN/WHERE IT'S APPLICABLE

PV is well suited for rural and urban off-grid applications and for grid-connected buildings with air-conditioning loads. The economic viability of PV depends on the distance from the grid, electrical load sizes, power line extension costs, and incentive programs offered by governmental entities or utilities.

PV applications include prime buildings, outbuildings, emergency telephones, irrigation pumps, fountains, lighting for parking lots, pathways, security, clearance, billboards, bus shelters or signs, and remote operation of gates, irrigation valves, traffic signals, radios, telemetry, or instrumentation.

Grid-connected PV systems are better suited for buildings with peak loads during summer cooling operation but are not as well suited for grid-connected buildings with peak wintertime loads.

Note that a portion of a PV electrical system is direct current, so appropriate fusing and breakers may not be readily available. A PV system is also not solely an electrical installation; other trades, such as roofing and light steel erectors, may be involved with a PV installation. When a PV system is installed on a roof or wall, it will likely result in envelope penetrations that will need to be sealed.

PRO

- Reduces greenhouse gas emissions.
- Reduces nonrenewable energy demand, with the ability to help offset demand on the electrical grid during critical peak cooling hours.

- Enhances green-image marketing.
- Lowers electricity consumption costs and may reduce peak electrical demand charges.
- Reduces utility infrastructure costs.
- Increases electrical reliability for the building owner. May be used as part of an emergency power backup system.

CON

- Relatively high initial capital costs.
- Requires energy storage in batteries or a connection to electrical utility grid.
- May encounter regulatory barriers.
- High-capacity systems require large-building envelope areas that are clear of protuberances and have uninterrupted access to sunshine.
- Capacity to supply peak electrical demand can be limited, depending on sunshine during peak hours.

KEY ELEMENTS OF COST

The following provides a possible breakdown of the various cost elements that might differentiate a PV system from a conventional one and an indication of whether the net cost for this system is likely to be lower (L), higher (H), or the same (S). This assessment is only a perception of what might be likely, but it obviously may not be correct in all situations. There is no substitute for a detailed cost analysis as part of the design process. The listings below may also provide some assistance in identifying the cost elements involved.

First Cost

- PV modules H
- Wiring and various electrical devices H
- Battery bank H
- Instrumentation H
- Connection cost (if grid-connected) H

Recurring Cost

- Electricity L

SOURCES OF FURTHER INFORMATION

California Energy Commission, Renewable Energy Program www.energy.ca.gov/renewables/.

Canadian Renewable Energy Network (CanREN) http://users.encs.concordia.ca/~raojw/crd/essay/essay002164.html.

Eiffert, P., and G. Kiss. Building-Integrated Photovoltaic Designs for Commercial and institutional Structures. Golden, CO: National Renewable Energy Laboratory. www.nrel.gov/docs/fy00osti/25272.pdf.

Natural Resources Canada, eco ENERGY for Renewable Power www.nrcan.gc.ca/ecoaction/14145.

North American Board of Certified Energy Practitioners (NABCEP) http://www.nabcep.org/

NRC. *Photovoltaic Systems Design Manual*. Ottawa, Ontario, Canada: Natural Resources Canada, Office of Coordination and Technical Information.

NRC. RETScreen (Renewable Energy Analysis Software). Natural Resources Canada, Energy Diversification Research Laboratory, Ottawa, Ontario, Canada. www.retscreen.net.

Photovoltaic Resource Site www.pvpower.com.

School of Photovoltaic and Renewable Energy Engineering, University of New South Wales www.pv.unsw.edu.au.

Solar Energy Industries Association www.seia.org.

Sustainable Sources www.greenbuilder.com.

ASHRAE GreenTip #10-4

Solar Protection

GENERAL DESCRIPTION

Shading the building's transparent surfaces from solar radiation is mandatory during summer and sometimes even necessary during winter. This way, it is possible to prevent solar heat gains when they are not needed and to control daylighting to minimize glare problems. Depending on the origin of solar radiation (i.e., direct, diffuse, reflected), it may be possible to select different shading elements that provide more effective solar control.

Depending on the specific application and type of problem, there may be different options for selecting the optimum shading device. The decision can be based on several criteria, from aesthetics to performance and effectiveness or cost. Different types of shading elements are suitable for a given application, result in varying levels of solar control effectiveness, and have a different impact on indoor daylight levels, natural ventilation, and overall indoor visual and thermal comfort conditions.

There are basically three main groups of solar control devices. The first group is external shading devices, which can be fixed and/or movable elements. They have the most apparent impact on the aesthetics of the building. If properly designed and accounted for, they can become an integral part of the building's architecture, as they are integrated into the building envelope. Fixed types are typically variations of a horizontal overhang and a vertical side fin, with different relative dimensions and geometry. When properly designed and sized, fixed external shading devices can be effective during summer, while during winter they allow the desirable direct solar gains through the openings. This is a direct positive outcome given the relative position of the sun and its daily movement in winter (when there is low solar elevation) and summer (when there is high solar elevation). Movable types are more flexible, since they can be adjusted and operated either manually or automatically for optimum results and typically include various types and shapes of awnings and louvers.

The second group is shading devices located between the window panes, which are usually adjustable and retractable louvers, roller blinds, screens, or films that are placed within the glazing. This type of a shading device is more suitable for solar control of scattered radiation or sky diffuse radiation. Given that the incident solar radiation is already absorbed by the glazing, thus increasing its temperature, one needs to take into account the heat transfer component to the indoor spaces.

The third group, which is internal shading devices, is very common because of indoor aesthetics. They offer privacy control and easy installation, accessibility, and maintenance. Although on the interior they are very practical and, most of the time, necessary, their overall thermal behavior needs to be carefully evaluated, since the incident solar radiation is trapped inside the space and will be absorbed and turn into heat if not properly controlled (i.e., reflecting solar radiation outward through the opening). Numerous types or combinations of the various shading devices are also possible, depending on the application.

HIGH-PERFORMANCE STRATEGIES

- Use natural shading. Deciduous plants, trees, and vines offer effective natural shading. It is critical for their year-round effectiveness not to obstruct solar radiation during winter (in order to increase passive solar gains). Plants also have a positive impact on the immediate environment surrounding the building (i.e., the microclimate) because of their evaporative cooling potential. However, the plants need some time to grow, may cause moisture problems if they are too close to opaque elements, and can suffer from various diseases. The view can be restricted and some plants, especially large leafless trees, can still obstruct solar radiation during winter and may reduce natural ventilation. In general, for deciduous plants, the shading effect is best for east and west orientations, along with southeast and southwest.
- Incorporate louvers into the design. These are also referred to as *venetian blinds* and can be placed externally (which is preferable) or internally (which allows for easier maintenance and

installation in existing buildings). The external louvers can be fixed in place with rotating or fixed tilt of the slats. The louvers can also be retractable. The slats can be flat or curved. Slats from semitransparent material allow for outdoor visibility. The louvers can be operated manually (i.e., slat tilt angle, up or down movement) or they can be electrically motor driven. Adjusting the tilt angle of the slats or raising/lowering the panel can change the conditions from maximum light and solar gains to complete shading. Louvers can also be used to properly control air movement during natural ventilation. Slat curvature can be used to redirect incident solar radiation before entering into the space. Slat material can have different reflective properties and can also be insulated. During winter, fully closed louvers with insulated slats can be used at night for providing additional thermal insulation at the openings.

- Incorporate awnings into the design. External or internal awnings can be fixed in place, operated manually, or driven electrically by a motor that can also be automated. Light-colored materials are preferable, as they allow for high solar surface reflectivity. Awnings are easily installed on any type and size of opening and may also be used for wind protection during winter to reduce infiltration and heat losses.

KEY ELEMENTS OF COST

Natural shading usually is a cost-effective strategy that reduces glare, and, depending on the external building façade, can improve aesthetics. Plants should be carefully selected to match local climatic conditions (in order to optimize watering needs).

External electrically driven and automated units have a higher cost and it takes money to maintain the motors, but they are more flexible and effective. Louvers are difficult to clean on a regular basis. Nonretractable louvers somewhat obstruct outward vision.

Awning fabric needs periodic replacement, depending on local wind conditions. Electrically driven and automated units have a higher cost and it takes money to maintain the motors.

SOURCES OF FURTHER INFORMATION

Argiriou, A., A. Dimoudi, C.A. Balaras, D. Mantas, E. Dascalaki, and I. Tselepidaki. 1996. *Passive Cooling of Buildings.* M. Santamouris and A. Asimakopoulos, eds. London: James & James Science Publishers Ltd.

Baker, N. 1995. *Light and Shade: Optimising Daylight Design.* European Directory of sustainable and energy efficient building. London: James & James Science Publishers Ltd.

LBNL. 2000. *Daylight in Buildings: A Source Book on Daylighting Systems and Components.* Berkeley, California: Lawrence Berkeley National Laboratory. http://gaia.lbl.gov/iea21/.

Givoni, B. 1994. *Passive and Low Energy Cooling of Buildings.* New York: Van Nostrand Reinhold.

Stack, A., J. Goulding, and J.O. Lewis. Shading systems: Solar shading for the European climates. DG TREN, Brussels, Belgium. http://www.romazoconsument.nl/uploads/Shading_systems_Solar_shading_for_the_European_climates_Es-so_.pdf

CHAPTER ELEVEN

Lighting Systems

OVERVIEW

This chapter is intended to familiarize the reader with the process of sustainable lighting design. Lighting has a significant impact on building loads and energy usage. In the United States, energy powering lighting accounts for roughly 20% of the total energy used in buildings. Percentages differ in other countries; for example, in Canada (where heating takes up a greater percentage of total energy use), only 4% of residential energy and about 12% of commercial energy is used for lighting. Lighting also adds heat to the conditioned space, which increases cooling loads and decreases heating loads. Therefore, it is important that the HVAC designer understand the chosen lighting systems and their effects on HVAC loads. Architectural configurations and interior design decisions dictate the potential to offset electrical lighting needs with daylighting. Consequently, the productive collaboration of all design professionals including the lighting designer; architect; interior designer; as well as the HVAC, energy, and electrical engineers can enhance the overall energy and the environmental performance of buildings. This chapter provides insight into the latest technologies or practices at the time of publication, but given the rapidly changing landscape of lighting, this chapter should just be used as a guideline. Because California is noted to be an example of stringent energy codes (Title 24), the reader is advised to look at the California Energy Design Resources pages on lighting, with the website reference listed at the end of this chapter.

ELECTRIC LIGHTING

Sustainable Lighting Design

Sustainable lighting design is a comprehensive concept that looks beyond just saving lighting energy to looking at total life cycle of the lightings systems and how to minimize their impact on the environment. Sustainable lighting solutions address the following:

- Decreased overall environmental impact
- Significant reduction in operating costs
- Safety and security in all parts of an operation
- Good indoor environmental quality (IEQ)
- Reduced carbon footprint
- Responsible disposal at end of life

Other environmental factors to consider when designing or selecting lighting systems and fixtures are as follows:

- Longer lamp life
- Reduced use of hazardous materials (for example, mercury content)
- Higher efficacy
- Lighting controls
- Proper disposal (recycling)
- Use of recycled materials
    - Preconsumer
    - Postconsumer
    - Packaging materials
- Transportation
    - Cube efficiency
    - Location of manufacturing and distribution centers

Energy-Effective Lighting Design Standards

The lighting design profession is increasingly focused on providing high-quality visual environments, using energy efficient strategies. Lighting technologies have become more efficient, and expectations for interior illuminance (measured in foot-candles [I-P] or lux [SI] at the task level) have moderated while still meeting the illuminance levels required for the tasks of activities in a given space. Therefore, it is now possible to design high-quality lighting at connected power levels that are much lower than 30 years ago. It is much easier to regulate lighting based on installed lighting power rather than actual illumination levels. A number of governmental entities have incorporated regulatory requirements regarding lighting, often focusing on restrictions limiting the lighting power density (LPD). The LPD is defined as the watts of power allocated for lighting per unit floor area (ft^2 [m^2]) in a building. Some have limited or banned outright the manufacture of lamps with low efficacies (lumen/watt ratio). For example, incandescent lamps that convert only one-tenth of their consumed energy to supply light have been banned in Europe and a phaseout has begun implementation in the United States.

Most energy codes currently have building LPD limits between 0.8 and 1.4 W/ft^2 (about 8.5 and 15 W/m^2), depending on the occupancy type of the space. All buildings can be efficiently and comfortably illuminated using carefully selected, standard lighting equipment. Successful lighting systems with LPDs of 0.7 to 1.0 W/ft^2 (8 to 11 W/m^2) can be applied to most building types by following basic application design criteria set forth by the Illuminating Engineering Society (IES) and lighting equipment manufacturers. For offices and schools in particular, lighting quality is an important consideration for good overall IEQ. Lighting quality is generally defined as being relatively uniform, visually comfortable, with low glare, and designed to appropriate illuminance levels for the tasks performed. This includes balancing both the horizontal and vertical brightness.

ANSI/ASHRAE/IES Standard 90.1-2016 lists LPDs for various types of buildings, but excludes some types of specialty lighting from a required maximum LPD (e.g., those used for museum displays, medical procedures, retail display, exit signs, performance theater). Users may use one of two methods to comply: (1) the space-by-space method, which assigns LPDs to individual space types based on the specific tasks which will occur within, or (2) the building area method, which averages all of a building's space types to a single building LPD. When using the space-by-space method, some spaces within a building are allowed to have much higher or much lower LPD levels than specified for the building through the building area method. The way the lighting systems are controlled also may be a factor in determining the final required LPD. ANSI/ASHRAE/USGBC/IES Standard 189.1-2017, *Standard for the Design of High-Performance Green Buildings Except Low-Rise Residential Buildings*, specifies reductions in LPDs from those specified in ANSI/ASHRAE/IES Standard 90.1-2016 for more aggressive energy savings targets and will be discussed later in this chapter.

The lighting designer will select the appropriate luminaires, sources, and control configurations to create appropriate illuminance levels (measured in footcandles or lux). These selections will be based primarily on the room's intended activity while conforming to the maximum LPD required by code. The fundamental guide used by lighting professionals is the tenth edition of the *IES Lighting Handbook* (IES 2011). This handbook comprehensively describes recommended design approaches, illuminance and uniformity levels, power density targets, daylighting integration strategies, luminaire efficiency, and related information necessary for appropriate lighting design.

Efficient Lighting Systems

The lighting industry has made great strides in energy efficiency in the last several decades and continues to undergo a major shift with the advent of high-efficiency lamp/ballast technologies and solid-state light-emitting diode (LED) technology. In general, a lighting designer should perform a detailed life-cycle cost analysis to ensure the investment makes sense.

Spacing measurements for lighting design are taken from the plan-view center of the luminaire. Luminaires should be mounted at least one-third of the indicated mounting distance away from any ceiling-high partition.

To allow integration with daylighting, luminaires in rows should ideally be parallel to windowed walls so that controls can be used to step lamps on or off based on the amount of daylight penetration into the room.

The lamp/ballast combination and luminaire type need to be carefully considered to provide visual acuity for the intended task, visual comfort, glare control, efficacy (lumen/watt output), and total system efficiency for any particular space type. The proper specification of lighting equipment is necessary to optimize lighting quality and minimize energy use.

The following luminaire types have cost effective energy alternatives and, therefore, should be avoided:

- Luminaires using Edison base (standard screw-in) sockets or halogen lamps using any sockets rated over 150 W
- Luminaires designed for incandescent or halogen low-voltage lamps exceeding 75 W
- Track lighting systems using any kind of incandescent or halogen
- Line-voltage monopoints allowing the later installation of incandescent or halogen track luminaires.

This chapter is not intended to be comprehensive summary of the lighting industry but rather to provide a straightforward and instructive approach to basic sustainable lighting design. Professional lighting design assistance may be needed to reach optimum performance in complex spaces and for special conditions.

In all cases, the reader should be certain to review the subsequent sections on other aspects of lighting for detailed information on product specifications and additional energy-saving alternatives.

Efficient Lamps and Ballasts

The lighting industry has made significant improvements in lamp and ballast technology over the last few decades. Recent improvements continue to permit quality lighting even at greatly reduced power levels. Proper specification of lighting equipment is an important part of actualizing these results.

Specifications. The following specifications are recommended to ensure the latest technology is being used:

- Ballasts for all fluorescent lamps and for high-intensity discharge (HID) lamps rated 150 W and less should be electronic. Harmonic distortion should be less than 20%.
- Fluorescent lighting systems should use high-efficiency electronic ballasts. Because instant-start ballasts are the most efficient and least costly, they should

be used in all longer duty cycle applications where the lights are turned on and off infrequently. Fluorescent systems controlled by motion sensors in spaces where the lights will be turned on and off frequently should have programmed-start ballasts.

- Designers are strongly encouraged to use ballasts with low-ballast-factors (BF < 0.80) wherever possible.
- Electronic dimming ballasts may be used as required by the function of the space, and are helpful also for achieving good daylighting control.
- Metal halide ballasts 150 W and less should be electronic. Metal halide systems greater than 150 W should use linear-reactor, pulse-start ballasts wherever 277 V power is available. For other voltages, pulse-start lamps and ballasts should be used.

The lamps specified for primary lighting systems should be both energy efficient and have long lamp life, representing excellent cost benefits. Lamps for other applications can be less efficient, have shorter lives, or both. Also, designers should strive to minimize the number of different types of lamps on a project to reduce maintenance costs and improve the efficiency of inventory management.

As previously mentioned, LED lamps have entered the market to replace many of the lamps considered to be the standard alternatives in the industry, and the technology is changing rapidly. Therefore, the reader is encouraged to seek out LED alternatives that produce high-quality light with both high efficacy and long life.

Lighting Power Density Determination Using ASHRAE Standards

ANSI/ASHRAE/USGBC/IES Standard 189.1-2017 specifies reductions in the LPD values given in ANSI/ASHRAE/IES Standard 90.1-2016 and these lower LPD values are encouraged to become the basis of design for high-performance buildings. Either of two methods are acceptable to use in determining the allowable LPD: building area method or the space-by-space method can be used to comply with this standard. Examples are given below for both methods for an office building.

The allowable LPD values specified in ANSI/ASHRAE/IES Standard 90.1 have been lowered in the past several versions of the standard to reflect lighting technology (and cost) changes. For example, under the building area method, the allowable Standard 90.1 LPD for an office building was listed as 0.90 W/ft^2 (9.65 W/m^2) in the 2010 version of the Standard, dropping to 0.82 W/ft^2 (8.8 W/m^2) in 2013 and then to 0.79 W/ft^2 (8.5 W/m^2) in 2016. The building area methods sets the maximum total wattage for installed lighting based on the total floor area. The LPD is then multiplied by the gross (interior) area of the building to arrive at the maximum allowable installed lighting wattage for the building. The Standard 90.1 values are code minimum; to conform to the more aggressive Standard 189.1-2017, this total must be further multiplied by the factors given in Table 7.4.6.1A of Standard 189.1.

For an office, that multiplying factor is 0.95, meaning that a high-performance green building would have a 5% lower overall LPD factor than a code minimum building.

Application of the space-by-space method for determinng LPD values in an office building are done for each different space type. Typical spaces found in office buildings were assumed for this example. A hardwired lighting system for a high-performance green building would be installed with the lighting power density (LPD) requirements shown in the last column of Table 11-1. The area of each space is multiplied by the LPD for that space to determine the maximum connected (hard-wired) lighting wattage. For example, for an open office area of 400 ft^2 (37.2 m^2), the maximum wattage for lighting would be 400 $ft^2 \times 0.69$ W/ft^2 = 276 W (37.2 $m^2 \times$ 7.42 W/m^2 = 276 W).

Additional lighting, such as that hardwired to a furniture system, may be acceptable, but should not be specified in a space unless a more complete analysis and design are undertaken. These systems need to be carefully coordinated with the permanent lighting systems of the building. An exception is portable plug-in lamps and manufacturer-installed luminaires on modular furniture, overhead cabinets, bins, or shelves.

The total amount of installed lighting is only part of the picture. It is important to note that there are additional requirements regarding lighting control that are not discussed in this chapter. A thorough review of appropriate LPD requirements for all types of buildings can be done by considering both latest versions of ANSI/ASHRAE/USGBC/IES Standard 189.1 and ANSI/ASHRAE/IES Standard 90.1.

Application Notes

Always check required illuminance levels using IES guidelines and LPD restrictions using Standards 90.1 and 189.1 or other applicable standards or codes to ensure both code compliance as well as following green design practices.

Open Offices. For general illumination in spaces with ceilings that are 9 ft (2.7 m) or higher, consider suspended lighting systems supplemented by task lights. Use high-performance luminaires and (when using fluorescent lighting) high-performance ballasts.

Private Offices. Where ceiling heights allow, suspended linear fluorescent or LED lighting should be a first consideration for private offices, although recessed troffers can also be used. Task lights can also be integrated where appropriate.

Executive Offices, Board, and Conference Rooms. Executive offices can be designed similarly to private offices. If desired, a premium approach using LED downlights, wallwashers, and/or accent lights can be used, but care should be taken not to exceed the required LPD.

Table 11-1: LPD Values for Office Building Example (Space-by-Space Method)

Type of Area	Connected LPD limit per Std. 90.1-2016		Percent Reduction of LPD per Std 189.1-2017	Connected LPD limit	
	(W/ft²)	(W/m²)		(W/ft²)	(W/m²)
Private Office Areas	0.93	10.00	95%	0.88	9.47
Open Office Areas	0.81	8.71	85%	0.69	7.42
Restrooms	0.85	9.15	100%	0.85	0.15
Conference/ Meeting/ Multipurpose Areas	1.07	11.48	90%	0.96	10.32
Lobby	1.00	10.76	95%	0.95	10.22
Elevator Lobby	0.69	7.42	85%	0.59	6.31
Hallways and Corridors	0.66	7.08	85%	0.56	6.01
Lounge/ Recreation Area	0.62	6.67	85%	0.53	5.67
Mechanical and Electrical Rooms	0.95	10.18	100%	0.95	10.18
Stairway	0.58	6.24	100%	0.58	6.24

Classrooms. Ideally, classroom lighting should be designed using direct/indirect lighting systems using high-efficiency T8 lamps and ballast combinations or LED-based fixtures. Care needs to be taken to ensure that all desks are evenly illuminated.

Corridors. Typically, corridors strategies include LED sconces, downlights, ceiling-mounted or close-to-ceiling decorative diffuse fixtures, or similar equipment. LED lamps will provide opportunities for changing lighting color as well as having the highest efficacy. Note that these luminaires may be connected to a generator and/or equipped with emergency battery backup when needed as an alternative to less attractive "bug eye" type emergency lighting.

Electrical/Mechanical Rooms. As seen in Table 11-1, Standard 90.1 allows a LPD totaling 0.95 W/ft^2 (10.18 W/m^2), but of this, 0.52 W/ft^2 (5.6 W/m^2) must be controlled separately from the base allowance of 0.43 W/ft^2 (4.58 W/m^2).

High-Bay Spaces. Industrial, grocery, and retail space without ceilings or with very high ceilings (usually 15 ft [4.6 m] and above) need special lighting fixtures. For mounting heights up to 20 to 25 ft (6.1 to 7.6 m), first consider LED or fluorescent industrial luminaires with T8 or T5 lamps.

For mounting heights above 25 ft (7.6 m), consider T-5HO high-bay luminaires. High-wattage metal halide should be reserved for very high mounting (above 50 ft [15.2 m]) and for special applications (e.g., sports lighting).

Other Applications. The following luminaire types are generally recommended for these areas:

- *Artwork, bulletin/display surfaces, etc.:* use LED wall washers or low-voltage monopoint LED lights. Make sure to pay attention to the color rendition index (CRI) of the specific lamps when illuminating art work. A higher CRI will render colors better.
- *Utility spaces (e.g., cable and equipment rooms):* use LED strip lights or surface luminaires.
- *Lobby spaces, cafeterias, and other public spaces:* use appropriate selections from among these luminaires as much as possible.

DAYLIGHT HARVESTING

General Description

Most buildings are designed to have some type of natural light that is transmitted through windows and/or skylights. The majority of commercial, industrial, and institutional buildings have windows and, in some cases, skylights, tubular daylighting devices, clerestories, or more extensive fenestration systems.

From an energy perspective, the optimal use of daylight is to reduce the load of the electric lighting system by dimming or switching off luminaires when natural light provides ample illuminance for the tasks performed in the space. It is important to note that the incoming light is only usable if it is controlled to

Image courtesy of Cannon Design.

Figure 11-1 Example of daylight harvesting rendering.

reduce glare and illuminates a space in a way that is comfortable to the occupant (including the consideration of thermal comfort). This process of reducing electric lighting in the presence of daylight, which is known as *daylight harvesting* (see Figure 11-1), is discussed in this section because of its significant energy saving potential. The prediction of daylight harvesting savings is a complicated process that involves a comprehensive understanding of the site, building orientation, weather conditions, materials, and system integration. There are added capital costs for daylight harvesting elements such as dimming ballasts, dimming controls, and photoelectric controls. It is important to justify these costs by accurately predicting the potential energy savings of daylight harvesting techniques. However, when the challenges of daylighting are appropriately addressed, significant energy savings are possible.

From a lighting design perspective, daylight can be treated similarly to any other light source, so it can be used to compose lighting design solutions that take illuminance, luminance, contrast, color rendition, and other lighting design elements into consideration. However, the lighting designer is challenged to deal with the fact that the light source varies daily and throughout the year. The designer can use blinds, shades, curtains, moveable shutters, light shelves, light conveyors, and other mechanical forms of attenuation and shielding to control daylight.

It is possible to simulate the performance of daylight to determine the amount, and to a certain extent, the quality of available daylight under varying conditions of season, time of day, and weather. However, this is exhaustive analytical work of a highly specialized nature, and it is recommended that appropriate experts perform such studies. In the meantime, some buildings can benefit tremendously from some simple daylight-harvesting considerations.

Basic Toplighting

Basic toplighting typically involves using simple skylights or tubular daylighting devices in the roof. This is not to say that other toplighting configurations (e.g., the clerestory, roof monitor, or sawtooth roof) are not viable options; indeed, they often have advantages over horizontal skylights. There are many architectural elements, including structure, waterproofing, and other details, that should be considered when exploring the appropriateness of skylights. When used in a manner similar to light fixtures and laid out to provide uniform illumination, toplights are an acceptable way to illuminate single-story, large spaces or the top floor of a multi-story building. Toplighting is best when a number of smaller skylights are used, in much the same way effective lighting systems have many light fixtures rather than one larger light source in the middle of the room. Skylights should be diffuse or prismatic rather than clear to avoid harsh glare from direct sunlight. When layouts adhere to these guidelines, skylights generally do not have to incorporate mechanical or electronic light control louvers, because the optimum size of the skylight is chosen for passive daylighting (i.e., typically, there are no active or moving elements needed to regulate the amount of interior light). If installed in an area requiring the ability to "black out" or darken the space (e.g., for showing movies or videoconferencing), some type of shading should be provided along with controls.

ANSI/ASHRAE/USGBC/IES Standard 189.1-2017 requires toplighting for buildings of three stories or less having a large enclosed space of 20,000 ft^2 (2000 m^2) located "directly under a roof with finished heights greater than 15 ft. (4 m) and that have a lighting power allowance for general lighting equal to or greater than 0.5 W/ft^2 (5.5 W/m^2)." Section 8.3.3.3 of that Standard details the minimum amount of toplighting required.

To determine the optimum size of skylights, one can download a program called SkyCalc from the Energy Design Resources website (reference URL at the end of this chapter). This program, which is optimized for California and the northwestern region of the United States, can be applied (with appropriate care) anywhere in North America. It takes into account location, utility rates, and other basic data, and yields recommended skylight area.

Note that the ideal amount of fenestrated roof is generally about 5%, but designers should always review governing standards and codes. All too often, architects, although well-intentioned, design skylights that are too big. Ideally, the optimized solution between light availability and overall energy use can be achieved.

When/Where Applicable

Daylight harvesting is best suited for skin-load-dominated (SLD) structures, especially single-story or narrow, multistory buildings. It is also an effective strategy for the top story of an internal-load-dominated (ILD) building. In structures where merchandise is sold, it has been found that sales increase for products displayed under skylights. Therefore, large-volume, single-story structures often used for big-box retail stores can still benefit from daylighting, because vertical glazing would be ineffective in bringing light to the deep interior.

Buildings serving several functions can benefit from daylighting, including the following types of buildings:

- Gyms
- Industrial work spaces
- Big box retail stores
- Grocery stores
- Exhibition halls
- Storage
- Warehouses
- Office buildings (if skin load dominated)
- Classrooms

It is essential to use automatic lighting controls programmed to dim or extinguish electric luminaires to efficiently integrate daylighting Without them, the energy savings will not be realized. (See the later section on Lighting Controls.)

Pros of Daylighting

- Daylight harvesting offers significantly reduced energy consumption (exceeding 60%) and reduced HVAC load (as long as solar gains do not outweigh electric lighting reductions).
- Daylight extends the electric lighting maintenance cycle (lamps can last two to three times as long in calendar years with proper selection of fluorescent lamp/ballast combinations).
- Daylight has been shown to lead to improved human factors and increased enjoyment of space.
- Merchandise daylit under skylights sells at a faster rate.
- Daylight assists in maintaining proper circadian rhythm, which is especially important to sufferers of seasonal affective disorder (SAD)

Cons of Daylighting

- Sunlight is intermittent and variable.

- Daylight harvesting requires intensive architectural, structural, and lighting design coordination.
- There is a potential for glare and, therefore, glare control devices are often required.
- The user must be educated on the proper use and maintenance of the components.
- There is no assurance that the design will meet exact project lighting requirements.
- There can be higher construction costs and higher maintenance.
- There is a risk of poor design or installation workmanship, which can result in roof leaks.
- Daylighting may not be suitable for uncommon room shapes, sizes, and/or finishes.
- There is net decreased roof insulation when skylights are installed.
- Even when using high-performance glazing, there will be significantly larger heat loss than that lost through a solid insulated wall in winter. Additionally, occupants in close proximity to glazing during colder periods may experience discomfort.
- Heat gain may occur at undesirable times.

LIGHT CONVEYORS (TUBULAR DAYLIGHTING DEVICES)

Light conveyors use a specialized technique whereby light from a source is transmitted some distance from the source to illuminate spaces, either along its length or some distance away. Some devices gather light and transmit it down a fiber-optic tube bundle to light an area at a distance; with sensors and lamps in the room it serves to seamlessly ensure constant illuminance. The source can be either natural light or an artificial source. More details on light conveyors are provided in GreenTip #11-1.

LIGHTING CONTROLS

General

While all modern energy codes require automatic shutoff controls for commercial buildings, implementing automatic controls in all building projects is a sound money- and energy-saving idea. There are two ways to reduce lighting energy use through controls:

- Turn lights off when not needed (which reduces hours).
- Reduce lighting power to minimum need (which reduces kilowatts).

With very few exceptions, each interior space enclosed by ceiling-high partitions must have a separate local switching or dimming device and some form of automated "off" control (e.g., occupancy-sensing, time-based scheduling, or other) to meet code. In addition, wherever possible, providing separate switching/dimming for lights in daylit zones can further reduce energy usage or provide occupant control in situations where overrides to provide more light are required. To comply with code requirements and ensure maximum energy savings, specify the most appropriate lighting control option(s) as outlined in Table 11-2 and described in the following paragraphs.

Control Options

Below are some considerations for useful lighting control system components that correlate with recommendation in Table 11-2:

1. *Ceiling-mounted motion sensor with transformer/relay, auxiliary relay, and series switch.* The sensor should be located to look down on the work area, so as to detect anything from a small hand motion to major movements. The sensor may be mounted to the upper wall if a ceiling location is not workable. More than one sensor can be used for a large room or a room with obstructions (e.g., a library or server room). In such situations, sensor coverage zones should be slightly overlapped to ensure comprehensive coverage. The main transformer relay should control the overhead lighting system (usually 277 volt) and the auxiliary relay should control at least one-half of a receptacle to switch task lights and other applicably controlled plug loads. Note that the light switch is in series so that it can only turn lights off in an occupied room; it cannot override the motion sensor's "off" control.
2. *Ceiling-mounted motion sensors connected to programmable time controller.* During programmed ON times, the lights remain on. During programmed OFF times, motion within the space initiates lights on for a time out period. The controller should be programmable according to the day of the week and should have an electronic calendar to permit programming holidays.
3. *Workstation motion sensors should be connected to a plug strip or task light with auxiliary receptacle.*
4. *Include one or more ceiling-mounted motion sensors with transformer/relay (minimum of two luminaires controlled.)*
5. *Programmable time controller with manual override switch(es) located in a protected or concealed location.* There are separate zones for retail and similar applications where displays can be controlled separately from general lighting. This may also control dimmers.
6. *Automatic daylighting sensor connected to fluorescent lighting with dimming ballasts or LED lighting in each luminaire in the daylit zone (in addition to any of the above).*
7. *Motion sensor connected to a high-low lighting system.*

When using controls such as motion sensors or daylight sensors, be very thorough and carefully read the manufacturer's literature. Different sensors work for different applications, and their sensing systems are optimized. For instance, avoid

Table 11-2: Recommended and Optional Lighting Controls

Type of Space	Minimum Recommended Control	Optional Control(s)
Private Office, Exam Room	1	2 2 + 4 1 + 8 2 + 4 + 8 2 + 8
Open Office	3	3 + 4 3 + 8 3 + 4 + 8
Conference Rooms, Teleconference Rooms, Boardrooms, Classrooms	2	1 2 + 8 1 + 8
Server Rooms, Computer Rooms, Other Clean Work Areas	5	
Toilet Rooms, Copy Rooms, Mail Rooms, Coffee Rooms	5 or 6	
Individual Toilets, Janitor Closets, Electrical Rooms, and Other Small Spaces	6	
Public Corridors	3	
Corridors, Hallways, Lobbies (Private Spaces Only)	3	3 + 8
Public Lobbies	7	7 + 8
Industrial Work Areas	7	7 + 8
Warehousing and Storage	9 (high-intensity discharge [HID] systems) 3 or 5 (fluorescent or LED systems)	3 or 5 + 8 (fluorescent)

Table 11-2: Recommended and Optional Lighting Controls (Continued)

Type of Space	Minimum Recommended Control	Optional Control(s)
Stores, Newsstands, Food Service	7	
Mechanical Rooms	Manual switching only	
Stairs	None	Motion sensors can be used to reduce light levels to minimum egress lighting levels only
Hotels		Standard 189.1-2017 requires automatic controls to turn power to lights off within 30 minutes of guests leaving their guest rooms.

wallbox motion sensors except in spaces where their sensing field is appropriate (e.g., small private offices, individual toilets, etc.). For spaces with small-motion work, a lookdown sensor (situated on the ceiling) generally works much better than a lookout sensor (situated on a wallbox). Consider the sensing technology when specifying sensors. Infrared, ultrasonic, and microphonic sensor technologies present opportunities to fine-tune lighting control system design and can enhance the overall effectiveness of the lighting system.

Applicability

The controls mentioned in the previous section are applicable to most commercial, institutional, and industrial buildings. Use common sense in special spaces, keeping in mind safety and security. Never switch path-of-egress lighting systems except with properly designed emergency transfer controls.

Pro

- There is low to moderate cost for most space types.
- There is virtually no maintenance.
- These controls will generally lower energy use.

Con

- If controls are not properly commissioned, unacceptable results may occur until they are fixed.

- Poorly installed and/or commissioned controls negatively affect system performance.
- Substitutions and value engineering can cause bad results.

COST CONSIDERATIONS

Lighting Systems

The systems described above are generally low-to-moderate cost lighting systems. On average, they also use low-maintenance lamps and ballasts. The combination of low first cost, low maintenance, and low energy use leads to lighting choices that are among the most economical available.

Users should consider life-cycle costs when evaluating the use of LEDs. Although high in first cost, LED lamps have significantly higher efficacies and longer life, which can offset first cost in the course of the life cycle. Note, however, that LEDs installed in areas where they will overheat will experience shorter life.

Daylight Harvesting

Daylight harvesting is a potentially complex undertaking in which the first cost of lighting remains the same, the cost of lighting controls increases, and the added cost of skylights and/or structural changes/complications are incurred as well. To be cost-effective, this needs to be offset by a combination of HVAC energy savings, lighting energy savings, HVAC system first-cost reduction, and perhaps savings from utility incentives or tax credits. Expect daylight-harvesting systems to yield a 4 to 5 year simple payback with a utility incentive, 6 to 8 year or more without.

Controls

The lighting control systems described previously are generally of low to moderate cost. However, using better quality sensors and separate transformer/relay packs with remote sensors costs much more than wallbox devices. Savings can range from modest to considerable, depending on the building and occupants.

REFERENCES AND RESOURCES

Published

ASHRAE. 2016. ANSI/ASHRAE/IES Standard 90.1-2016, *Energy Standard for Buildings Except Low-Rise Residential Buildings*. 2016. Atlanta: ASHRAE.

ASHRAE. 2017. ANSI/ASHRAE/USGBC/IES Standard 189.1-2017, *Standard for the Design of High Performance Green Buildings*. Atlanta: ASHRAE.

Benya, J. and Peter Schwartz. 2010. *Advanced Lighting Guidelines*. Fair Oaks, CA: New Buildings Institute. http://newbuildings.org/advanced-lighting-guidelines-0.

LBNL. 2000. Daylight in Buildings: A Source Book on Daylighting Systems and Components. Berkeley, CA: Lawrence Berkeley National Laboratory. http://gaia.lbl.gov/iea21/IES. 2011.

Lighting Handbook. New York: Illuminating Engineering Society of North America.

Online

Advanced Lighting Guidelines Online
www.advancedbuildings.net.

Architectural Energy Corporation
www.archenergy.com.

BetterBricks
www.betterbricks.com.

California Energy Commission
www.energy.ca.gov.

Collaborative for High Performance Schools
www.chps.net.

Energy Design Resources—Lighting Design
https://energydesignresources.com/technology/lighting-design.aspx/.

Energy Design Resources—SkyCalc
https://energydesignresources.com/resources/software-tools/skycalc.aspx/.

Energy Trust of Oregon
www.energytrust.org.

EIA, U.S. Energy Information Administration
www.eia.gov.

Lighting Research Center at Rensselaer Polytechnic Institute
www.lrc.rpi.edu/programs/NLPIP/publications.asp.

Natural Resources Canada
http://oee.nrcan.gc.ca/node/17725#bb.

New Buildings Institute
www.newbuildings.org.

RealWinWin, Inc.
www.realwinwin.com.

Rising Sun Enterprises
www.rselight.com.

Savings by Design
www.savingsbydesign.com.

U.S. Department of Energy
www.energy.gov.

ASHRAE GreenTip #11-1

Light Conveyors (Tubular Daylighting Devices)

GENERAL DESCRIPTION

A light conveyor is pipe or duct with reflective sides that transmits artificial or natural light along its length. There are three types of such light-directing devices: plastic square or round duct with curves or bends that transmits light along its length; a relatively straight tube having an opaque, highly reflective interior coating that transmits daylight or high-intensity electric light to an interior space; and a bundle of multiple fiber-optic tubes that transmit light from the outside (horizontally or vertically) to the interior.

The first type is a square duct or round pipe made of plastic. Based on how the inside of the duct or pipe is cut and treated, light entering one end of the pipe is reflected off these configurations (similar to the way light is refracted through a prism) and transmitted through. The reflected light continues to travel down the pipe and out the other end. Because some light is absorbed and escapes along the length of the pipe (i.e., it is lost), the maximum distance that light can be piped into a building is generally about 90 ft (27.4 m). There are a few installations where sun-tracking mirrors concentrate and direct natural light into a light pipe. In most applications, however, a high-intensity electric light is used as the light source. Having the electric light separate from the space where the light is delivered isolates the heat, noise, and electromagnetic field of the light source from building occupants. In addition, the placement of the light source in a maintenance room separate from building occupants simplifies replacement of the light source.

The second type of light-directing device is a straight tube with a highly reflective interior coating. The device is mounted on a building roof and has a clear plastic dome at the top end of the tube and a translucent plastic diffusing dome at the bottom end. The tube is typically 12 to 16 in. (300 to 400 mm) in diameter. Natural light enters the top dome, is reflected down the tube, and is then diffused throughout the building interior. The light output is limited by the amount of daylight falling on the exterior dome.

A second type of tubular daylighting device uses the technology of fiber optics to transmit light into interior zones of a building. The most complex designs incorporate sensors, controls, and LED lamps to maintain relatively constant light output.

WHEN/WHERE IT'S APPLICABLE

Light-conveyor systems are best suited to building applications where there is a need to isolate electric lights from the interior space (e.g., operating rooms or theaters) or where electric light replacement is difficult (for example, above swimming pools or in roadway tunnels). For the latter, reflective-tube system, each device can light only a small area (10 ft^2 [1 m^2]) and is best suited to small interior spaces with access to the roof, such as interior bathrooms and hallways.

PRO

- A light conveyor transports natural light into building interiors.
- The first type of light conveyor isolates the electric light source from the lighted space.
- The first type of light conveyor reduces lighting glare.
- It lowers lighting maintenance costs.

CON

- A light conveyor may have greater capital (first) cost than traditional electric lighting.
- The tube type may increase roof heat loss.
- The tube type runs the risk of poor installation, resulting in leaks.
- The effectiveness may not be worth the additional cost.

KEY ELEMENTS OF COST

Because of the specialized nature of these techniques, it is difficult to address specific cost elements. As an alternative to conventional electric lighting techniques, it could add to or reduce the overall cost of a lighting system—and the energy costs required—depending on specific project conditions. A designer should not incorporate any such system without thoroughly investigating its benefits and applicability, and should preferably observe such a system in actual use.

SOURCES OF FURTHER INFORMATION

Laouadi, A. and H.H. Saber. 2014. Performance of Tubular Daylighting Devices. National Research Council Canada, Publication No. 82. http://www.nrc-cnrc.gc.ca/ctu-sc/ctu_sc_n82.

Mayhoub, M.S., and D.J. Carter. 2011. Hybrid lighting systems: Performance and Design. Lighting Research and Technology. 44(3):261-276.

CHAPTER TWELVE

WATER EFFICIENCY

Water efficiency and conservation continues to become a critical factor in green-building design. In 2015, water use in residential and commercial buildings was estimated at 45.6 billion gal (173 million m^3) per day, nearly 13% of total fresh and saline water withdrawals in the United States. Between 1985 and 2005, total water use increased less than 3%, while water use in the buildings sector increased 27% (DOE 2012). Efficient operational practices provide opportunities to save significant amounts of water. The reduction in energy use through efficiency measures, the desire for lower operating costs, and the expectation of increased government regulation and standards are expected to drive adoption of water-efficient products and methods.

The areas of focus for this GreenGuide chapter are the HVAC systems, water heating, plumbing design and potential water recovery and reuse systems. In a typical commercial building with a water-cooled chiller, the HVAC system can account for approximately one third of water consumption. Therefore, minimizing the water needed to operate HVAC systems while taking care not to significantly increase energy usage should be a major consideration in green-building design. Figure 12-1 shows the rough level breakdown of water use within commercial buildings.

Plumbing systems are normally not considered within the purview of the HVAC&R designer's expertise. Nevertheless, designers, in practice, must work closely in putting together a functional building mechanical system. Indeed, frequently the designing firm for the HVAC and plumbing systems are one and the same. For detailed design guidelines and information on these systems, please refer to the American Society of Plumbing Engineers (ASPE).

Recently, ASHRAE partnered with the American Water Works Association (AWWA), U.S. Green Building Council (USGBC), and ASPE to develop a new proposed standard titled Standard 191, *Standard for the Efficient Use of Water in Building, Site and Mechanical Systems.* The target date for the final standard to be published is sometime in 2017.

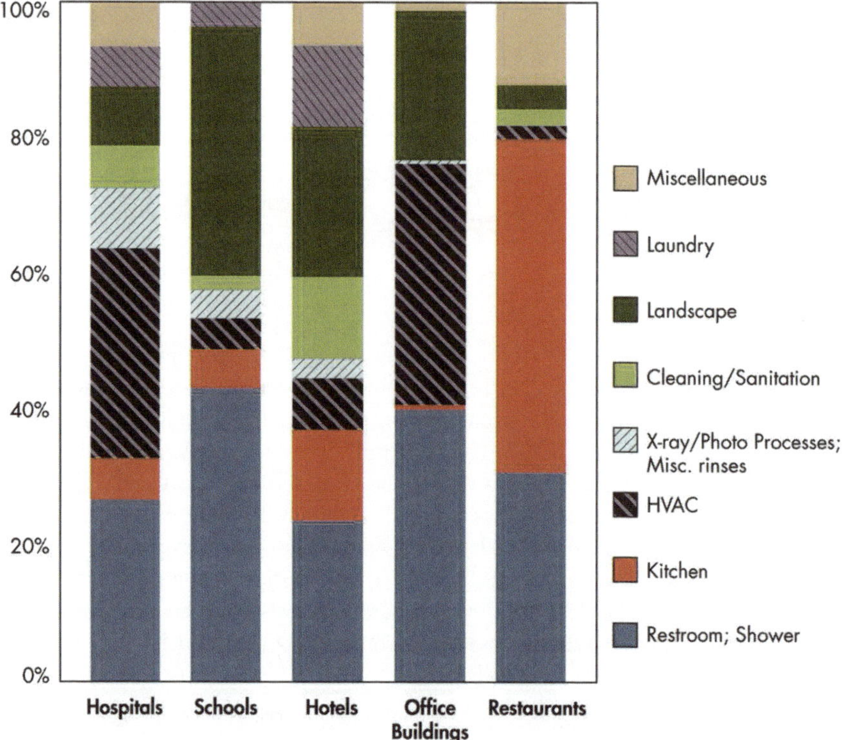

Image courtesy of EPA Water Sense

Figure 12-1 Water uses in commercial buildings.

In green-building design, it is important for the practitioners of each design discipline to be familiar with what the other disciplines may bring to an effective green design. This holds true as well with plumbing design. The editors of this guide have chosen to include a discussion of some key aspects of plumbing design that impact green design, including several ASHRAE GreenTips. Several of these GreenTips may have an impact in other areas as well. For instance, point-of-use hot-water heaters (GreenTip #12-1) not only save heating energy and distribution energy, but they also often result in the use of less water.

THE ENERGY-WATER BALANCE

According to the World Health Organization, approximately 663 million people still do not have access to improved water supply sources. The amount of water in its various forms on the planet is finite. The amount of fresh water, of a quality suitable for the purposes for which it may be used, is not uniformly distributed (e.g.,

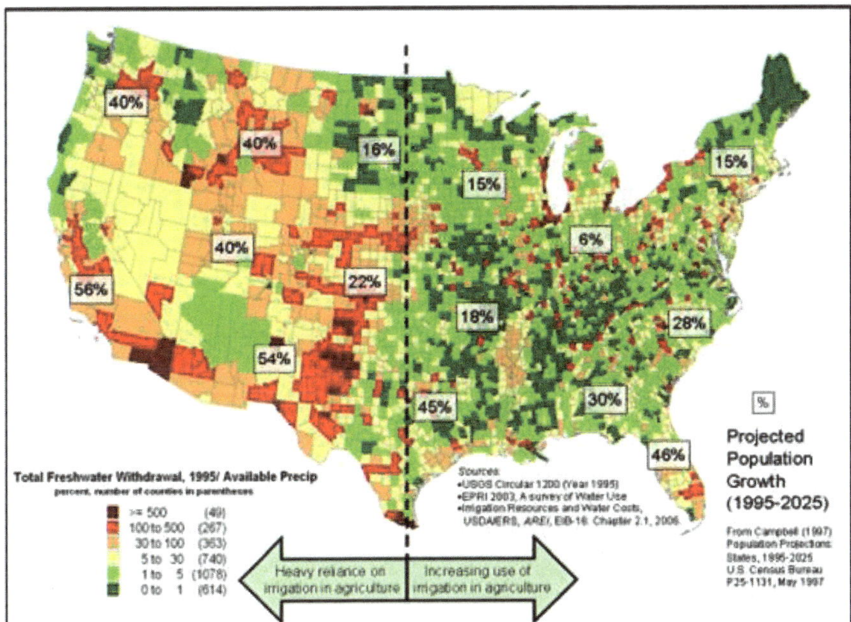

Image courtesy of Universities Council on Water Resources Journal of Contemporary Water Research and Ed/Sandia National Laboratories.

Figure 12-2 United States water availability map.

20% of the world's fresh water is in the U.S. Great Lakes, while elsewhere it is often nonexistent or in meager supply). In addition, the distribution of water relative to the population growth is not always the most even, as seen with respect to the United States in Figure 12-2. Nevertheless, water must be allocated somehow to the world's populated lands, many of which are undergoing rapid development. In short, it is becoming more and more difficult to provide for the adequate and equitable distribution of the world's water supply to those who need it.

The continued security and economic health of the global population depends on a sustainable supply of both water and energy. These two critical resources are inextricably linked. A nation's ability to continue providing both clean and affordable energy and water is being seriously challenged by myriad issues, including those stated above.

Energy production requires a reliable, abundant, and predictable source of water. Electricity production from fossil fuels and nuclear energy is the largest user of water in the United States at 45% of total water withdrawals, requiring 161 billion gallons (609 million m^3) of water per day in 2010 (Maupin et al. 2010). Although some water consumed by thermoelectric power plants using once-through cooling systems is returned to waterways, it is done so at a higher temperature leading to a greater evaporation rate. On the other side, the water use cycle involves considerable energy expenditures; energy is needed to extract, treat, and transport potable water great distances, and to collect and treat wastewater. For reference, estimates of water-related energy use range from 4% to 13% of U.S. electricity use (Copeland 2014).

It is estimated that the operation of residential and commercial buildings in the United States requires 13% of the nation's water use (Maupin et al. 2010) and 40% of its energy consumption (EIA 2009). It is clear that water is a precious resource, essential at every building site. In areas where water supply is not plentiful, an engineer focusing on green design should take into consideration the total energy as well as water consumption required to operate the building or facility. For example, when considering a cooling plant, a life-cycle analysis should be performed to compare an air-cooled chilled-water plant to a water-cooled chilled-water plant. The total embodied energy of the air-cooled plant may end up being approximately the same as that of a water-cooled plant if the water-cooled plant was consuming desalinated water that was delivered to the site. Therefore, while many of the measures to protect and preserve the world's freshwater supplies are beyond the design engineer's purview, there are a number of simple things related to building sites that can be done as part of a green design effort.

COOLING TOWER SYSTEMS

Because cooling towers are commonly used in larger-scale building systems and they consume water through evaporation, drift, and blowdown losses, the following section provides details on cooling tower operations, particularly as they affect green-building design and water efficiency.

Cooling towers are efficient devices for removing heat from water through evaporation and can do so close to the ambient wet-bulb temperature. The wet-bulb temperature is always lower than the dry-bulb; thus, water cooling allows more efficient condenser operation (as much as a 50% energy savings) than air cooling. However, cooling towers use a significant amount of water. A typical tower operation will evaporate 3 gpm of water per 100 tons of cooling capacity (3.2 L/min of water per 100 kW of cooling capacity). In addition to evaporation, some recirculated water must be bled from the system to prevent soluble and semisoluble minerals from reaching too high a concentration. This bleed or blowdown is usually drained to the sanitary plumbing system and ends up being treated further downstream at a sewage waste treatment plant. It is possible to maximize both the energy and the water efficiency of the cooling tower system by appropriately addressing splash-out and spill issues, limiting microbiological activity, treating the water with chemical-free systems, and using filtration systems with low water usage for the blowdown requirement.

It is important, from a sustainability perspective, that a proper tower design is selected. The tower that is selected must be energy efficient as defined below and designed to minimize splash losses, spills, drift, and algae growth. For more information, consult Chapter 40 in the 2016 *ASHRAE Handbook—HVAC Systems and Equipment* (ASHRAE 2016).

Drift

To promote efficient heat transfer through evaporation, cooling towers force intimate contact between outdoor air and warm water. During this heat exchange process dust, pollen, and gas in the air become entrained in the water, while some lower-vapor pressure components in the water (e.g., bromine or chlorine), as well as entrained water drops, migrate to the air. Airborne dust and pollen that are captured by the water can promote biological growth in the tower.

Small water droplets entrained and carried out with the air passing through the tower are called *drift*. Drift is always present when operating a cooling tower. Because drift is generated by small droplets of the cooling tower water, it contains all of the dissolved minerals, microbes, and water treatment chemicals in the tower water. Drift is a source of PM_{10} emissions (i.e., particulate matter that is less than 10 μm in diameter) and, since it can contain biological growth suspended in the water, it is a suspected vector in *Legionella* transmission.

Drift is usually reported as a percentage of the recirculating water flow rate, though it is more accurately described in terms of the parts per million (ppm) of the air passing through the tower. Tower designs use drift eliminators to capture virtually all of this entrained water. A typical value of drift for high-efficiency eliminator performance in modern cooling towers is 0.001 to 0.005% of the recirculating water rate.

It should further be noted that since drift is liquid water being aspirated from the cooling tower, is it simply one component of tower blowdown. Because of this, the drift rate has no impact on total water consumption since the amount of drift will result in a commensurate reduction in blowdown.

Water Consumption of Typical Cooling Tower

In times of increasing pressures on water supplies, establishing methods to reduce water consumption in cooling towers is significant for building systems. Consider as an example a 400 ton (1407 kW of refrigeration) rated cooling system that circulates approximately 1200 gpm (4542 L/min) of water through the tower and chiller at full cooling load. At nominal rates, 12 gpm (45.4 L/min), or about 1% of the total system flow, is lost to evaporation, while 0.06 gpm (0.23 L/min) is lost to drift (0.005%). At four cycles of concentration, 4 gpm (15.1 L/min) is intentionally bled from the system.

Table 12-1 shows monthly evaporation, blowdown, and drift estimates for operating this 400 ton rated tower at an assumed 75 hr/week or 300 hr/month. In this

Table 12-1: Water Consumption from Example Cooling Tower Operation

	Flow at Rated Capacity	Total/Month (Assuming 60% Average Capacity)
Evaporation	12 gpm (45.4 L/min)	129,600 gal (490,536 L)
Drift	0.012 gpm (0.05 L/min) (0.001%) 0.060 gpm (0.23 L/min) (0.005%)	130 gal (491 L) 648 gal (2453 L)
Bleed at Four Cycles	4 gpm (15.1 L/min)	43,200 gal (163,512 L)

table, it is assumed that the tower operates at an average of 60% of its rated capacity, or 240 tons cooling for 300 hours per month. These values also assume that the water flow rate is reduced proportionally using a variable-speed pumping system.

Depending on the water treatment system used, this cooling tower could potentially be sending a blowdown waste stream potentially containing heavy metals, phosphates, and biocides to a publicly owned treatment works (POTW) system. Most POTW systems are designed to handle only organic waste; much of these cooling tower chemicals will pass through the system untreated or will be released later as gaseous emissions at the POTW. In addition, harmful air emissions may result from the water treatment chemicals used, again depending on the type of water treatment.

Over the lifetime of the building, these releases could be among the more significant impacts on the local environment that the building will cause. This example highlights the magnitude of what can happen if this issue is not addressed.

Green Choices—Water Treatment

The water in the evaporative cooling loop must be treated to minimize biological growth, scaling, and corrosion. Typically, a combination of biocides, corrosion inhibitors, and scale inhibitors are added to the system.

Corrosion inhibitors are usually phosphate- or nitrogen-based (e.g., fertilizers) or molybdenum- or zinc-based (e.g., heavy metals). These inhibitors are more effective when added in combinations. These materials have low vapor pressure and are not used by the system. The inhibitors simply need to remain in the solution at the proper concentration to maintain a protective film on the metal components. Their only loss is through bleed and drift.

Most scaling inhibition is done by polymer-based chemicals, organic phosphorous compounds (e.g., phosphonates), or by acid addition. The acid reacts with the alkalinity in the water to release CO_2 and is consumed. The polymer and phosphonate scale inhibitors, on the other hand, remain in the solution to delay scaling; their major loss is through bleed and drift. Some polymers are designed to be easily broken down by bacteria in the environment, while others are not.

Many biocides are available. A typical system will use an oxidizing biocide (e.g., bromine or chlorine) maintained at a constant level and slug feed a nonoxidizing biocide once a week. Chlorine and bromine have a high vapor pressure in water. Much of the chlorine and bromine added to the tower is stripped from the water into the air and a small quantity reacts with organics in the tower. Drift and bleed will contain all of the nonoxidizing biocides small quantity of oxidizing biocides, and the reaction products of the biocides.

There are additional ways to treat the system with fewer negative impacts on the environment. Besides being rapidly stripped into the air, chlorine and bromine may react with organic molecules to produce very hazardous products. Other oxidizing biocides (e.g., hydrogen peroxide), on the other hand, do not have this issue. Further chemical additions can be added only when needed if continuous monitoring of the cooling system is in place. This technique can yield equivalent performance with fewer added chemicals. For reference, the U.S. Environmental Protection Agency (EPA) website on green chemistry (www.epa.gov/greenchemistry/) contains criteria on how to evaluate the life-cycle environmental impact of a particular chemical.

Nonchemical water treatment has the potential to be a powerful method for water treatment. By not adding chemicals to the cooling tower water system, there are substantially reduced levels of toxic chemicals stored and handled on site, a reduced risk from spills or leaks of cooling tower water, the bleed water will be better to send to the local POTW, and any concern associated with drift will be reduced. Additionally, instead of being sent to the POTW, bleed water could be used on site for irrigation or other needs where potable water is not required. There are several nonchemical technologies available; however, their successful use depends on the water chemistry, operating procedures, and degree of pollution of the specific system. Examples include those that use ozone, UV, pulsed electric fields, mechanical agitation, ultrasound, and hydrodynamic cavitation. Each of these technologies has developed a widespread following.

ASHRAE has conducted several research projects that investigated water treatment options and their effectiveness. For example, the effectiveness of scale prevention of some of these nonchemical water treatment technologies was assessed in ASHRAE Research Project RP-1155 (Cho et al., 2003). Others, such as ASHRAE Research Project RP-1361 (Vidic et al. 2010) raised questions regarding the overall efficacy of alternative methods. Technologies for water treatment continue to be developed and the reader is encouraged to investigate thoroughly any option chosen and evaluate all the trade-offs.

Green Choices—Tower Design and Selection

All cooling tower designs are efficient at removing heat from water through evaporation; however, not all designs perform as well environmentally. Some tower designs are more prone to splashout, spills, drift, and algae growth than others. Splashout involves tower water splashing from the tower and often happens when circulating tower water with the fans off while there are strong winds. Crossflow towers are more prone to this issue since, in the no-fan condition, some water will fall outside of the fill.

Spills from the cold-water basin can occur when all of the water in the piping drains into the basin at shutdown and overflows. Attention to proper water levels will prevent this. Some tower designs use hot-water basins or pray header pipes to distribute water at the top of the tower. The hot-water basin design can overflow if the nozzles clog; however, a spray header pipe never overflows.

Algae are a nuisance in basins and can contribute to microbial growth. Algae control requires harsher chemical treatment than typical biological control and thus are particularly problematic. Algae are plants and need sunlight to grow. Therefore, tower designs that are light-tight rather than open completely eliminates algae as an issue in cooling towers.

The amount of drift varies extensively in tower design. Some tower designs have very little drift, and the less the drift from a tower, the lower the amount of water containing minerals, water treatment chemicals, and microbes that will be released into the surrounding environment.

Green Choices—Filtration for Water Cooled Systems

A critical consideration with respect to green operation of water-cooled systems is to design, install, and maintain a filtration system capable of continually removing visible solids (40 μm and larger) from the water-cooled system. Keeping the system free from dirt, corrosion by-products, and scale maintains optimal heat transfer efficiency and reduces under-deposit corrosion. Filtration systems that use centrifugal action and require no media or moving parts have the advantage of reduced maintenance, no disposal of used media, better water efficiency, and offer minimal or zero liquid loss.

The filtration system and sweeper piping should be able to provide adequate pressure and flow (constant 20 psi at the nozzles, 1 gpm/ft^2 [138 kPa at the nozzles, 40.8 L/min per m^2]) to remove visible solids from the water and provide for efficient handling of the solids. It is not necessary to remove particles smaller than 40 μm, because the extra energy and water loss to achieve this level of filtration for a water-cooled system is wasteful.

Green Choices—Operation and Maintenance

An often overlooked method to minimize environmental impact is maintenance. Cooling towers operate outdoors under changing conditions. Wind damage to inlet air louvers, excessive airborne contamination, clogging of water distribution nozzles, and mechanical problems can best be prevented and quickly corrected with periodic inspections and maintenance.

DOMESTIC WATER HEATING

While space heating and cooling is the largest consumer of site energy in residences, the production of hot water comes second at 17.7% (EIA 2009). Many techniques can be used to minimize the energy required for domestic water heating. First, the use of water-conserving plumbing fixtures is recommended as it will in turn conserve energy required to heat water. Guidance is provided by the U.S. EPA's WaterSense program, ANSI/ASHRAE/USGBC/IES Standard 189.1, and this chapter about incorporating these fixtures into building design. Further, minimizing hot-water distribution heat losses by reducing premise plumbing lengths and insulated exposed pipe is another effective measure. GreenTip #12-1 discusses the use of point-of-use water heaters.

An important energy-reducing technique is the use of water heaters with higher uniform energy factors (UEFs). The UEF is a measure of efficiency as defined by U.S. regulations on residential and commercial water heaters (DOE 2014). Gas and electric storage water heaters, are currently the dominant technology in the water heater market. In recent decades, advanced water heaters have been developed that achieve UEFs beyond those dictated by the minimum efficiency standard, in part facilitated by the EPA ENERGY STAR® program launched in 2009. Such technologies include tankless water heaters, electric heat pump water heaters, condensing gas water heaters, and direct-contact water heaters, among others.

PLUMBING

Water-Conserving Plumbing Fixtures

Water conservation strategies save building owners money when it comes to both consumption and demand charges. Further, municipal water and wastewater treatment plants save on operating and capital costs for new facilities. As a general rule, water conservation strategies can yield significant cost savings when properly applied.

The Energy Policy Act of 1992 set reasonable standards for the technologies then available. Now, there are plumbing fixtures and equipment capable of significant reduction in water usage. For example, a rest stop in Minnesota that was equipped with ultralow-flow toilets and waterless urinals has recorded a 62% reduction in water usage.

The minimum water usage standards were established by the Energy Policy Act of 1992 (EPACT) for typical fixture types. These values would be used to calculate the baseline water consumption on a LEED project. ANSI/ASHRAE/USGBC/IES Standard 189.1 requires more efficient fixtures in the design, equivalent to the levels set under the U.S. Environmental Protection Agency's WaterSense program (http://epa.gov/watersense/). A good comparison of the various water efficiency requirements has been published by the Alliance for Water Efficiency and included in the references list at the end of the chapter.

Applicable state and local codes should be checked prior to design, as they have approved fixture lists; some code officials have not approved the waterless urinal and low-flush toilet technologies. Waterless urinals and low-flow lavatory fixtures can have a rapid payback period. Toilet technology continues to evolve rapidly, so be sure to obtain test data and references before specifying. Some units work very well, while others perform marginally.

Options that should be considered in the design of water-conserving systems include the following:

- Infrared faucet sensors
- Delayed-action shutoff or automatic mechanical shutoff valves (metering faucets at 0.25 gal/cycle [0.95 L/cycle])
- High efficiency (WaterSense compliant or equivalent) toilets
- Lavatory faucets with flow restrictors
- Low-flow kitchen faucets
- Domestic dishwashers that use 10 gal (38 L)/cycle or less
- Commercial dishwashers (conveyor type) that use 120 gal (455 L)/h
- Waterless urinals
- Closed cooling towers (to eliminate drift) and filters for cleaning the water

A comparison of water consumption for a typical office building using the baseline Energy Policy Act specifications compared to the more efficient WaterSense program values is given in the Digging Deeper sidebar at the end of this chapter.

WATER RECOVERY AND REUSE

For HVAC&R engineers looking for water recovery and reuse opportunities, condensate collection should be considered on most projects. Factors that determine whether condensate collection should be considered include location (i.e., whether building is in a more humid climate), building type (particularly relating to amount of outdoor air required), the size, number and accessibility of air handlers that condition outdoor air, and location of potential uses for the condensate. Location determines both the potential to collect a significant amount of condensate as well as the value of the water to the local community.

ANSI/ASHRAE/USGBC/IES Standard 189.1 requires that condensate collection be done on all air-conditioning units with a cooling capacity greater than 65,000 Btu/h (19 kW) and in regions where the ambient mean coincident wet-bulb temperature at 1% design cooling conditions is 72°F (22°C) or greater.

Building or space occupancy type determines the amount of outdoor air required, and thus, the amount of moisture in the incoming air. As the air passes across cold cooling coils, if the coil surface temperature is less than the dew point of the airstream passing by, then the potential exists for water condensation on the coils. (See the Digging Deeper sidebar titled "How Much Water Will Collect at Design Conditions?") Water that condenses on the coils will collect and drop to the drain pan below. The actual amount of water collected depends on parameters such as the absolute humidity level, total airflow, and coil bypass ratio. A major source of the moisture being condensed is from outdoor air brought in through outdoor air intakes or through infiltration exchange with the outdoors. A building or space that requires a lot of outdoor air on an ongoing basis (e.g., a laboratory) is an ideal candidate for condensate collection. Other obvious candidates include spaces with indoor water features, natatoriums, gymnasiums, and shower rooms, although these may face special challenges. Dedicated outdoor air systems, or DOAS units, are also prime candidates for consideration. More information on the subject of condensate collection can be found in GreenTip #12-2.

Rainwater Harvesting

GreenTip #12-3 also deals with a strategy for conserving potable water, though the water source differs.

Graywater Systems

See GreenTip #12-4, which deals with sanitary wastewater and a strategy to conserve potable water.

REFERENCES AND RESOURCES

Published

ACEEE. 2011. Emerging Hot Water Technologies and Practices for Energy Efficiency as of 2011. http://aceee.org/research-report/a112.

ASHRAE. 2017. ANSI/ASHRAE/USGBC/IES Standard 189.1-2017, *Standard for the Design of High-Performance Green Buildings Except Low-Rise Residential Buildings*. Atlanta: ASHRAE. (Will be incorporated into the 2018 version of the *International Green Construction Code* [IgCC].)

ASHRAE. 2016. ANSI/ASHRAE/IES Standard 90.1-2016, *Energy Standard for Buildings Except Low-Rise Residential Buildings*. Atlanta: ASHRAE.

ASHRAE. 2016. *ASHRAE Handbook—HVAC Systems and Equipment*. Chapter 40, Cooling Towers. Atlanta: ASHRAE.

Digging Deeper

EXAMPLE CALCULATION TO COMPUTE A BASELINE PREDICTED WATER CONSUMPTION FOR A BUILDING

Often, the design project team may want or need to compute the estimated water consumption of a building and compare that to some baseline level of performance. For example, this type of comparison is necessary for a building project working to demonstrate the minimum prerequisite water efficiency for a LEED project and to determine water savings credit points. This procedure involves estimating the water consumption using flow rate or usage rate values specified for the baseline performance in the LEED program description. These baseline levels correspond generally to the 1992 EPACT values listed in Table 12-2, with one exception for residential faucets.

It is fairly easy to determine the values to use for both the baseline level and the proposed building fixture flow or usage rates; these are taken from the table listed in the corresponding LEED program description and the specifications for the building project. Total water consumption is estimated based on occupancy expectations and thus requires some estimates or assumptions on occupancy levels and patterns. This is because it is the number of people in a building, not necessarily the number of fixtures in the building, that determines how much water is consumed.

One common method for determining water consumption is to base it on the number of full-time occupants in the building on daily basis. For this example, consider an office building that contains space for 100 employees, with a 50/50 breakdown of males and females. We assume that 10% of the employees are not at this office on any given normal workday, that this facility has 14 outside visitors a day (7 male, 7 female), and that each visitor stays an average of two hours per visit. On weekends and holidays, also assume that on average 5% of the employees are in the office for a total of four hours for each person each day, with no visitors. Basing the full-time occupancy estimate on an 8 h workday, the full-time equivalent occupancy of this building therefore is estimated as follows.

Weekdays

$$FTE_{male} = \frac{100-10}{2} \times \frac{8}{8} + \frac{14}{2} \times \frac{2}{8} = 46.75 \approx 47 \text{ people}$$

Because we have assumed an equal distribution of males and females, this is also the female FTE for weekdays. In this example, we are also rounding up the population to a whole number for each case.

Weekends, holidays

$$FTE_{male} = FTE_{female} = \frac{5}{2} \times \frac{4}{8} = 2.5 \approx 3 \text{ people}$$

Next, an assumption is needed for how many times per day each occupant uses the facilities. We will assume that each person visits them three times during each full, 8 h day. During each visit we assume that the males use the water closet once and a urinal twice. For each visit, regardless of gender, we assume that the lavatory sink is used a total of 15 s. We will also assume that this building contains a break room with a sink where the water is run a total of 30 min per day on weekdays and 5 min per day on weekends and holidays.

We now have all the information needed to compute an estimate of the water consumption, which is illustrated in the following tables.

Weekdays:

Fixture Type	No. Occupants or Devices	Flow/ Flush Rate	Units	Daily Use per Person	Duration, min	Daily Water Volume, gal (L)
W.C. Female	47	1.6 (6.1)	gpf (Lpf)	3		226 (854)
W.C. Male	47	1.6 (6.1)	gpf (Lpf)	1		75 (284)
Urinal	47	1 (3.8)	gpf (Lpf)	2		94 (356)
Lavatory Faucet	94	0.5 (0.03)	gpm (L/s)	3	0.25	35 (133)
Kitchen Sink	1	2.2 (0.14)	gpm (L/s)		30	66 (250)

Weekends, holidays:

Fixture Type	No. Occupants or Devices	Rate	Units	Daily Use per Person	Duration, min	Daily Water Volume, gal (L)
W.C. Female	3	1.6 (6.1)	gpf (Lpf)	3		14 (53)
W.C. Male	3	1.6 (6.1)	gpf (Lpf)	1		5 (18)
Urinal	3	1 (3.8)	gpf (Lpf)	2		6 (23)
Lavatory Faucet	6	0.5 (0.03)	gpm (L/s)	3	0.25	2 (9)
Kitchen Sink	1	2.2 (0.14)	gpm (L/s)		5	11 (42)

Digging Deeper

These give an estimated total water consumption of 496 gal (1878 L) per day during each weekday and 38 gal (144 L) on weekends and holidays.

The final step in the calculation is to apply this to the number of days per year for each day type, which gives us an estimated total annual water consumption of 133,010 gal (503,498 L) per year.

Day Type	No. Days per Year	Daily Use, gal (L)	Total, gal (L)
Weekday	260	496 (1878)	119,808 (455,780)
Weekend	105	38 (146)	4037(15,283)
Total			133,010 (503,498)

A similar procedure would be conducted for estimating the water consumption in the proposed building design that includes water conservation features. Note that these numbers provide an estimate for the amount of potable water used for the restroom and breakroom sink usage only. It does not include water use for other processes, such as a cooling tower, which would need to be factored in depending on the particular building situation. If alternative sources of water were used to supplement, for example, the toilet flushing, then this would have to be factored into the water use per fixture, based on specific information for that project.

Cho, Y.I., S. Lee, and W. Kim. 2003. Physical water treatment for the mitigation of mineral fouling in cooling-tower water applications. *ASHRAE Transactions* 109(1):346–57.

Copeland, C. 2014. Energy-water nexus: The water sector's energy use. Washington, DC: Congressional Research Service.

DOE. 2012. 2011 *Buildings Energy Data Book*. Chapter 8. Washington, D.C.: U.S. Department of Energy.

DOE. 2014. Energy Conservation Program for Consumer Products and Certain Commercial and Industrial Equipment: Test Procedures for Residential and Commercial Water Heaters; Final Rule. 10 CFR Parts 429, 430, and 431. Federal Register. Vol. 79, No. 133: 40542–88.

EIA. 2009. Residential Energy Consumption Survey. Washington, D.C.: U.S. Department of Energy. http://www.eia.gov/consumption/residential/index.php.

Maupin, M.A., Kenny, J.F., Hutson, S.S., Lovelace, J.K., Barber, N.L. and Linsey, K.S. 2014. Estimated use of water in the United States in 2010. *U.S. Geological Survey Circular* 1405, 56. https://pubs.usgs.gov/circ/1405/pdf/circ1405.pdf.

Vidic, R.D., S.M. Duda, and J.E. Stout. 2010. ASHRAE RP-1361. Biological control in cooling water systems using non-chemical treatment devices. Pittsburgh: University of Pittsburgh.
www.prowaterfl.net/wp-content/uploads/2012/08/Non-Chemical-Biological-Control.pdf.

Online

Alliance for Water Efficiency, "National Efficiency Standards and Specifications for Residential and Commercial Water Using Fixtures and Appliances". http://www.allianceforwaterefficiency.org/uploadedFiles/CodeAndStandards/US-Water-Product-Standard-Matrix-March-2010.pdf.

American Society of Plumbing Engineers
www.aspe.org.

ASHRAE Standard Project Committee 191
http://spc191.ashraepcs.org/.

U.S. Environmental Protection Agency, Green Chemistry
www.epa.gov/greenchemistry/.

U.S. Environmental Protection Agency, WaterSense Program.
www.epa.gov/watersense/.

U.S. Green Building Council
www.usgbc.org.

U.S. Geological Survey
www.usgs.gov.

ASHRAE GreenTip #12-1

Point-of-Use Domestic Hot-Water Heaters

GENERAL DESCRIPTION

As implied by the title, point-of-use domestic hot-water heaters provide small quantities of hot water at the point of use, without tie-in to a central hot water source. A cold-water line from a central source must still be connected as well as the energy source for heating the water, which could be electricity or gas.

There is some variation in types. Typically, the device may be truly instantaneous (e.g., lavatories), or it may have a small amount of storage capacity. With the instantaneous type, the heating source is sized such that it can heat a normal-use flow of water up to the desired hot-water temperature (e.g., 120°F [49°C]). When a small tank (usually 3 to 10 gal [11 to 38 L]) is incorporated in the device, an electric heating coil can be built into the tank and can be sized somewhat smaller because of the small amount of stored water available. The device is usually installed under the counter of the sink or bank of sinks. Slightly larger-sized units are available that can provide instantaneous heating for a residential house without requiring a storage tank and its associated thermal losses.

A similar type of device boosts the water supply (which is cold water) up to near boiling temperature (about 190°F [88°C]). This is typically used, for example, to make a cup of coffee or tea without having to brew it separately in a coffeepot or teapot.

WHEN/WHERE IT'S APPLICABLE

These devices are applicable wherever there is a need for a hot-water supply that is low in quantity and relatively infrequently used and is excessively inconvenient or costly to run a hot-water line (with perhaps a recirculation line as well) from a central hot-water source. Typically, these are installed in lavatories or washrooms that are isolated or remote, or both. However, they can be used in any situation where there is a hot-water need but where it would be too inconvenient and costly to tie in to a central source. (There must, of course, be a source of incoming water and a source of electricity.)

PRO

- A point-of-use device is a simple and direct way to provide small amounts of domestic hot water per use.
- Long pipe runs—and, in some cases, a central hot-water heating source—can be avoided.
- Energy is saved by avoiding heat loss from hot-water pipes and, if not needed, from a central water heater.
- In most cases, where applicable, it has a lower first cost.
- It is convenient—especially as a source of 190°F to 210°F (88°C to 99°C) water supply.
- When installed in multiple locations, central equipment failure does not knock out all user locations.
- It may save floor space in the central equipment room if no central heater is required.
- Water is saved by not having to run the faucet until the water warms up.

CON

- This is a more expensive source of heating energy (though cost may be trivial if usage is low and may be exceeded by heat losses saved from a central heating method).
- Water impurities can cause caking and premature failure of electrical heating coils.
- It cannot handle changed demand for large hot-water quantities or too-frequent use.
- Maintenance is less convenient (when required), since it is not centralized.
- A temperature and pressure relief valve and floor drain may be required by some code jurisdictions.

KEY ELEMENTS OF COST

The following provides a possible breakdown of the various cost elements that might differentiate a point-of-use domestic hot-water heater from a conventional one and an indication of whether the net incremental cost for the system is likely to be lower (L), higher (H), or the same (S). This assessment is only a perception of what might be likely, but it obviously may not be correct in all situations. There is no substitute for a detailed cost analysis as part of the design process. The listings below may also provide some assistance in identifying the cost elements involved.

First Cost

- Point-of-use water-heater equipment H
- Domestic hot-water piping to central source (including insulation thereof) L
- Central water heater (if not required) and associated fuel and flue gas connections L
- Electrical connection H
- Temperature and pressure relief valve and floor drain (when required by code jurisdiction) H

Recurring Cost

- Energy to heat water to appropriate temperatures H
- Energy lost from piping not installed L
- Maintenance/repairs, including replacement H

SOURCES OF FURTHER INFORMATION

American Society of Plumbing Engineers www.aspe.org.

ASPE. 1998. *Domestic Water Heating Design Manual*. Chicago: American Society of Plumbing Engineers.

Fagan, D. 2001. A comparison of storage-type and instantaneous heaters for commercial use. *Heating/Piping/Air Conditioning Engineering*, April.

ASHRAE GreenTip #12-2

Air-Handling Unit (AHU) Condensate Capture and Reuse

GENERAL DESCRIPTION

As air passes across cold cooling coils in an AHU, if the coil surface temperature is less than the dew point of the airstream passing by, then the potential exists for water condensation on the coils. Water that condenses on the coils will collect and drop to the drain pan below. In conventional building systems, this water typically drains, unused, to the sewer system or elsewhere, but in some situations it can be worthwhile to capture and reuse it.

Condensate collection can either be designed into new construction or retrofitted into existing buildings. While the former is preferable (due to lower costs and fewer complications), the latter presents the highest potential, since existing buildings comprise about 98% of the building stock.

The best end use for the collected water will depend on the particular circumstances of the location. In a building with its own chiller and cooling tower, the most logical choice is to route collected condensate to the cooling tower sump to reduce the need to use fresh water for makeup. In most cases, peak condensate production will occur at the same times as peak makeup water demands, creating an elegant synergy. This is also the simplest retrofit, involving reasonably inexpensive equipment and piping; water can be routed directly to the tower with no need for treatment.

Condensate collection can also be integrated with a rainwater collection system, a scheme often referred to as *rainwater plus*. This will usually involve a storage tank or cistern and can require considerably more expense and engineering than using the condensate in a cooling tower. Depending on the intended use (e.g., irrigation, ornamental fountains, or other internal uses including toilet flushing), different amounts of further treatment will be required. In all cases, local building codes must be followed.

WHEN/WHERE IT'S APPLICABLE

Factors that determine whether condensate collection should be considered include location (e.g., climate), building type (particularly relating to amount of outdoor air required), the size, number, and accessibility of air handlers that condition outdoor air, location of potential uses for the condensate, etc. Location determines both the potential to collect a significant amount of condensate and the value of the water to the local community. In periods of drought, the actual value of a unit of water to the local society and economy may be worth much more than the rate currently paid to the local utility.

A building or space that requires a lot of outdoor air on an ongoing basis (e.g., laboratories) is an ideal candidate for condensate collection. Other obvious candidates include spaces with indoor water features, natatoriums, gymnasiums, or locker rooms. DOAS units are also prime candidates for consideration, since they are typically designed for optimal latent load removal.

Using typical meteorological year data, assumptions about the air-handling system, and the following equation, it is possible to estimate the amount of condensate that can be collected annually in a particular location.

Condensate collection is required by ANSI/ASHRAE/USGBC/IES Standard 189.1 for air-conditioning units that are above a certain cooling capacity and in more humid climates.

LESSONS LEARNED

Attention must be paid to the cleanliness of the water and the system components. For example, any external condensate collection pan should be covered to prevent foreign particles from getting into the system. The potential is also there for biological growth and contamination. Also, there may be an increase in corrosion potential in the cooling tower loop, if that is where the condensate is sent.

Sweating on the outside of the condensate piping can be an issue, particularly in semiconditioned mechanical rooms, so all lines (as well as perhaps the collection basin itself) should be insulated. If the condensate line is tied into rain downspouts, you may want to consider running a smaller pipe or tube inside of the downspout for the condensate to avoid moisture buildup on the outside downspout surface.

The dimensions of the U-trap in the existing condensate drain pipe between the AHU and the floor drain should be maintained when connecting the drain pipe to the external collection pan. It is also highly recommended that a condensate flow meter be installed and that it be located to facilitate easy reading. The additional cost of the meter is worthwhile because of the good feedback on functionality and the education potential it provides. It's also a good way to verify water and cost savings.

PRO

- If condensate is routed to a cooling tower, demand for makeup water will reduce and so will the need for treatment chemicals. Blowdown frequency should decrease, and sewer costs could be reduced with appropriate metering.
- Cool condensate routed to a cooling tower will provide residual free cooling for condenser water.
- Incorporating condensate collection into a rainwater collection and storage system can reduce the cistern size requirement by providing a supplemental water source during long periods between rain events.

CON

- Complications arise when dealing with district cooling systems with satellite chillers, because it is possible to produce condensate in an AHU while the chiller and cooling tower (for that particular building) are idle. This leads to the risk that treatment chemicals in the sump will be diluted and needlessly washed away via the overflow drain.
- In general, less-efficient systems have higher condensate production potential. Enthalpy wheels, for example, will greatly improve system efficiency but will dramatically reduce condensate production. A building with 100% outdoor air supply that is overpressurized will produce more condensate but will waste energy. Energy efficiency should always take precedence over water production.

KEY ELEMENTS OF COST

The following capital cost issues list the various cost elements associated with either building condensate collection into new construction or retrofitting an existing building. This assessment is only a perception of what might be likely, but it obviously may not be correct in all situations. There is no substitute for a detailed cost analysis as part of the design process.

Pipe. Depending on the distance between the AHU and the end use, and whether a storage tank is involved, the material and labor costs of the pipe installation are likely to be the most expensive part of the system. For new construction the additional cost should be minimal, since a condensate drain pipe would need to be furnished, regardless of the end use.

Storage. If condensate is to be stored for later use, a cistern or storage tank can represent a considerable part of the system cost. Additional costs will be incurred for system design (i.e., tank sizing) and tank site selection and installation. Finally, treatment of stored water prior to end use, if necessary, will add equipment, design, and maintenance costs.

Metering. A totalizing meter is a relatively inexpensive but important component of a condensate collection system. Once in place, a condensate meter will help verify payback on the investment in the system.

SOURCES OF FURTHER INFORMATION

Guz, K. 2005. Condensate water recovery. *ASHRAE Journal* 47(6):54–56.

Lawrence, T., J. Perry, an P. Dempsey. 2010. Making every drop count: Retrofitting condensate collection on HVAC air-handling units. *ASHRAE Journal* 52(1): 48–54.

Lawrence, T,M., J. Perry and T. Alsen. 2012. AHU condensate collection economics. *ASHRAE Journal* 54(5):12–17.

HOW MUCH WATER WILL COLLECT AT DESIGN CONDITIONS?

For simplicity, consider the process of a unit conditioning 100% outdoor air (where the unit is a dedicated outdoor air system). The psychrometric chart represents a path of outdoor air as it passes across the cooling coil for the 0.4% cooling design condition in Athens, Georgia. Assuming a supply air condition of 55°F (12.8°C) and 85% relative humidity (wet-bulb temperature = 52.5°F [11.4°C]), the humidity ratio changes across the coil from 0.0141 to 0.0078 lb/lb$_{air}$ (kg/kg$_{air}$). The difference in absolute humidity ω between the incoming outdoor air and supply air leaving the unit represents the amount of condensation that occurs. Thus, for every lb (kg) of air supplied by the unit, 0.0141 – 0.0078 or 0.0063 lb (0.00286 kg) of water are condensed.

The total amount of condensate expected is determined by the equation below:

$$\text{Condensate} = \text{Airflow} \times \text{Density} \times 60\frac{\text{min}}{\text{h}}\ (\text{I-P})\ \text{or}\ 3600\frac{\text{s}}{\text{h}}\ (\text{SI}) \times \Delta\omega$$

Assuming 1000 cfm (472 L/s) of outdoor air is being conditioned, the total amount of condensate expected would be the following:

$$\text{Condensate} = 1000\frac{\text{ft}^3}{\text{min}} \times \frac{\text{lb}}{13.133\ \text{ft}^3} \times 60\frac{\text{min}}{\text{h}} \times (0.0141 - 0.0078)\frac{\text{lb}_{water}}{\text{lb}_{dry\ air}} \quad (\text{I-P})$$
$$= 28.8\frac{\text{lb}}{\text{h}}$$

$$\text{Condensate} = 0.472\frac{\text{m}^3}{\text{s}} \times \frac{\text{kg}}{0.820\ \text{m}^3} \times 3600\frac{\text{s}}{\text{h}} \times (0.0141 - 0.0078)\frac{\text{kg}_{water}}{\text{kg}_{dry\ air}} \quad (\text{SI})$$
$$= 13.05\frac{\text{kg}}{\text{h}}$$

This is approximately 3.5 gal (13.1 L)/h at the cooling design condition.

Similar calculations can be run for any locality, and the result can vary widely depending on the climate. For example, when the calculation is run for other representative cities the condensate yields are as follows:

- Boston, MA (90.8°F [32.6°C] dry-bulb/73.3°F [22.8°C] mean coincident wet-bulb [MCWB] temperature)
  = 3.2 gal (12.1 L)/h

Digging Deeper

Digging Deeper

- Sacramento, CA (100.4°F [37.9°C] dry-bulb/70.7°F [21.4°C] MCWB) = 0.8 gal (3.1 L)/h
- Denver, CO (94.3°F [34.5°C] dry-bulb/60.3°F [15.6°C] MCWB) = no condensate collected

Interestingly, the total annual rainfall is only a partial indicator of how much condensate might be collected, as shown in Table 12-2. The two comparative eastern and western U.S. cities have similar rainfall totals, but they vary significantly in terms of the total amount of condensate collection potential.

What to do with Collected Water

The best end use for the collected water will depend on the particular circumstances of the location. In many locations the primary use of city water is for makeup water in cooling towers. Therefore, it may be the most logical choice to collected condensate to its cooling tower sump. In most cases, peak condensate production will occur at the same times as peak makeup water demands, creating an elegant feedback loop. This is also the simplest retrofit, involving reasonably inexpensive equipment and piping; water can be routed directly to the tower with no need for treatment.

Complications arise when dealing with district cooling systems with satellite chillers, because it is possible to produce condensate in an air-handling unit while the chiller and cooling tower for that particular building are idle. While it is no tragedy that condensate sent to the cooling tower will simply overflow to the sewer (where it would have gone prior to retrofit), there is

Table 12-2: Annual Condensate Collection Compared to Total Annual Rainfall

	Annual Condensate for Continuous Outdoor Air, gal/cfm (L/L/s)	Average Annual Rainfall, in. (m)
Athens, GA	12.5 (100.4)	47.8 (1.21)
Boston, MA	4.5 (36.1)	42.5 (1.08)
Sacramento, CA	1.3 (10.4)	17.9 (0.45)
Denver, CO	0.5 (4)	15.4 (0.39)

the risk that treatment chemicals in the sump will be diluted and needlessly washed away. In this scenario, care should be taken to prioritize condensate retrofits in buildings with baseline chiller plants.

Condensate collection can also be integrated into a rainwater collection system, a scheme often referred to as *rainwater plus*. This will usually involve a storage tank or cistern and can require considerably more expense and engineering than using the condensate in a cooling tower. Depending on the intended use (e.g., for irrigation, fountains, toilet flushing, or potable water), different amounts of further treatment will be required. In all cases, local building codes must be followed.

ASHRAE GreenTip #12-3

Rainwater Harvesting

GENERAL DESCRIPTION

Rainwater harvesting is a simple technology that has been around for thousands of years. Harvested rainwater can stand alone or augment other water sources. Systems can be as basic as a rain barrel under a downspout or as complex as a pumped and filtered graywater system providing landscape irrigation, cooling tower makeup, and/or building waste conveyance.

Systems are generally composed of five or fewer basic components: a catchment area, a means of conveyance from the catchment, storage (optional), water treatment (optional), and a conveyance system to the end use.

The catchment area can be any impermeable area from which water can be harvested. Typically this is the roof, but paved areas (e.g., patios, entries, and parking lots) may also be considered. Roofing materials that are metal, clay, or concrete-based are preferable for roofs planned for rainwater harvesting compared to those with potential contaminants, such as asphalt or those with lead-containing materials. Similarly, care should be given when considering a parking lot for catchment due to oils and residues that can be present.

Conveyance to the storage will be gravity-fed, like any stormwater piping system. The only difference is that now the rainwater is being diverted for reuse purposes.

Commercial systems will require a means of storage. Cisterns can be located outside the building (e.g., above-grade or buried) or placed on the lower levels of the building. The storage tank should have an overflow device piped to the storm system and a potable water makeup if the end-use need is ever greater than the harvested volume.

Depending on the catchment source and the end use, the level of treatment will vary. For simple site irrigation, filtration can be achieved through a series of graded screens and paper filters. If the water is to be used for waste conveyance, then an additional sand filter may be appropriate. Parking lot catchments may require an oil separator. The local code authority will likely decide acceptable water standards, and, in turn, filtration and chemical polishing will be a dictated parameter, not a design choice.

Distribution can be via gravity or pump depending on the proximity of the storage tank and the end use.

WHEN/WHERE IT'S APPLICABLE

If the building design is to include a graywater system, condensate collection, or landscape irrigation—and space for storage can be found—rainwater harvesting is a simple addition to those systems.

When a desire exists to limit potable water demand and use, depending on the end-use requirement and the anticipated annual rainfall in a region, harvesting can be provided as a standalone system or to augment a conventional makeup water system. Sites with significant precipitation volumes may determine that reuse of these volumes is more cost effective than creating stormwater systems or on-site treatment facilities.

Rainwater harvesting is most attractive where municipal water supply is either nonexistent or unreliable, hence its popularity in rural regions and developing countries.

PRO

- Rainwater harvesting reduces a building's potable water use, and reduces demand on the municipal water supply, lowering costs associated with water.
- Rainwater is soft and does not cause scale buildup in piping, equipment, and appliances. It could extend the life of systems.
- It can reduce or eliminate the need for stormwater treatment or conveyance systems.

CON

- There is added first cost associated with the cisterns and the treatment system.
- There are additional materials and their associated embodied energy costs.

- The storage vessels must be accommodated. Small sites or projects with limited space allocated for utilities would be bad candidates.
- Costs include maintenance of the system (e.g., maintaining the catchments, conveyance, cisterns, and treatment systems).
- At the time of the writing of this document, there is no U.S. guideline on rainwater harvesting. However, the International Code Council (ICC) and the Canadian Standards Association (CSA) are working on a standard for rainwater harvesting (CSA/ICC B805) that should be published in late 2017. The local health code authority has jurisdiction, potentially making a particular site infeasible due to backflow prevention requirements, special separators, or additional treatment.

KEY ELEMENTS OF COST

The following provides a possible breakdown of the various cost elements that might differentiate a building utilizing rainwater harvesting from one that does not and an indication of whether the net cost is likely to be lower (L), higher (H), or the same (S). This assessment is only a perception of what might be likely, but it obviously may not be correct in all situations. There is no substitute for a detailed cost analysis as part of the design process. The listings below may also provide some assistance in identifying the cost elements involved.

First Cost

- Catchment area — S
- Conveyance systems — S
- Storage tank — H
- Water treatment — S/H
- Distribution system — S/H
- Design fees — H

Recurring Cost

- Cost of potable water — L
- Maintenance of system — H
- Training of building operators — H

- Orientation of building occupants S
- Commissioning cost H

SOURCES OF FURTHER INFORMATION

CSA/ICC B805-201X. "Rainwater Harvesting Systems." (should be released in 2017).

American Society of Plumbing Engineers www.aspe.org.

Irrigation Association www.irrigation.org.

Texas Manual on Rainwater Harvesting www.twdb.texas.gov/publications/brochures/conservation/doc/RainwaterHarvestingManual_3rdedition.pdf.

USGBC. 2009. *LEED 2009 Green Building Design and Construction Reference Guide*. Washington, D.C.: U.S. Green Building Council.

Waterfall, P.H. 2006. Harvesting Rainwater for Landscape Use. https://extension.arizona.edu/sites/extension.arizona.edu/files/pubs/az1344.pdf.

Environmental Protection Agency, Water Sense Program http://epa.gov/watersense/.

ASHRAE GreenTip #12-4

Graywater Systems

GENERAL DESCRIPTION

Graywater is generally wastewater from lavatories, showers, bathtubs, and sinks that is not used for food preparation. Graywater is further distinguished from blackwater, which is wastewater from toilets and sinks that contains organic or toxic matter. Local health code departments have regulations that specifically define the two kinds of waste streams in their respective jurisdictions.

Where allowed by local code, separate blackwater and graywater waste collection systems can be installed. The blackwater system would be treated as a typical waste stream and piped to the water treatment system or local sewer district. However, the graywater would be recycled by collecting, storing (optional), and then distributing it via a dedicated piping system to toilets, landscape irrigation, or any other function that does not require potable water.

Typically, for a commercial graywater system (e.g., for toilet flushing in a hotel), a means of short-term on-site storage, or, more appropriately, a surge tank, is required. Graywater can only be held for a short period of time before it naturally becomes blackwater. Often some treatment of the graywater is done, such as with a bleach solution or other means, to prolong storage time. The surge tank would be provided with an overflow to the blackwater waste system and a potable makeup line for when the end-use need exceeds stored capacity.

Distribution would be accomplished via a pressurized piping system requiring pumps and some low level of filtration. Usually, there is a requirement for the graywater system to be a supplemental system. Therefore, systems will still need to be connected to the municipal or localized well service.

Plumbing codes require that a colorant be added to graywater used within buildings, such as for toilet flushing, to help distinguish it as nonpotable water.

WHEN/WHERE IT'S APPLICABLE

Careful consideration should be given before pursuing a graywater system. While a graywater system can be applied in any facility that has a nonpotable water demand and a usable waste stream, the additional piping and energy required to provide and operate such a system may outweigh any benefits. Such a system is best applied where the ratio of demand for nonpotable water to potable water is relatively high and consistent, as in restaurants, laundries, and hotels.

Some facilities have a more reliable graywater volume than others. For example, a school would have substantially less graywater in the summer months. This may not be a problem if the graywater was being used for flushing, since it can be assumed that toilet use would vary with occupancy. However, it would be detrimental if graywater were being used for landscape irrigation.

PRO

- A graywater system reduces a building's potable water use, in turn reducing demand on the municipal water supply and lowering costs associated with water.
- It reduces a building's overall wastewater generation, thus becoming less taxing on the existing sewage systems.

CON

- There is an added first cost associated with the additional piping, pumping, filtration, and surge tank required.
- There are additional materials and their associated embodied energy costs.
- There is negative public perception of graywater reuse and health concerns regarding ingestion of nonpotable water.
- Costs include maintenance of the system, including the pumps, filters, and surge tank.
- Local health code authority has jurisdiction, potentially making a particular site infeasible due to that authority's definition of blackwater versus graywater.

KEY ELEMENTS OF COST

The following provides a possible breakdown of the various cost elements that might differentiate a building using a graywater system from one that does not and an indication of whether the net incremental cost is likely to be lower (L), higher (H), or the same (S). This assessment is only a perception of what might be likely, but it obviously may not be correct in all situations. There is no substitute for a detailed cost analysis as part of the design process. The listings below may also provide some assistance in identifying the cost elements involved.

First Cost

- Collection systems H
- Surge tank H
- Water treatment H
- Distribution system H
- Design fees H

Recurring Cost

- Cost of potable water L
- Cost related to sewer discharge L
- Maintenance of system H
- Training of building operators H
- Orientation of building occupants S
- Commissioning cost H

SOURCES OF FURTHER INFORMATION

Advanced Buildings Technologies and Practices www.advancedbuildings.org.

American Society of Plumbing Engineers www.aspe.org.

Del Porto, D., and C. Steinfeld. 1999. *The Composting Toilet System Book*. New Bedford, MA: The Center for Ecological Pollution Prevention.

Ludwig, A. 1997. *Builder's Greywater Guide* and *Create an Oasis with Greywater*. Oasis Design.

USGBC. LEED Green Building Design and Construction Reference Guide. Washington, DC: U.S. Green Building Council.

CHAPTER THIRTEEN

Smart Building Systems

Society has entered the age of Big Data and the Internet of Things (IoT). Using smart devices and the ever-expanding wireless network, mankind is intimately connected to the world in which we (meaning humans and their devices) operate. Portable access to information allows us to make informed decisions more quickly and accurately than ever before.

Along these lines, building operations and maintenance is being transformed by this newfound connectivity. At the device level, smart thermostats—capable of remote adjustment and monitoring via any number of graphical interfaces (smartphone, tablets, computers)—and IP-addressable rooftop units—which facilitate remote troubleshooting and control—are examples of how systems infrastructure is leveraging technology to improve control, comfort, and maintenance efficiency. At a building- or portfolio-scale, open architecture controls and cloud-based software platforms now facilitate integration, aggregation and analysis of building data and control across systems. The built environment is rapidly evolving from closed systems to those that are interconnected and accessible.

Smart buildings stand apart in their quest to optimize building performance by continually turning data into information and knowledge, and ultimately into action. This chapter highlights components and strategies often used in smart building systems, and is divided into nine sections as follows:

- Integrated building automation systems. A smart approach to operations and maintenance integrates building infrastructure systems into a common operations interface, facilitates data visualization and sharing across systems and serves as a repository of maintenance information and procedures.
- Detailing the design of a controls system. Clear definition with respect to design intent and system functionality affords the contractor, commissioning authority and facilities staff a roadmap for installing, configuring and operating the building.
- Automated fault detection and diagnosis. With the objective to maximize ongoing building operational performance, automated fault detection and diag-

nosis (AFDD) serves as a database overlay designed to uncover, report, characterize, and oftentimes correct, system faults.

- Smart hardware. Hardware in smart buildings often possess intelligence to diagnose defective devices and recalibrate and/or fix those devices.
- Controls systems and building occupants. Through the use of smart applications and graphical user interfaces, building occupants can both directly control and solicit feedback from buildings, a process critical to ongoing optimization.
- Controls systems for energy and water efficiency. Through both stand-alone and integrated systems strategies, smart controls have the capability of significant reductions in building energy and water usage in a typical commercial building.
- Controls systems and indoor environmental quality (IEQ). In most commercial buildings, controls play a crucial role in monitoring and regulating space temperature, humidity, ventilation, acoustics, and light levels.
- Controls system commissioning. Commissioning occurs from design through postoccupancy and helps ensure that control systems—both stand-alone and interconnected—are installed and configured correctly.
- The interaction of a smart building with the coming smart grid. With smart grid technology being deployed by utilities worldwide, new concepts and technologies are needed for a smooth interaction between the electric grid and building automation systems.

INTEGRATED BUILDING AUTOMATION SYSTEMS

Building operation is a systematic process comprehensively influenced by building technology, cultural concept, occupant interactions with building equipment and systems, social equity, and other factors, among which occupant behavior plays an extremely important role. The occupants' and operator's expectation of comfort and/or satisfaction in the built environment oftentimes drives building controls.

The integration of building systems data and control provides a framework from which to build a more efficient, higher-performing building. A smart building systems approach to facilities management and control aggregates monitoring and control into a common system that provides the visualization and operational feedback that allows facilities staff to better understand the building's dynamics. With knowledge and actionable information in hand, building staff can make informed decisions on how to efficiently position, operate, maintain and optimize their assets over time.

The heart of a smart building system is an open architecture or integrated systems approach to building management that allows the dynamic features of the modern building—information technology (IT), lighting and lighting controls, mechanical/HVAC, plumbing, power/power monitoring, security, audio-visual, and conveyance—to integrate to a common front-end. As specified in the commercial buildings Master Format Division 25 for Integrated Automation, (CSI 2016) this front end serves as both the:

- Single point of operations, controls and maintenance information for the building's dynamic systems.
- Data aggregation, analytics, visualization, and reporting platform for sustainable operation and maintenance.

Understanding that the facilities staff truly drive long-term operational performance of the building, a key part of smart building system conceptualization is educating the facilities staff on their options, gaining an understanding of the facilities staff's knowledge and resources and, ultimately, defining the owner's project requirements (OPR) (refer to Chapter 4 for more information). With the OPR in hand, a collaborative, performance-based process facilitates the design and specification of systems and components and the careful evaluation and selection of integration contractors aligned to deliver an optimal outcome.

However, a well-intentioned design devoid of proper functionality and utilization will never maximize its value. Coordinating, specifying, and commissioning the connectivity, installation, and integration of otherwise disparate systems becomes critical to the utility of a smart building system. This cohesive process is led by a dedicated, multidisciplinary team. In addition to coordinating the design effort as it relates to systems integration, the smart building systems team specifies the metering, monitoring, training, and ongoing commissioning requirements of the system. Defining these components helps ensure that the owner has easily-accessible tools to drive operations and optimization.

DETAILING THE DESIGN OF A CONTROL SYSTEM

Lack of clarity with respect to design or control intent oftentimes results in poor building performance. To ensure the contractor and commissioning authority understand how the building systems should function, the following control-related elements should be included in the contract documents. While most of this section focuses on controls for HVAC systems, the process described can and should be applied when developing controls designs for other systems (e.g., lighting). This list of items is not comprehensive, but it provides an idea of the type of issues that should be addressed.

Provide Detailed Control Descriptions. One of the most prevalent reasons control systems fail to perform as intended is that insufficient forethought is given to the sequence of operation prior to the contractor programming setup. The designers, and later, the control contractor's programmer often do not think the sequence out and consider how it will (or will not) function during all possible modes and scenarios of weather, loads, staging, and interactions. This issue can be mitigated by the following:

- Ensure the designer provides flow diagrams of the major controlled systems, showing interfaces and control authorities between local and central control.

- Ensure the designer includes detailed sequences of operation that include a brief system narrative, points list, alarms, what initially starts equipment, staging, failure and standby functions, power outage response and reset requirements, interlocks to other systems, control authorities with local (packaged) controls, trending requirements, and energy efficiency strategies with given set points.
- Develop a graphical test simulation of the control program(s) to ensure the mechanical equipment sequences on and off as load increases and decreases, according to the sequences of operation.

Match Control Strategies to Operator Capabilities. If the operators do not understand the features or sequences sufficiently, and there is not a qualified controls technician maintaining the system, the advanced features or sequences that have problems will likely be overridden or disabled. Designers and design reviewers should make sure the complexity of control schemes matches the expected level of technical expertise of the operators. It is critical that operator training be conducted in a timely manner, including follow-up sessions as needed.

Strategies Relying on Drift-Prone Sensors. Where smart hardware is not applied, control sensor and loop recalibrations are a necessary function for maintaining high-performing systems. The OPR document should define the operator training and skillsets needed to maintain the system functioning at design efficiency. Major control strategies that depend on sensors that are known to drift should be avoided or, if called for, the necessary training and recalibration programs must be institutionalized into the building maintenance culture. For example, consider the case of a chiller staging sequence using supply and return temperatures and flowmeter(s). The sensors may drift over time, and typical accepted error ranges in these types of sensors will yield inaccurate load calculations that may disrupt proper staging. This strategy can result in high overall efficiency, but it requires a regular calibration and maintenance checks to maintain this high efficiency.

Requirements for System Architecture Rationale. It is important to ensure that, in the requirements for the controls submittal, the controls contractor is mandated to provide calculations and rationale for the number and layout of the primary (peer-to-peer) and local (application-specific) controllers in relation to the total number of points and other network traffic. It should also be required that the contractor describe how many points can be reasonably trended without appreciably affecting point value refresh rates, and describe the impacts on network speed that alternative layouts would have.

The building automation system (BAS) performance requirements should be defined in the contract documents (i.e., specifications such as those set forth in ASHRAE Guideline 13-2015, *Specifying Direct Digital Control Systems* [ASHRAE 2015b]). The performance requirements are defined during the predesign phase and should be contained in the OPR document in accordance with ASHRAE Standard 202-2013.

Requirements for Clear Control Sequences. Ensure that the requirements for the control drawing submittals in the specifications include statements requiring the following:

- A brief overview narrative of the system, generally describing its purpose, components, and function (i.e., the design intent for the controls)
- All interactions and interlocks with other systems
- Detailed delineation of interaction between any localized controls and the BAS, listing what points only the BAS monitors and what BAS points are control points and are adjustable
- Start-up, warm-up, cooldown, occupied, and unoccupied operating modes, plus power failure recovery and alarm sequences
- Capacity control sequences and equipment staging
- Initial and recommended values for all adjustable settings, set points, and parameters that are typically set or adjusted by operating staff and any other control settings or fixed values, delays, etc., that will be useful during testing and operating the equipment
- Rough zone analysis requirements to assure reset strategies are effective
- Energy mapping requirements to help operators see the building efficiency at a glance
- System override abilities and requirements
- Description of building isolation areas for off-hours operation
- Front-end graphics requirements including summary screens by system type, zone, and plant

To facilitate review and referencing in testing procedures, all sequences should be written in short statements, each with a number for easy reference.

Requirements for Clear Control Drawings. The specifications must ensure the control drawing submittal requirements include at least the following:

- The control drawings must contain graphic schematic depictions of all systems showing each component (e.g., valves, dampers, actuators, coils, filters, fans, pumps, speed controllers, piping, ducting, etc.), each monitored or control point and sensor, and all interlocks to other equipment. Drawings may include fan and pump flow rates as well as horsepower.
- The schematics should include the system and component layout of any equipment that the control system monitors, enables, or controls, even if the equipment is primarily controlled by packaged controls.

- A full points list should be provided, including point abbreviation key, point type, system point with which it is associated, point description, units, panel ID, and field device.
- Network architecture drawing should include all controllers, workstations, printers, and other devices in a riser format and including protocols and speeds for all trunks. Include the network buses with the bus speeds.
- Sketches of all graphics screens for review and approval must be included.

Include Commissioning Engagement in Design Fees

In estimating the fees for the design process, owners and the design team should include sufficient allowance to be fully engaged in the commissioning process during design. This includes responding to design review comments and incorporating commissioning requirements into the project specifications.

AUTOMATED FAULT DETECTION AND DIAGNOSIS

Three factors that are critical to sustaining the efficiency level of a new building are (1) a well-designed measurement and verification (M&V) process, (2) implementing a commissioning program that will evaluate the function of all key systems and equipment on a regular basis, and (3) good operator training on the control system functions (see previous section, Match Control Strategies to Operator Capabilities). The first two factors allow building operators to monitor performance on a regular basis and to intervene when problems are detected. As discussed in this chapter, the control system is essential in implementing a strong M&V process. Good training can help ensure that all of the capabilities of the control system, including those related to M&V, are used to their full potential over the lifetime of the building. To aid building operators in maintaining an improved level of efficiency, the field of fault detection and diagnosis (FDD) was developed before the development of automated fault detection and diagnosis (AFDD). FDD is a non-software driven means of detecting and correcting operational faults. This method often uses in-field or handheld equipment that maintenance personnel can use on site to assess equipment performance under different conditions.

The objective of automated fault detection and diagnosis (AFDD) is the early detection and correction of system faults and inefficiencies. In this two-step process, first the fault in equipment operation is detected, then second, the actual cause of the fault is isolated. Enabling this practice is monitoring and metering infrastructure that continuously trends data across building systems, using algorithms that range from simple to complex to process and interpret the collected data. At the equipment level, this data originates from a set or subset of sensor readings taken from sensors installed on the HVAC equipment and its components, such as temperatures, pressures, humidity levels, energy demands, or other related values. At the building level, this may also include data such as indoor and outdoor temperatures, smart meters, and other data enabled through the growing sensing and communication capabilities of the wireless sensors and IoT. This measured data is combined with known information the equipment, system and/or building characteristics. AFDD software then analyzes the combined data set using a host of rules and algorithms, and flags instances of nonconformance. These instances are then typically prioritized based on another set of rules and/or evaluated by building operations staff to determine the underlying cause and appropriate corrective action.

AFDD methods may be divided into three main categories, including qualitative model-based methods, quantitative model-based methods, and process-history based methods. The two model-based methods use a priori knowledge to identify actual operational data and model result predictions. Qualitative methods trigger a fault when measured data violate a set of rules describing the proper operation of a building, system, or equipment. These rules are determined based on expert knowledge, process history data, and quantitative models. Quantitative methods use a mathematical model based on the underlying physical relationship and characteristics of the equipment, system, or building performance, which is then compared to the actual performance, triggering a fault when the actual and predicted values do not match. The data-based methods (process-history) use limited to no knowledge of the physical system being analyzed, focusing on data previously collected to train the systems. This approach typically uses statistics-based black box methods, or grey-box models where the model parameters are based on physical principles, to relate measured input sensor data, measured output data, and/or predicted data, triggering a fault if the relationship between these values is violated. Compared to the first two methods, this method requires less knowledge of the building, system, and/or equipment characteristics but also requires a significant amount of historical data to use as training data to develop the input/output sensor relationships that are comprehensive enough to cover a large range of operational conditions. This is important for this method, as these methods typically do not extrapolate well outside the bounds of the training data.

AFDD can be implemented for building equipment in a number of different ways. The first is through embedded AFDD developed by the manufacturer and specific to a piece of equipment. This method includes equipment-specific values and algorithms that cannot be applied elsewhere. AFDD can also be implemented using non-manufacturer-specific methods that can be added to existing equipment through a standalone or online software as a service (SaaS) or cloud-based model. Similar to the embedded FDD, this provides continuous monitoring and feedback on the performance of a piece of equipment using a set of sophisticated algorithms and expert rules. These bottom-up methods focus on specific pieces of equipment. Conversely, top-down methods are also used, which focus on detecting the presence of a fault using building- or system-level data rather than equipment-level data.

AFDD methods have been applied to all types of equipment through significant research efforts in recent years, including air handlers, packaged units, VAV terminal units, boilers, chillers, and economizers. Common HVAC&R faults can include stuck or leaking valves or air dampers, actuator failures, leaky ducts, clogged air filters, incorrect sensor calibrations, low refrigerant, and restricted airflow. They can also be used in a broader sense to identify system faults such as equipment not operating in accordance with the specified schedule and/or associated design parameters.

Moving forward, the IoT has a strong potential to further improve the AFDD methods that currently exist through providing more data, that can in turn, better inform AFDD algorithms, enabling smarter building operations that can more quickly identify issues and resolve them. More IoT devices continue to be connected at a rapidly expanding rate, while at the same time, sensors are becoming smaller, lower-cost, and more readily available, and improved security and standard protocols for communications are currently under development. These devices, such as networks of low-cost sensors, and portable devices that can and already do collect data, such as smart phones, can provide additional data feeds to AFDD, potentially at low to no cost. Data fusion methods have been proposed that use large amounts of data that come from many different types of sensors that alone would not be sufficient, but because of the insights that can be determined from multiple different overlapping types of sensors (e.g., occupant presence from a low-cost infrared (IR) camera and from a low-cost CO_2 sensor), enabling a check of accuracy of the input data into AFDD and improved performance and reliability of the resulting output from these methods.

SMART HARDWARE

Smart operation of buildings depends on quality input data from a range of sensors and information collected about a building and its systems, as discussed in previous sections of this chapter. However, if the sensors that are relied upon for intelligent operation are providing inaccurate information, this can have a negative impact on the building performance. Typically sensors need calibrating every few years, depending on the manufacturer, sensor quality, sensor type, and how they are used. The types of sensor technologies currently used today for the collection of common building system operations must generally be checked every 1 to 3 years.

In recent years, modern sensors have become more “plug-and-play” solutions that require minimal setup to use. Information that was previously required from a manufacturer, such as calibration coefficients to ensure the correct interpretation of a sensor output signal, are now integrated into the sensors themselves. This makes for a more user-friendly set of sensors that are easier to install and used by a wide variety of end users and technical skill levels. However, it also reduces awareness of the needs for calibration and the limitations of these sensors by end users, who may or may not be aware of calibration methods. Therefore, smart hardware with improved accuracy and reliability for longer periods of time is of interest to the building community who continuously rely on the values of the sensor outputs. The accurate calibration of sensors is also of interest to the manufacturers of the sensors themselves to ensure a level of accuracy for their sensors at a low cost.

In an effort to improve sensor reliability and reduce frequent maintenance needs, new sensors are being developed that self-calibrate. While complete self-calibration is impossible, a strong reduction in the amount of calibration effort needed is now possible, enabling longer time periods between manual calibration and/or replacement of sensors. This can be accomplished in a number of ways, most of which involve the use of multiple sensors. Methods include co-integration of other sensors that enable self-calibration through methods such as cross-sensitivity compensation, differential sensing, and background calibration, and co-integration of actuators to generate a calibration signal. Cross-sensitivity compensation uses two sensors, including a main sensor that is sensitive to the property being measured and also an interfering quantity from the second sensor. Assuming the sensitivity of the sensor to the interference is consistent over time, this can be used to correct the value of the property being measured by the main sensor. Differential sensing includes two identical sensors that measure the same value but with the opposite signs, where the error can be compensated for by taking the difference between the outputs. Background calibration involves the comparison of the output values of two different types of sensors that measure the same parameter, enabling compensation for their differences. The final method uses the integration of actuators that generate a calibration signal that is used to compare the sensor's output to a reference output, which then is used to correct the sensor's output.

CONTROL SYSTEMS AND BUILDING OCCUPANTS

Human perceptions of thermal and visual comfort vary considerably among individuals. For this reason, a good design principle to keep in mind is that building occupants should be given as much control over their thermal and visual environments as possible. In fact, some definitions for smart buildings emphasize that 'smartness' is achieved through the balancing of the ability of the occupants to control their environment and the ability of the building systems to manage their energy consumption efficiently (Buckman et al. 2014). Room thermostats, operable windows, dimming switches, and adjustable blinds are all means of giving people this control. Intelligently integrating these manually operable controls into the BAS can contribute significantly toward optimizing both IEQ and energy efficiency.

Among the six driving factors of building performance, occupant behavior plays a key role (Hong et. al. 2016, Yan et. al. 2015). This is because of the following:

- Significant interactions between occupants and building systems. People spend more than 90% of their time in buildings (Robinson and Godbey 2010). The occupants' expectation of comfort and satisfaction in the built environment drives the occupant to perform various controls, such as adjusting the thermostat in spaces, opening windows or adjusting vents for ventilation, turning on lights and adjusting the window blinds for visual comfort compatible with the tasks being performed, and consuming domestic water.

- Strong coupling between occupant behavior and building performance. Various occupant behaviors have different impacts on the built environment (e.g. indoor temperature, humidity level, lighting, CO_2, etc.) and energy and water consumption. Building performance also has economic, physiological and psychological impacts on occupancy expectations.

Occupant behavior is also a key factor in evaluating various technologies in building design and retrofit. Many case studies demonstrate that occupant behavior has great influence on the adaptability and implementation of a building technology. Correctly understanding the interactions between occupant behavior and building technologies is critical to improve building design and operations. Different occupant behavior needs different technical solutions, and different technical solutions may affect or change occupant behavior in buildings. The efficiency level of building technologies is important, and equally important is the effective interaction between occupants and the building equipment and systems to meet occupants' comfort and health needs.

Most current building control systems use static and homogeneous human profiles, ignoring diverse and dynamic human comfort needs and thus leading to energy waste and human dissatisfaction in buildings. Smart buildings challenge this convention through the collection of operations data and transfer of information and control to building occupants. This occurs by way of smart devices, operations dashboards, and occupant-comfort-centric controls, which have opened a pathway for improved occupant satisfaction and building operational efficiency.

One such smart control strategy is the integrated-building-human-cyber systems technology (Figure 13-1), which integrates wireless sensing, data collection, analytics, machine learning of human behavior, and comfort feedback to optimally manage energy end uses in buildings.

CONTROL SYSTEMS AND ENERGY AND WATER EFFICIENCY

Control systems are at the core of building performance. When they work well, the indoor environment promotes productivity with the appropriate lighting, comfort, and ventilation people need to carry out their tasks effectively and efficiently. When they break down, the results are higher utility bills, loss of productivity, and occupant discomfort. In modern buildings, control systems operate lights, chilled- or hot-water plants, ventilation, space temperature and humidity control, plumbing systems, electrical systems, life-safety systems, and other building systems. These control systems can assist in conserving resources through the scheduling, staging, modulation, and optimization of equipment to meet the needs of the occupants and

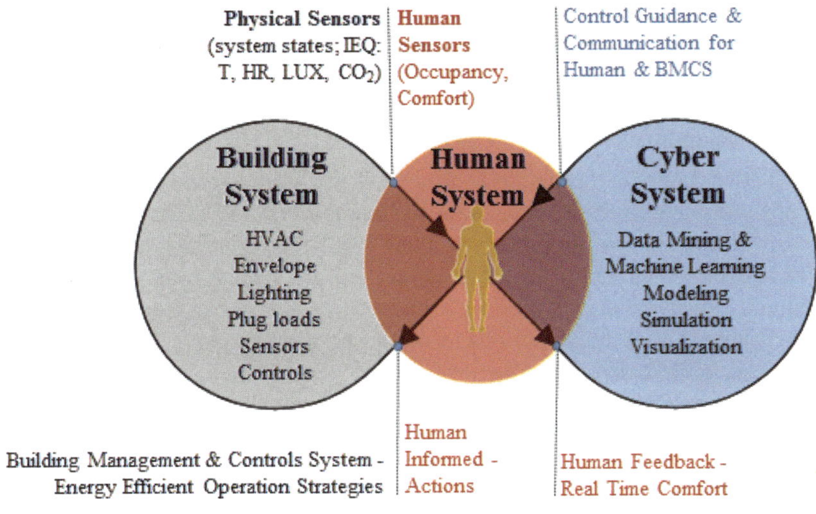

Source: LBNL (Tianzhen Hong)

Figure 13-1 Integrated building-human-cyber systems.

systems that they are designed to serve. This chapter contains a number of recommendations on how building and system controls can be used to support a high-performance, sustainable design.

The potential for energy and water savings through proper design, installation and operation of building automation systems is significant. Controls can save energy through a variety of methods, including the following, as discussed in ANSI/ASHRAE/IESNA Standards 90.1-2016 and/or ANSI/ASHRAE/USGBC/IES 189.1-2017:

- Reduction of equipment runtime. Examples include scheduled control of lighting and air-conditioning systems inside buildings and photoelectric controls for site (exterior) lighting.
- Daylight harvesting control with dimming or stepped lighting in spaces with access to natural light.
- Static pressure reset of fans (and/or pumps) with variable-frequency drives (VFDs).
- Resetting temperature set points for chilled and condenser water supply, hot-water temperature, and/or coil discharge air temperature.
- Variable-speed control of pumps, fans, and compressors. This is another common method employed for cooling systems.
- Demand-controlled ventilation. Modulation of system and zone ventilation airflow rates based on occupancy or occupancy level.

Control systems can deliver building water efficiency in three main areas: landscape irrigation, leak detection, and cooling tower make-up. Additionally, controls can be used to regulate and monitor on-site wastewater treatment plants where those systems are in place. Chapter 12 of this guide provides additional details on efficient and sustainable water system design, operation and maintenance.

Integrating landscape irrigation controls into the smart controls system provides a number of advantages. Among these are the ability to adjust schedules and set points from a single location, the ability to perform remote diagnostics, and the ability to track system performance and water use. Integrating water meters enables continuous measurement of water consumption. Water consumption data can be analyzed during unoccupied periods to determine whether leaks are present in the water distribution system. The judicious placement of submeters can allow building maintenance staff to find the system or location in which the leak or leaks are present. Continuous water meter data can also be used to identify processes and areas of high water use and guide postconstruction water conservation efforts.

Smart irrigation controllers using embedded sensors in the ground can make a significant contribution to reducing the annual water use of a building. (Note that this is a requirement for landscaping irrigation in ANSI/ASHRAE/USGBC/IES Standard 189.1-2017, *Standard for the Design of High-Performance Green Buildings Except Low-Rise Residential Buildings* [ASHRAE 2017a].) The simplest irrigation controls are based on time clocks that open valves for a set duration, for a set number of periods per week. Unless the duration and frequency of watering is adjusted throughout the irrigation season, the use of time clock controllers often results in excessive water use. To automatically address seasonally varying irrigation requirements, some time clock controllers allow for 365 day programming.

An improvement on the time clock controller is to add moisture sensors, which can enable the system to bypass a watering period (if ground moisture levels are above a set point). Still more sophisticated controllers can gather data about local weather conditions—either directly via sensors, indirectly via a remote weather station, or by direct input from a remote weather station—and use that data to adjust the amount of water delivered to the landscape.

Chapter 12 covers HVAC water use reduction in detail. Installing controls to continuously monitor condenser water quality in the interest of reducing the frequency of chemical treatment and cooling tower blowdown can reduce potable water consumption. Independently submetering cooling tower make-up water and blowdown should also be considered as a means to reduce sanitary/sewer bills.

Finally, some advanced green buildings feature on-site water and wastewater treatment plants. Such plants generally include pumps, fans, and sensors to monitor characteristics of the treated water (e.g., dissolved oxygen levels and total suspended solids). Depending on the goals and complexity of the water treatment plant, an industrial supervisory control and data acquisition system may be necessary to ensure maximization of throughput, efficient energy use, and code compliance.

CONTROL SYSTEMS AND INDOOR ENVIRONMENTAL QUALITY (IEQ)

Factors regulated by building control systems that impact IEQ include operative temperature, relative humidity, outdoor airflow rates, and light levels. The first two factors are addressed in ANSI/ASHRAE Standard 55, *Thermal Environmental Conditions for Human Occupancy* (ASHRAE 2017b), the third factor is addressed in ANSI/ASHRAE Standard 62.1, *Ventilation for Acceptable Indoor Air Quality* (ASHRAE 2016), and the fourth factor is addressed in Chapter 10 of the IES Lighting Handbook (IES 2011). Chapter 8 (Indoor Environmental Quality) of this guide delves into the design and implementation of each of these factors.

Thermal Comfort

Standard 55 specifies, based on occupant activity level, clothing level, air temperature, radiant temperature, air velocity and relative humidity. One of the most common controls to all types of buildings is the thermostat, which adjusts the dry-bulb temperature of a space, therefore adjusting this comfort condition. Thermostats modulate the HVAC system operation to meet the desired temperature within a generally narrow deadband range above or below this set point. While they are less common than thermostats and temperature sensors, stand-alone humidistats and humidity sensors can be used to control room humidity levels. Standard 55 specifies a maximum humidity ratio of 0.012. The standard does not specify a minimum humidity ratio for thermal comfort (although it did in earlier versions) but nonthermal comfort factors (e.g., skin drying) may be used to establish such a minimum.

Air Quality and Ventilation

Procedures for determining minimum outdoor airflow rates are described in Standard 62.1. Building controls play a critical role in ensuring that minimum out door airflow rates are achieved in three areas: VAV system control, mixed-mode ventilation systems, and dynamic reset of outdoor air intake flows.

It is difficult to ensure minimum outdoor airflow rates are met by VAV systems over the entire range of operating conditions in the absence of controls designed, installed, and maintained specifically for that purpose. One means of achieving such control is to measure the supply airflow rate and CO_2 concentrations in the occupied zones, and adjust the outdoor air intake on a continuous basis based on these input values. Another potential method is measuring the outdoor air intake flow. Control systems based on CO_2 levels, as well as those that measure outdoor air intake flow, all have their potential errors and design difficulties. In addition, like any measurement device, maintenance such as calibration must be done as recommended by the manufacturer to ensure accuracy of that information. The BAS can determine if the outdoor airflow rate is sufficient based on this information and can adjust the mixed-air dampers accordingly. The preferred method of controlling the outdoor air intake flow rate is to control the pressure in the static pressure in the mixing plenum. This will indirectly control the needed variations in outdoor airflow. If airflows are also measured at each VAV box, this control routine can be further improved upon. As described in Standard 62.1, zone air distribution effectiveness can change when the temperature of the supply air changes. Therefore, it may be necessary for the control system to reset the minimum outdoor airflow after a seasonal switchover of supply air temperature.

Mixed-mode ventilation refers to the combination of mechanical ventilation and operable windows that provide natural ventilation. During some times of the year, if the windows are sized, located, and operated (by automatic control) properly, it is not necessary to provide mechanical ventilation when the windows are open. Perhaps the most straightforward way to control mixed-mode ventilation systems is to use the output of a window switch to shut down the terminal unit (e.g., the VAV box) when the window is open. This is most applicable for single-occupant zones or areas with a relatively small number of occupants. When the occupants decide that it is preferable to shut the window, the mechanical HVAC system is brought back online. It may be necessary to install an alarm or override based on space temperature, wind annoyance, wet weather, or to provide freeze protection.

Lighting Levels

The BAS can maintain desired light levels by either adjusting electric light output or controlling the amount of daylight entering the building. One or more photocells may be used to measure the light level in the occupied space and the output used to brighten or dim electric lights. Photocell output may be used to switch between multiple light levels, raise or lower window blinds, or adjust louvers to keep daylighting at comfortable levels and eliminate glare. Equations that calculate the sun position angle for various wall or roof orientations can be input into the BAS and used to vary the tilt angle of window blinds, to minimize glare and maximize daylighting. Chapter 11, Lighting Systems, provides insight into designing and operating lighting and lighting controls for efficiency and visual comfort.

CONTROL SYSTEMS COMMISSIONING

Improper control systems operation can lead to 15% to 30% excess energy consumption in commercial buildings (Katipamula and Brambley 2005a, 2005b). Defined in Chapter 4, the commissioning process serves as a first and ongoing means of ensuring that building systems operate as intended. Coupled with AFDD, an ongoing commissioning program provides early fault detection, diagnosis, and correction. The commissioning process is defined in ASHRAE Standard 202-2013 as follows:

> *A quality focused process for enhancing the delivery of a project. The process focuses upon verifying and documenting that the facility and all of its systems and assemblies are planned, designed, installed, tested, operated, and maintained to meet the Owner's Project Requirements.*

The full process from project planning through occupancy and operations is explained in ASHRAE Standard 202-2013 and in Chapter 43 of the *ASHRAE Handbook—HVAC Applications* (ASHRAE 2015a). Additional information relative to design reviews and commissioning for designers and commissioning providers can be found on the Energy Design Resources website (www.energydesignresources.com/publication/gd/).

This section covers salient elements of applying the commissioning process to controls in new construction and will focus on what the commissioning provider can do during the design phase to facilitate a successful commissioning program. For commissioning of existing buildings, the reader is referred to sources of information, such as U.S. Green Building Council's (USGBC) LEED® program for Existing Buildings (see "References and Resources" section at the end of the chapter for website).

Conduct and Participate in Design Reviews

During the design reviews, the commissioning provider should see that the control-related elements are included in the contract documents. The section called "Detailing the Design of a Control System" provides a list of items that should appear in the contract documents that should be reviewed by the commissioning provider.

Specify a Systems Manual

Ensure that the commissioning scope for the commissioning provider, contractor, and designer includes a systems manual that, among other things, includes the following:

- Narratives explaining all energy-efficiency features and strategies
- Set point and parameter table that indicates the impacts of changing the values
- Recalibration and retesting frequency table
- Suggested smart alerts in the control system to send alerts on malfunctioning sensors and actuators
- List of standard trend logs to view to verify proper performance

The building operators also need to be trained on the systems manual and its contents in order to properly operate and maintain the system.

INTERACTION OF A SMART BUILDING WITH THE SMART GRID

Electric utilities have been developing the tools and procedures for smart grid applications; however, there is a growing need for developing hardware and software that will allow smart buildings to interact smoothly with the coming smart grid. Prior research has demonstrated that information gathering, modeling, and more intelligent controls can help reduce energy consumption within individual buildings and the electrical grid, but the coming smart grid will require a new and more complex integration of information gathering, decision, and control. Technology innovations are leading to the potential for a number of items interacting with a smart grid that may or may not be controlled through the normal building automation systems.

The Open Automated Demand Response (OpenADR) Standard will help pave the way for building energy management and automation systems to communicate with the electric utility provider and implement demand response measures. The development of the OpenADR Standard during the past decade was the result of national research labs and companies, with Lawrence Berkeley National Laboratory (LBNL) providing a significant, early role. The OpenADR Alliance was formed to support the development and deployment of commercial demand response through adoption of the OpenADR Standard. We are just now beginning to see products enter the market that will allow demand-response programs to be implemented in commercial buildings, and this area will continue to grow in importance in society and with green buildings in particular. More information on OpenADR can be found in the OpenADR primer (OpenADR Alliance).

The way smart buildings interact with the Smart Grid will see even more rapid changes in the coming years as technology advancements and distributed energy resources will modify traditional demand response value streams. The result will be the need for a multitude of demand response strategies to take form. LBNL has identified these strategies as shape, shift, shed, and shimmy. These strategies are beginning to take form by the integration of demand response into wholesale markets where it can be dispatched consistent with locational marginal prices, the use of demand response aggregators by utility programs to deliver tailored options that work for customers with unique needs, utility tariff restructuring, and offering incentives based on relative locational value (LBNL 2015).

REFERENCES AND RESOURCES

Published

ASHRAE. 2013. ANSI/ASHRAE Standard 202, *Commissioning Process for Buildings and Systems*. Atlanta: ASHRAE.

ASHRAE. 2015a. *ASHRAE Handbook—HVAC Applications*. Atlanta: ASHRAE.

ASHRAE. 2015b. ASHRAE Guideline 13-2015, *Specifying Direct Digital Control Systems*. Atlanta: ASHRAE.

ASHRAE. 2015c. ASHRAE Guideline 14-2015, *Measurement of Energy and Demand Savings*. Atlanta: ASHRAE.

ASHRAE. 2016. ANSI/ASHRAE Standard 62.1-2016 *Ventilation for Acceptable Indoor Air Quality*. Atlanta: ASHRAE.

ASHRAE. 2016. ANSI/ASHRAE/IES Standard 90.1-2016, *Energy Standard for Buildings Except Low-Rise Residential Buildings*. Atlanta: ASHRAE.

ASHRAE. 2017a. ANSI/ASHRAE/USGBC/IES Standard 189.1-2017, *Standard for the Design of High-Performance Green Buildings Except Low-Rise Residential Buildings*. Atlanta: ASHRAE. (Will be incorporated into the 2018 version of the *International Green Construction Code* [IgCC].)

ASHRAE. 2017b. ANSI/ASHRAE Standard 55-2017, *Thermal Environmental Conditions for Human Occupancy*. Atlanta: ASHRAE.

Barwig, F.E., J.M. House, C.J. Klaassen, M.M. Ardehali, and T.F. Smith. n.d. The National Building Controls Information Program. Washington, D.C.: American Council for an Energy-Efficient Economy. http://aceee.org/files/proceedings/2002/data/papers/SS02_Panel3_Paper01.pdf.

Buckman, A.H., M. Mayfield, and S.B.M. Beck. 2014. What is a Smart Building?, *Smart Sustain. Built Environ*. 3(2):92–109.

CSI. 2016. MasterFormat Specifications. Alexandria, VA: Construction Specifications Institute. www.csiresources.org/practice/standards/masterformat.

DOE. 2005. Energy impact of commercial building controls and performance diagnostics: Market characterization, energy impact of building faults and energy savings potential. Washington, D.C.: U.S. Department of Energy.

Hong T., S. Taylor-Lange, S. D'Oca, D. Yan, S. Corgnati. 2016. Advances in Research and Applications of Energy-Related Occupant Behavior in Buildings. *Energy and Buildings* 116: 694–702.

IES. 2011. *Lighting Handbook*. New York: Illuminating Engineering Society of North America.

Robinson, J., and G. Godbey. 2010. Time for life: The surprising ways Americans use their time. University Park, PA. The Pennsylvania State University Press.

USGBC. 2009a. *LEED® 2009 Green Building Operations and Maintenance*. Washington, D.C.: U.S. Green Building Council.

USGBC. 2009b. *LEED® 2009 for New Construction and Major Renovations*. Washington, D.C.: U.S. Green Building Council.

Yan D., L. O'Brien, T. Hong, X. Feng, B. Gunay, F. Tahmasebi, A. Madhavi. 2015. Occupant behavior modeling for building performance simulation: current state and future challenges. *Energy and Buildings* 107:264–78.

Online

DDC Online
www.ddc-online.org.

Energy Design Resources
www.energydesignresources.com/publication/gd/.

International Performance Measurement and Verification Protocol
www.evo-world.org.

LBNL. 2015. California Demand Response Potential Study: Charting California's Demand Response Future.
www.cpuc.ca.gov/General.aspx?id=10622.

OpenADR Alliance
www.openadr.org.

OpenADR Primer
www.openadr.org/assets/docs/openadr_primer.pdf.

U.S. Green Building Council
www.usgbc.org.

CHAPTER FOURTEEN

Completing Design and Documentation for Construction

DRAWINGS/DOCUMENTATION STAGE

Once the project has reached the working drawing/construction document stage, the green design concepts and resulting configurations should be well set, and the task of incorporating them into the documents that contractors will use to build the project should be relatively routine. However, quality control at this stage is especially important in green design projects.

Many firms have a routine procedure to review the documents for quality before they are released to contractors (to ensure that concepts are adequately depicted and described and to catch errors and omissions). This process should also include a green design concept review, preferably by one or more design team members that were in on the early stages of the project. This is particularly true if those preparing the construction documents were not part of that process. This is not the time to allow an excellent green design to become diluted or slip away.

SPECIFYING MATERIALS/EQUIPMENT

Green-Building Materials

Sources for guidance on selecting and specifying materials for a green project are as follows:

- Athena Institute, www.athenasmi.ca
- BuildingGreen, www.buildinggreen.com

Also see the Digging Deeper sidebar titled "One Design Firm's Materials Specification Checklist."

Controlling Construction Quality

It is far easier to control construction quality in the design and specification stage of a project than during its construction. During preparation of the final construction drawings and specifications, it is critical to be diligent in spelling out the quality expected in the field to carry through the green design concepts developed throughout the early design stages. Some further thoughts on this subject, many applicable to the design phase, are covered in Chapter 17 of this guide.

COST ESTIMATING AND BUDGET RECONCILIATION

It is very important to have the cost estimator involved right from the start of the project to ensure that the project budget reflects the decisions made by the rest of the project team throughout the integrated design process.

The chapters, case studies, resources, and cost data included in the latest edition of the RSMeans' publication *Green Building: Project Planning & Cost Estimating,* third edition (RSMeans 2010) are good resources that include information on the following:

- Green -building approaches, materials, systems, and standards
- Evaluating the cost versus value of green products over their life cycle
- Specifying green-building projects—complete with a list of often-specified products/materials and a sample spec
- Low-cost green strategies—and special economic incentives and funding
- Deconstruction—featured in a new chapter on this key element of sustainable building

This publication has been completely updated with the latest in green-building technologies, design concepts, standards, and costs.

BIDDING

The following is an excerpt from the Harvard University Office for Sustainability, Green Building Resource, Design Phase Guide website (www.energyandfacilities.harvard.edu/green-building-resource):

> *"The expectations of general and subcontractors on Leadership in Energy and Environmental Design (LEED®) and other green projects are different than those associated with conventional projects. Practices such as recycling, erosion, and sedimentation control, indoor air quality (IAQ), and filling out LEED (and other rating system) submittals require special attention, and should be communicated to the team early in the construction process."*

Include sustainability language in the Request for Proposals and Owner's Project Requirements document. Consider green-building expertise and LEED (or other green rating system) project experience as key criteria in the selection process. Ask to see evidence of experience with LEED, material tracking, construction and demolition waste management, and plans for IAQ during construction."

The contractor should be required to have a LEED coordinator (if the project is going to LEED) and a commissioning coordinator. Having experienced people involved in these tasks is critical to success. Also, the contractor should report on regional materials at a minimum, but reuse and/or rapidly renewable materials as well.

According to the same website, the contractor is responsible for conveying the following to subcontractors: workplace practice expectations for recycling and erosion and sedimentation control, and construction IAQ practices. The Owner's Sustainability Representative (OSR) is responsible for coordinating regular LEED meetings to track the LEED documentation process and to collect submittal and audit requirements from contractors. Also, a submittal review process should be established among the commissioning authority, the architect, and OSR.

MANAGING RISK

Green Design Documentation Issues

Reviewing green design work done by others allows engineers to see how built projects have created new opportunities and can guide efforts to sell and provide green engineering services. Performing sustainable design can yield important benefits to design engineers individually and to their firms. Improved client service, more repeat work, improved market position, enhanced public relations, and better employee satisfaction and retention are among the many benefits that numerous architectural firms and several pioneering engineering firms have derived from informing their practices with sustainable design expertise.

On the other hand, failure to address client concerns about green design can harm reputation. Disproportionate start-up costs, risk, or other perceived obstacles (e.g., the educational investment and commitment required to change engineering thinking and culture) may be perceived as being associated with undertaking sustainable design. Managing each of these issues, communicating frankly, and crafting creative solutions, as required, can reduce exposure to these potentially negative issues.

However, when documenting the project requirements for bidding by contractors and subcontractors, it is important that the engineer not guarantee that the results predicted by building simulation modeling will actually be achieved. It is important that the engineer not take a disproportionate share of the risk in incorporating newer technologies into a building design project. Sustaining fair and practical business practices is critical to being a successful green engineer.

Contract Provisions

The following is an excerpt from the online essay, "Green Buildings and Risk," by Tim Corbett (www.greenbiz.com/blog/2007/11/26/managing-risk-green-building-projects):

"A good contract is the best line of defense when it comes to mitigating your risk. The contract is an excellent method for defining your scope of services (e.g., what will be provided, when, and what will not). Contracts are also an excellent method for qualifying clients and managing and establishing expectations. Contracts should address the following:

- *New and innovative products and technology may be used; they may lack proven history of successful application. Owner understands and agrees that project objectives may not be realized. A caveat is that the decision to try new, unproved products or technologies needs to be a decision by the owner and the design team together to make sure the owner is informed of the risks and opportunities.*
- *Ordinary skill and care will be used to achieve project objectives; however there is no warranty or guarantee the project will achieve LEED certification.*
- *Verify the level of investigation and analysis that will be performed for new material and technologies, with no expressed or implied warranty or guarantees of results.*
- *Client agrees to measure the potential risks related to incorporating the innovation product and/or systems and accepts the risks.*
- *Limit your exposure to consequential damages by including appropriate language in your contract."*

PROJECT DELIVERY METHODS AND CONTRACTOR SELECTION

Successful projects depend upon the entire team of players involved: architect, engineers, program managers, construction managers, owner's representatives, facilities personnel, building users, and contractors. It is assumed that all parties will be ethical, reliable, diligent, and experienced. There are a number of project delivery methods that could be used to deliver the design and construction of a project. The major methods, to be briefly discussed here, are the construction manager approach, the design/bid/build approach (D/B/B), the design/build (D/B) approach, and the public-private partnership (P3) approach. It is important that the project delivery method be chosen early in the project for the same reasons that it is important when considering green design options. The discussion of the three methods below is intended only to relate to the effect this decision will have on the success of the optimized design and operation of the building.

Construction Manager

The construction manager method is the process undertaken by public and private owners in which a firm with extensive experience in construction management and general contracting is hired during the design phase of the project to assess project capital costs and constructibility issues. This is especially important when considering design alternatives in an effort to deliver a high-performance building to the client. The initial design process often includes a project definition stage, or programming, in which the owner works with the design professionals, the CxA, and the construction manager to define the specific scope of the project. The design professional uses this information to prepare a set of bidding documents that the construction man-ager uses to obtain bids from qualified subcontractors. The lowest responsible price is usually selected and the contractor then constructs the project.

- *Advantages:* Budget control, buy-in of green concepts by contractors.
- *Disadvantages:* Perceived lack of competitive general contractor and subcontractor pricing, innovative systems could be shelved due to overly conservative first-cost estimates.

D/B/B Method

D/B/B construction is the traditional process undertaken by public and private owners. The initial design process often includes a project definition stage, or programming, in which the owner works with the design professionals to define the specific scope of the project. The design professional uses this information to prepare a set of bidding documents. The bid documents are then available for qualified contractors and subcontractors to prepare pricing. The lowest responsible price is usually selected and the contractor then constructs the project.

- *Advantage:* Usually results in the lowest first cost at the outset of the construction of a project.
- *Disadvantages:* No contractor buy-in to green process and concepts, prequalification of contractors is difficult to do well.

D/B Method

D/B construction is typically a response to a request for proposal (RFP) developed by an owner. The RFP is usually a document that defines the general scope of the project and then solicits price proposals to accomplish this work. The work effort to prepare the specific design of the project is to be included in the D/B offering. The D/B team usually consists of an architect, engineers for the various disciplines involved, a general contractor, and the trade subcontractors. This entire team should be in place until the project is turned over to the owner.

As the D/B team develops the design, it must respond to the premises defined in the original scope of work.

- *Advantages:* Can result in lowest first cost, agreement by design/construction team with regard to design and building operation concepts.
- *Disadvantages:* Can result in uneven distribution of risk among team members, can result in loss of design team members as owner-advisors.

Public-Private Partnerships (P3)

The use of P3 is well established in Canada, and is beginning to gain favor in the United States and other countries. P3 are playing a bigger role in building and capital projects across all areas of government, including power generation, energy delivery, water and wastewater facilities, waste disposal, transportation, communications, education and health facilities, and public service buildings. Recently, its popularity has led to P3 being used more frequently in smaller-scale developments across Canada, such as schools, courthouses, and hospitals.

Bridging documents are provided to guide an integrated and facilitated P3 process by

- providing narratives that describe, in conceptual terms, the requirements for the new facility;
- outlining general performance and prescriptive requirements, operations requirements, programmatic space requirements and room data sheets, space standards, adjacency requirements, conceptual blocking and stacking;
- defining business terms; and
- proposing criteria for evaluation of submitted proposals.

The P3 process allows market forces to play out, with the opportunity for the government entity to decide later on funding options. Projects are delivered quickly. Competition on a wide variety of services is balanced by an integrated approach, while risk transfer from the public to private sector in the areas of pricing, compliance, and operations benefits the stakeholders of the government entity.

Information from this section was taken from website resources listed in the references section of this chapter, specifically those by Pierce (2003) and the California Debt and Investment Advisory Commission (2008).

Factors in Choosing an Approach

In all of the previously described scenarios, the team or contractors should be prequalified to perform the work prior to a request for pricing. It's important to confirm that the team or contractors fulfill the following requirements:

- Experience in similar work
- Record of past performance by responsible references
- Financial capability to complete the project
- Experienced staff available to work on the project

A record of exemplary performance and fiscal capability may be more important than experience in similar work. In any case, it is valuable to preselect the teams or contractors from whom you request pricing. Successful projects occur due to careful planning and implementation, including the following actions:

- Specify whether a shortage or unavailability of green materials must be treated as a compensable or noncompensable delay under the contract documents.
- A monthly progress report should be submitted to indicate project progress along with the required documentation such as photographs, periodic waste records, and periodic material purchase records.

REFERENCES AND RESOURCES

Published

Corbett, T. *Green Buildings and Risk.* http://www.greenbiz.com/blog/2007/11/26/managing-risk-green-building-projects.

EPA. Toxicity Characteristic Leaching Procedure (TCLP) Test, 40 CFR Part 260. Washington, D.C.: U.S. Environmental Protection Agency.

EPA. Universal Waste Rule, 40 CFR Part 260. Washington, D.C.: U.S. Environmental Protection Agency.

RSMeans. 2010. *Green building: Project planning and cost estimating*, third ed. Hoboken, NJ: John Wiley & Sons.

Online

Athena Institute
www.athenasmi.ca.

BuildingGreen
www.buildinggreen.com.

California Debt and Investment Advisory Commission. 2008. *Public-private partnerships: A guide to selecting a private partner*. CDIAC #08-02.
www.treasurer.ca.gov/cdiac/publications/p3.pdf.

Harvard University Office for Sustainability, Green Building Resource, Design Phase Guide
www.energyandfacilities.harvard.edu/green-building-resource.

Pierce, D.G. 2003. Public-private partnerships: Demystifying the process. Mondaq: http://www.mondaq.com/canada/article.asp?articleid=23737.

Digging Deeper

ONE DESIGN FIRM'S MATERIALS SPECIFICATION CHECKLIST

Choose at least one portion of materials/products that have the following characteristics:

- Local and/or indigenous, reducing the environmental impacts resulting from transportation and supporting the local economy
- Extracted, harvested, recovered, manufactured regionally within a radius of 500 miles (805 km)
- Low-embodied energy
- Reused, recycled, and/or recyclable, reducing the impacts resulting from extraction of new resources
- Salvaged building materials (e.g., lumber, millwork, plumbing fixtures, hardware, etc.)
- Postconsumer recycled content material and/or postindustrial recycled content (recovered) material
- Nonhazardous to recycle, compost, or dispose of
- Renewable and sustainably harvested (no old-growth timber), preferably with minimal associated environmental burdens (reducing the use and depletion of finite raw and long-cycle renewable materials)
- Rapidly renewable building materials, including any nonwood materials that are typically harvested within a 10 yr or shorter cycle
- Nontoxic/nonpolluting in manufacture, use, and disposal
- Use finished materials, products, and furnishings that are free of known, probable, and suspected carcinogens, mutagens, teratogens, persistent toxic organic pollutants, and toxic heavy metals pursuant to the U.S. Environmental Protection Agency's (EPA) Toxicity Characteristic Leaching Procedure Test, 40 CFR Part 260. (Based on the EPA's Universal Waste Rule and Part 260, building owners and their contractors must use the toxicity characteristic leaching procedure to determine if they are generators of hazardous waste and subject to EPA and state hazardous waste regulations when disposing of mercury lamps and other waste products.)
- Specify lead-free solder for copper water supply tubing; do not use plastic for supply water.
- Avoid plastic foam insulation.
- Specify natural, nontoxic, low-volatility organic compound–emitting, nonsolvent-based finishes, paints, stains, and adhesives.

- When specifying painting, remember the following:
  - Consider surfaces that don't require painting.
  - Choose paints that have been independently certified (e.g., Green Seal).
- Choose latex over oil-based paint.
- Increase direct-to-outdoors ventilation when painting.
- Dispose of oil-based paints like hazardous waste.
- Recycle latex paint or save for touch-ups.

Avoid the use of materials containing or produced with ozone-depleting chlorofluorocarbons (CFCs) or hydrochlorofluorocarbons (HCFCs) (often embedded in foam insulation and refrigeration/cooling systems).

Avoid CFC-based refrigerants in new base-building HVAC systems. When reusing existing base-building HVAC equipment, complete a comprehensive CFC phaseout conversion.

Commit to working with manufacturing teams who provide performance/service contracts of integrated systems for service delivery, product longevity, adaptability, and/or recycling.

Favor suppliers who will minimize packaging and will take back excess packaging (e.g., pallets, crates, cardboard, and excess building materials).

Maximize the use of materials that retain a high value in future life cycles

Specify products or systems that extend manufacturer responsibility through a lease or take-back program that ensures future reuse or recycling into a product of similar or higher value.

To guide material selection choices during the design process, educate designers and elicit life-cycle information from manufacturers to encourage selection of building materials and furnishings with favorable life-cycle performance. Select materials from manufacturers that are committed to improving their overall environmental performance at their manufacturing facilities and their suppliers' facilities.

CHAPTER FIFTEEN

CONSTRUCTION

To minimize gaps between design intent and what is actually built, the green-conscious engineer should provide a level of construction administration that is beyond the norm. That engineer (or an independent agent) may also provide mechanical/electrical system commissioning. (See Chapter 4.)

SITE PLANNING AND DEVELOPMENT

Green-building design includes the design of, and implementation of site accommodations to accomplish pollution reduction and water and energy conservation. While not generally the role of a mechanical or electrical engineer or a typical ASHRAE member regarding site-specific development, it is important that as a project team member, understanding and participation with the decision process is important.

THE ENGINEER'S ROLE IN CONSTRUCTION QUALITY

In construction, the engineer's key roles encompass shop drawing review, review of equipment substitutions, handling of change order and value engineering requests, and site visits and inspections. Codes and jurisdictional reporting measures are evolving and, in addition to Leadership in Energy and Environmental Design (LEED®) or other green design certification processes, the engineer may need to document the compliance with specific measures after the design or maybe even before construction is complete.

Shop Drawing Review

There should be a thorough review of shop drawing submittals. The specifications should require their timely submission from the contractor to allow time for proper review without delaying the project. The purpose of the review is to ensure that the contractor is correctly interpreting the specifications and drawings and is not missing any important details. Areas for green-building emphasis include checking motor power and efficiency ratings, checking air filter details, checking for proper clearances around equipment for servicing, and checking anything else relevant on the project concerning green design elements. Thorough review of all control sequences submitted by the building automation system subcontractor is critical, as sophisticated energy operational sequences used for energy performance can be missed.

Value Engineering

Review value engineering (VE) suggestions should be carefully studied for their impact on the project's green design goals. (VE is often offered under the assumption that first-cost savings are paramount to the owner and project team.)

The impact on life-cycle costs and predicted building operation and performance must also be clearly understood, as, in some cases, VE modifications not only adversely impact energy performance, but can impact code acceptance of previously approved design.

Site Visits/Observations

Work with the commissioning authority to review their site observations and address any design-based questions that arise during the commissioning process. In some jurisdictions, the requirement falls on the engineer to certify that intent conveyed in design is implemented in the installation.

CONSTRUCTION PRACTICES AND METHODS

In the design phase, the specifications should prescribe that certain construction methods and procedures must be followed to ensure a fully realized green project. This would include topics such as reduced site disturbance, handling of construction waste, control of rainwater runoff, and indoor air quality (IAQ) management during construction. This section includes a brief discussion of several items that are key to the construction and delivery of a green building.

Indoor Air Quality During Construction

It is recommended that the engineer and architect work with the construction manager or general contractor and associated subcontractors to develop an IAQ construction management plan that the contractor would be required to follow. The Sheet Metal and Air Conditioning National Association's *IAQ Guideline for Occupied Buildings Under Construction* (SMACNA) is cited by U.S. Green Building Council (USGBC) as a source document. The contractor should particularly protect installed or stored absorptive materials (e.g., insulation or gypsum wallboard) from water damage or other contamination. Water damage is especially insidious if materials get wet, are installed wet, and are then covered up. If air-handling units (AHUs) run during construction, the construction manager and associated subcontractor should be required to protect AHU components and the duct distribution system from dirt and debris, clean any components and ductwork that are damaged, and replace filters before occupancy. (ANSI/ASHRAE Standard 52.2-2012, *Method of Testing General Ventilation Air-Cleaning Devices for Removal Efficiency by Particle Size* [ASHRAE 2012] deals with this subject.) In addition, for renovation projects within an occupied building, isolation of the construction space from the occupied space is essential to avoid exposing occupants to airborne particulates or volatile organic compounds (VOC) associated with material manufacture or installation.

Even if your project is not going to pursue certification in a green-building rating program (such as LEED) or is not subject to a green-building code (such as ANSI/ASHRAE/USGBC/IES Standard 189.1 or the *International Green Construction Code*), it is a good idea to also conduct a building flush out or indoor air quality test after construction is complete and before occupancy begins. Established criteria for indoor air quality construction management practices can be found in Section 10 of Standard 189.1.

The construction team should also consider the issue of whether permanent HVAC systems should be operated during construction. This practice is not allowed in high-performance, green-building situations (such as a project needing to comply with Standard 189.1), and the use of permanent HVAC is not recommended for any green-building project if at all possible.

Construction Waste

One of the more positive side benefits from an increased adoption of green building practices has been a better awareness of the need to minimize waste sent to a landfill from new construction or renovation projects. Specific criteria again are contained in the LEED program descriptions and in Section 10 of Standard 189.1. These criteria include minimum percentages of the total waste being diverted from landfills via recycling or reuse, as well as (in Standard 189.1) a limit on the total waste generated based on the building size. Many waste haulers now will gladly cooperate with a waste minimization program that includes sorting of waste for reuse or recycling, as this can reduce the tipping fees that they will have to pay and can provide another revenue stream from the sale of recyclable materials, such as metals.

The ability to readily recycle or reuse materials from demolition and construction can depend on the location of the building project; some areas have better established recycling programs than others. As an engineer, contractor, or other professional on the project, you can help encourage the development of a waste minimization and recycling program as part of the building project.

COMMISSIONING DURING CONSTRUCTION

Once construction starts, the commissioning process enters a new phase. Depending on the commissioning plan and the activities that were identified in the plan, the commissioning team will be active during construction with tasks such as verification that equipment is being installed properly. It is during construction that the building envelope commissioning is done, and there is a growing recognition of the importance and value of commissioning the building envelope. The chapter on commissioning (Chapter 4) contains a thorough discussion on the commissioning process.

MOVING INTO OCCUPANCY AND OPERATION

Plans for Operation

An increased focus is being placed on the need to have a good operation and maintenance program in place in order for the benefits from energy and water efficiency and other green features of a new building project to continue long after occupancy takes place. Standard 189.1, Section 10 requires that plans for operation be developed as part of the construction documents. Although responsibility for final approval and implementation of these plans obviously is with the owner and the building operations staff, the project design team may be asked to help prepare these plans.

REFERENCES AND RESOURCES

Published

ASHRAE. 2012. ANSI/ASHRAE Standard 52.2-2012, *Method of Testing General Ventilation Air-Cleaning Devices for Removal Efficiency by Particle Size.* Atlanta: ASHRAE.

ASHRAE. 2017. ANSI/ASHRAE/USGBC/IES Standard 189.1-2017, *Standard for the Design of High-Performance Green Buildings Except Low-Rise Residential Buildings*. Atlanta: ASHRAE. (Will be incorporated into the 2018 version of the *International Green Construction Code* [IgCC].)

EPA. *Storm water management for construction activities*. EPA document No. EPA-8320R-92-005. Washington, D.C.: U.S. Environmental Protection Agency.

SMACNA. 2007. *IAQ Guidelines for Occupied Buildings Under Construction*. Chantilly, VA: Sheet Metal and Air Conditioning Contractors' National Association, Inc.

Online

Construction Management Software
www.softwareadvice.com/construction/project-management-software-comparison.

CONSTRUCTION FACTORS TO CONSIDER IN A GREEN DESIGN

Determine locations for construction vehicle parking, temporary piling of topsoil, and building material storage in order to minimize soil compaction and other site impacts.
Control erosion to reduce negative impacts on water and air quality.
Design a site sediment and erosion control plan that conforms to best management practices specified in the U.S. Environmental Protection Agency's *Storm Water Management for Construction Activities* (EPA) or local erosion and sedimentation control standards and codes, whichever is more stringent.
Conserve existing natural areas and restore damaged areas to provide habitat and promote biodiversity.
Schedule construction carefully to minimize impacts.
Avoid leaving disturbed soil exposed for extended periods.
Fill trenches quickly to minimize damage to severed tree roots.
Avoid building when the ground is saturated and easily damaged.
Estimate the amount of material needed to avoid excess.
Design to accommodate standard lumber and drywall sizes.
Assess the construction site waste stream to determine which materials can be reduced, reused, and recycled.
Conduct a waste audit, quantifying material diversion by weight.
Recycle and/or salvage construction and demolition debris.
Specify materials that minimize waste and reduce shipping impacts through bulk packaging, dry-mix shipping, reused bulk packaging, recycled-content packaging, or elimination of packaging.
Develop and implement a waste management plan, quantifying material diversion by weight.
Research markets in area for salvaged materials.
Establish on-site construction material recycling areas and recycle and/or salvage construction, demolition, and land clearing waste.
Contract with licensed haulers and processors of recyclables.
Require subcontractors to be responsible for their waste (including lunch wastes); create incentives for minimizing waste.
Educate employees and subcontractors.
Monitor and evaluate waste/recycling program.
Review the IAQ Construction Management Plan to ensure that all subcontractors understand the process and goals desired.

CHAPTER SIXTEEN

Operation, Maintenance, and Performance Evaluation

The marketplace has broadly embraced green building and sustainability as a philosophical pathway for the design, construction and operation of buildings which perform "better" than what we have traditionally done. The definition of "better" is still somewhat ambiguous. It can go from some level better than a similar building designed in compliance with ASHRAE Standard 90.1 to a net zero energy building and even more aggressively towards regenerative buildings. The initial wave of this uptake—the bleeding edge, so to speak—was aspirational in feeling and voluntary in adoption. People built green buildings because they wanted to. As the group of first adopters grew, the use of green building certifications and/or sustainability design concepts became more market demand driven as people began to perceive enhanced or increased value for green buildings. Now, green buildings and sustainability initiatives are routinely mandated by local ordinances, corporate organizational imperatives, and national building program requirements. But along with the increasing acceptance, marketplace insistence and compliance pulls, there is an increasing refinement of the definition of a green building and expectations of end users for real performance improvements.

The plans for operation as specified in Standard 189.1 are focused mostly on maintaining equipment and systems that are included in the building design to comply with specific high-performance requirements set by the standard. These plans can, however, be used as guidelines for outlining how a green building should be operated in general. For example, modern control and automation systems provide an opportunity for monitoring the performance of systems as they relate to energy and water consumption, indoor air quality (ventilation rates and perhaps CO_2 levels), and thermal comfort. A well-thought-out and written plan for operation would include the planned response to take when temperature excursions outside the design or set point range occur or if CO_2 levels exceed a predetermined value in a zone. It is also recommended that the building operations team include an evaluation of energy consumption in a more detailed fashion than just the aggregated monthly utility bill. This evaluation should at least include monitoring of the total energy consumption for key systems such as the HVAC or lighting, and preferably would be done on a finer level of detail to include the energy consumption of key equipment, such as a chiller.

This chapter outlines key aspects associated with the operation of buildings, be they designated green, sustainable, high performance, etc. Noting that the operation of the building should be taken into consideration during the design phase (such as required with the "plans for operation" in Standard 189.1), this section discusses strategies for improving the operation, maintenance, and performance of buildings long after construction is over and occupancy begins. Additional aspects particular to existing buildings are discussed in Chapter 19 of this guide.

ENERGY MODELING

At its industry outset many years ago, computerized energy modeling for greenfield projects was done to provide a comparative tool to help ascertain which systems or design solutions offered a more energy efficient result from the options available. The difference between the initial or baseline design model and design options in subsequent models was what was important. If System or Design Option A is the baseline, how much more (or less) energy efficient is Design Option B than Option A? And Option C than Option B? And so on, and so on. Hence, even if model A—the baseline—was off, the comparative aspects of A to B and B to C would be valid since all comparisons would be measured off the same baseline. While it was certainly the intent and hope that the model performance would be close to the actual building's performance, it was understood that the model was not going to be a predictor of actual, absolute energy use.

On existing buildings, this construct was a little different. When used on existing operations, energy models were often extensively tailored so that the output matched as closely as possible to the energy performance of an existing building, generally based on the building's energy bills. In that instance, the model could then be reasonably well used to predict the impact of proposed energy efficiency or conservation measures (ECMs) on the building's future energy use. Those were not simple or quick models to construct.

Today, with the intense focus on (and intent to measure) actual performance, project teams are looking for more predictive accuracy in even greenfield, baseline models. Major design, construction, and capital investment decisions are made based on the baseline, and any selected options driven by that baseline. This expectation puts a lot of pressure on the modeler to get it right—especially the baseline model. Incumbent with that pressure is an increased risk if they get it wrong. The advent of building performance labeling, both as-designed and in-operation (as used) marks a bright line from which the model's quality can be evaluated.

Given this increased importance and concomitant risk, is there a heightened duty of care expected of energy modelers? While the model derived decisions are not often public health, safety, and welfare decisions, they can certainly have fiduciary impacts. Does any energy model associated duty of care approach the level associated with the designer of record's duty of care? Can you get insurance to address faulty energy decisions? The increased use of, and reliance upon energy models along with the increasing need and requirement to measure, verify and report on in-operation performance is likely to create new considerations of and focus on the role of energy modeling in design, and the need for more affirmative risk mitigation tools.

ON CONSTRUCTION

High-performance buildings create an expectation of high-quality construction. You have to build the project as designed and if the design calls for certain equipment, materials and techniques, the contractor must comply with the requirements of the drawings and specifications as they are almost always a part of the contract. But what about substitutions or alternate techniques? Most specifications allow for multiple possible manufacturers or suppliers of equipment and/or materials (three possible providers is a commonly set minimum). Are material and equipment compatibility issues which might arise when there is a substitution or alternative technique employed under the province of the contractor? Or is the designer responsible for including in his or her specification discordant or incompatible materials or failing to execute their veto power during the shop drawing review period?

A contractor has an obligation to comply with the drawings and specifications (as reasonably interpreted); this is true irrespective of it being a green building or otherwise. Many of the risk aspects of green buildings are no different than other buildings—defective workmanship, latent defects, and schedule/cost delivery issues. With the exception of latent defect issues, contractors are typically held to a one year warranty on their overall work. However, the expectation of high performance over time and the movement to measure and quantify performance can create some more oblique risk profiles—a construction defect which impacts the building's performance may not manifest itself in such a way as to be physically evident (i.e., install less roof insulation than specified). In the past, such defects might never really be caught by anyone. But measured performance might reveal these types of shortfalls which do not necessarily look like substandard construction work but results in less than expected performance. What type of liability would a contractor have for these types of performance issues? What would be a reasonable cure?

BUILDING PERFORMANCE LABELING

Most design and/or construction contracts (with owners) make no mention or reference to performance at all and many contractor contracts do not extend warranty beyond one year after completion. Most design professionals' contracts make no mention of performance as a metric of contract compliance and their professional liability insurance (E&O) generally excludes coverage of warranties or guarantees of performance.

With the development and rollout of ASHRAE's Building EQ, there now exists a quantitative method of comparing a building's energy performance from both an as-designed and in-operation perspective and rating a building's energy performance against other similar facilities. If there is a performance label (as-designed and in-operation) affixed to the building, it is reasonable to expect that in some instances, the label and the actual performance diverge. ASHRAE Building EQ was described in detail earlier in Chapter 2 of this guide.

Does the application of the Building EQ label create a heightened expectation of transparency of a building's performance? Is that coincident with a heightened obligation (or expectation) to disclose performance problems? Are the problems system or component issues?

FILLING THE GAP: MEASUREMENT AND VERIFICATION

One of the key trends in the progression of green buildings is the growing importance of measurement of performance and verification that the measured performance meets or exceeds that predicted during the design—that it does what you said it would do. Or that the performance is at least consistent with the performance of similar buildings operating in similar circumstances. This desire to improve the predicted building performance means a more rigorous scrutiny of energy modeling techniques is now necessary.

Most current green-building certification programs require some measurement and verification (M&V) reporting over some time period, typically in excess of one year. The act of measuring parameters and the act of verifying that said measurements are reasonably accurate and truthful creates the potential for there to be a gap between what was expected and what is. And with that gap grows the need for the affirmative construction of a bridge between design, construction, and operations or the seeds of misunderstanding will surely spring forth a tangled and thorny weed.

ANSI/ASHRAE/IES Standard 189.1 requires the installation of measuring devices to record such things as the energy use of key systems, water consumption, and the flow rate for incoming outdoor air (at each air intake point). These provide the owner and operator with tools for monitoring in a little more detail than say the simple energy monthly billing statement. By recording the performance of a building from these and perhaps other devices installed, a baseline performance standard can be established for a particular building. This can then serve as a reference point to compare future operations. While Standard 189.1 requires energy consumption data only aggregated by system type (e.g., HVAC or lighting), it is possible to have the building automation and energy management system to tap into data available from devices such as variable frequency drives. This can allow for an even more detailed and disaggregated set of system measured performance data that will allow for verification of performance for that particular piece of equipment.

When developing a M&V plan for a particular building, the plan must be tailored to the specific purpose in mind. A plan intended to only implement a process that monitors system operations and highlights when potential problems crop up may well have a different set of measurements and frequency of data collected compared to a plan designed to verify the effectiveness of a particular retrofit or system modification. There are several good resources to refer to in developing a measurement and verification program, although they are oriented toward the verification of energy (and water) savings specifically from conservation projects. These include the following:

- ASHRAE Guideline 14-2014, *Measurement of Energy, Demand and Water Savings*
- M&V Guidelines: Measurement and Verification for Performance-Based Contracts (U.S. Department of Energy)
- International Performance Measurement and Verification Protocol

Reference citations for each of these are provided at the end of this chapter.

OCCUPANT SURVEYS

Occupant surveys can alert owners to specific problems with IEQ, which could potentially have implications for occupant health and productivity. Occupants are an important and often underutilized source of information about IEQ and its effect on comfort, satisfaction, and productivity. Poor thermal comfort is a common occupant complaint in buildings. Relying on complaint logs only provides an indication of local, personal, or sporadic dissatisfaction. Surveys are much more effective, as they provide a systematic mechanism for occupants to provide feedback about a broader range of aspects of the indoor environment. They are an important tool for identifying the specific nature and location of any problems, guiding postoccupancy retrocommissioning and corrective actions, and helping owners decide how to prioritize their investments in building improvements. Everyone in the building process benefits from learning how a building actually performs in practice. Occupant surveys are an informative and critical link in closing the feedback loop and allowing building owners, facility managers, and the design team to understand more completely how the building design and operation is affecting occupant productivity and well-being.

A survey should be designed so that participation is voluntary, occupants' responses remain anonymous, and results are reported only in aggregate. The survey should ask occupants for general location information only, so one can identify if problems are occurring in particular zones of the building. After asking about basic demographics and workstation characteristics, a survey should then ideally address a wide variety of IEQ features, including thermal comfort, air quality, lighting, acoustics, office layout and furnishings, and building cleanliness and maintenance. A common form of satisfaction question asks occupants to respond on a seven-point satisfaction scale ranging from very dissatisfied, to neutral, to very satisfied. Ideally, occupants who are dissatisfied with a particular aspect of their environment are presented with follow-up questions that allow them to more specifically identify the nature and potential source of their dissatisfaction. This is important for providing diagnostic information, and helping the building operators become more informed about how to respond.

Surveys can be administered in a variety of ways, but once the method is selected it should be consistently applied and available for all normal occupants of the building. Surveys can be administered directly, either by phone or in person, although this is very time-intensive and raises potential issues about privacy and accuracy of results. Web-based surveys are becoming a more common alternative to the traditional paper-based surveys and offer many advantages. They can be far less expensive to administer to a large number of people or to multiple buildings. The cost and potential errors of manually entering data from a paper survey are not present. They allow for more interactive branching features that provide diagnostic information, while keeping the survey to a reasonable length. Lastly, they offer the potential for automated reporting so that building owners and professionals can get quick access to the survey results.

Sources of sample surveys include but are not limited to the following:

- Center for the Built Environment (CBE) Occupant IEQ Survey
- Usable Buildings Trust

See the "References and Resources" section at the end of the chapter for more information.

ON OPERATION AND MAINTENANCE

Keep in mind that a building's performance is also very dependent upon how that building is operated and maintained. The fuel efficiency while driving a car is greatly impacted by how it is driven and how well it is maintained.

High-performance buildings frequently have fairly sophisticated controls and often some new or newer technologies and/or materials. There may also be some innovative system design features or system configurations that are key to achieving the high-performance objectives. The proper use and upkeep of those systems is essential to realizing and sustaining the target performance. The impact of O&M on performance should be obvious. However, some of those areas of impact may be a bit more opaque to the casual observer.

Given all of the above, the objective of high performance makes demands on a facility operations and maintenance program. The key ingredients of making such a program work are to: hire quality personnel, train them well, provide quality tools, and keep the systems, controls, and devices in good working order, including calibration.

PERSONNEL MANAGEMENT

One area that can impact system performance is the schedule stability of the operating staff. Frequent rotations or changes in work shifts may negatively affect the familiarity of operating personnel with how the system(s) respond to varied inputs especially across seasons. Frequent changes in outside service providers or changes in their internal organization point of contact can also create knowledge gaps which can hamper the ability of the operating team to respond effectively to changes in operational needs or environment.

In the evaluation of a building's performance, and the ongoing measurement and verification effort, what is the owner's role in establishing and ensuring that the day-to-day operations remain true to the design intent and that the facility and systems are maintained in such a manner as to support the high-performance objective? If there is such obligation, how does that affect any evaluation of (and potential disputes about) ongoing resource (energy, water) consumption that may arise?

A number of questions can arise when a building fails to perform as predicted. If a building is not performing as expected or predicted, what are "reasonable" efforts to correct the situation? Who should be responsible for those corrections? How do you distinguish between design, construction and operational impacts or issues that affect performance? To what extent are the design and construction project participants entitled to inspect and/or audit operations or maintenance practices relative to building performance?

Appropriate Staffing Levels

Tempting as it is to short staff on what is definitely an overhead cost function, some careful consideration of the comparative costs should help properly prioritize things. If you have a property with, say, 200,000 ft^2 (or 20,000 m^2) of leaseable space, how much does one additional staff member cost compared to a couple of additional hours or days of having the building off-line? Evaluate staffing levels not just in the context of their direct cost, but also in the context of costs avoided.

Appropriate Staff Expertise and Training

Related to staffing levels is the issue of staff expertise and training. You are not just putting a body in a seat and expecting that to work well. While it seems obvious when discussed at conference room tables, all too often, echelons far away from the on-the-ground impacts of their decisions make personnel decisions based strictly on monetary concerns with an eye on reducing staff costs wherever possible. This may not be the best metric to use in certain situations. It may be prudent to retain some specialized expertise to deal with specialized systems, sensitivities, or circumstances.

Along with the idea of retaining appropriate staff comes the idea of providing appropriate training. If you can't find the appropriate personnel, you can choose to "grow your own." Staff training, especially for complex building systems, is a smart way to keep systems in good shape and to get the best value from your operation. One approach to help reduce the cost of bringing in specialized training (or sending everybody out of town for three days) is to try to pool the training program with other owners. Bundling a program in this way can help to spread the costs over a larger audience. If you have made the investment in to a high-performance building, maintaining a high-performance team is a rational way to protect that investment.

Appropriate Tools and Information

Now that you have the right number of the right people who know the right things, you have to make sure they have the right tools. System data collection, reporting, trending, archiving, and feedback has to be appropriate for the objectives at hand. If one of the operating objectives is to optimize energy and IEQ performance, then historical operating data (trend logs) correlated with occupancy and weather data will be essential. It is not uncommon for building information systems to be set up without any significant trending, archiving or performance comparison protocols. If you do not know it is broken, you are unlikely to fix it. And do not be deluded into thinking that all problems can be resolved with software. Regular calibration of sensors and devices and keeping some reasonable inventory of spare parts is a good idea too.

Moving to the Future: Industry Trends

The drive towards high performance for the building along with the drive for higher performance in project delivery is converging in several ways. A crucial advance in the design/construct world is the advent of building information modeling (BIM). BIM has been around for more than 20 years, but in the initial years it was primarily used in industrial and large scale infrastructure projects. The massive drop in the cost of computing power and shifts to the Software as Service (SaS) approach have made BIM a much more reasonable tool for commercial and institutional projects. BIM, when done well helps to foster a more collaborative design and construction process. Couple this with the inherently more collaborative nature of high-performance buildings and there is a de facto push towards more collaborative projects.

The integration and adaptation of computer driven technologies across the entire platform of capital project execution—design, construction, and operations have been underway for some time now. But often these various technology platforms were set up in silos—they did not talk among themselves with the exception of a limited number of narrowly configured interconnections.

As the capability of computing power has increased exponentially and the cost of same has dropped precipitously, there is an increasing convergence in communication protocols which has enhanced the ability to transfer data across platforms and to work with that data in meaningful ways. One of most useful methods to use that data is to develop rapid failure mode recognition and response (fault detection and diagnosis) capabilities; thus creating more proactive maintenance models and the development of computerized maintenance management systems. These types of initiatives are already in place in the marketplace. Proactive maintenance management programs (PMM) incorporating fault detection and diagnosis (FDD) technologies have been developed using computerized maintenance management systems (CMMS) based on computer-based control systems. Some aspects of this type of instantaneous feedback are being directly incorporated into the manufacturer's equipment control systems. Some versions are being applied at the system level as sort of an overlay or umbrella that ties things together.

Computerized maintenance management systems are a very powerful tool that can be used to enhance facility operations and maintenance; and from there performance. CMMS tools can be used to capture data provided by FDD systems and provide a key platform for the development of proactive maintenance management programs. The increasing use of BIM has enhanced the capacity of CMMSs by making it possible to transfer information directly from the design process to an integrated facility model to be used in facility management. An industry effort to further enhance the value of BIM in the downstream operations and management of a facility is the deepening development and application of construction-operations building information exchange (COBie) configured data formats.

COBie is a performance-based protocol or specification for recording, compiling, transferring and delivering facility asset information. That information takes two primary forms: equipment and spaces. COBie establishes performance criteria organize the plethora of information collected in design specifications and equipment schedules. Shop drawing submissions, as-built documents and key space information can be incorporated and transferred using COBie compliant protocols. COBie compatible data formats can also be used to input BIM element data, thus allowing a smooth transfer of information from the initial BIM development to construction execution to maintenance management to operations optimization. A compatible information capture and transfer protocol makes the promise of fully integrating facility management goals into the design and construction process much more attainable.

One thing that should be kept in mind is that none of these technologies or concepts—proactive maintenance management, fault detection and diagnosis, computerized maintenance management systems, or construction-operations building information exchanges—are expressly or exclusively a green or high-performance building solution, just as BIM is not necessarily a green or high-performance building solution. However, to that extent that such tools improve the information flows and enhance the ability to use that information to the operator's advantage, they can be crucial to the ongoing achievement of high performance.

REFERENCES AND RESOURCES

Performance Measurement and Verification

ASHRAE. 2014. ASHRAE Guideline 14-2002, *Measurement of Energy, Demand and Water Savings*. Atlanta: ASHRAE.

The International Measurement & Verification Protocol (IPMVP). http://evo-world.org/en/.

FEMP. 2016. M&V Guidelines: Measurement and Verification for Performance Contracts, Version 4.0. Washington, D.C.: U.S. Department of Energy.

CBE. Occupant Indoor Environmental Quality Survey. Berkeley, CA: University of California, Berkeley. www.cbe.berkeley.edu/research/survey.htm.

Operation and Maintenance of High-Performance Buildings

ASHRAE. 2012. Guideline 32-2012, *Sustainable, High-Performance Operations and Maintenance*. Atlanta: ASHRAE.

ASHRAE. 2017. ANSI/ASHRAE/USGBC/IES Standard 189.1-2017, *Standard for the Design of High-Performance Green Buildings Except Low-Rise Residential Buildings*. Atlanta: ASHRAE. (Will be incorporated into the 2018 version of the *International Green Construction Code* [IgCC].)

ISO. 2012. ISO 13823-2008, General Principles on the Design of Structures for Durability. Geneva: International Standards Organization.

ISO. 2014. ISO 15686-1 through 15686-11, *Building Construction—Service Life Planning*. Geneva: International Standards Organization.

ISO. 2016a. ISO/DIS 18480-1, *Facility management—Part 1: Terms and definitions*. Geneva: International Standards Organization.

ISO. 2016b. ISO/DIS 18480-1, *Facility management—Part 2: Guidance on strategic sourcing and the development of agreements*. Geneva: International Standards Organization.

Commissioning

ASHRAE. 2007. ASHRAE Guideline 1.1, *HVAC&R Technical Requirements for the Commissioning Process*. Atlanta: ASHRAE.

ASHRAE. 2013. ASHRAE Standard 202, *Commissioning Process for Buildings and Systems*. Atlanta: ASHRAE.

ASHRAE. Guideline 1.4-2014, *Systems Manual Preparation for the Commissioning Process*. Atlanta: ASHRAE.

ASHRAE. 2015. Guideline 0.2-2015, *Commissioning Process for Existing Buildings and Systems*. Atlanta: ASHRAE.

Computerized Maintenance Management Systems (CMMS)

Bagadia, K. 2006. *Computerized maintenance management systems made easy: How to evaluate, select and manage CMMS.* New York: McGraw-Hill.

ISO. 2014. ISO 55000\1\2 (2014), *Asset Management*. Geneva: International Standards Organization.

Construction-Operations Building Information Exchange (COBie)

BSI. 2008. BS 1192:2007+A2:2016, *Collaborative production of architectural, engineering and construction information*. London: British Standards Institution.

Section 3: Postdesign—Construction and Beyond

CHAPTER SEVENTEEN

Residential Applications

A residence is a person's home. For many people the image of home is a house where the person lives or where they lived as a child. Many people live in homes where the building is shared by other people or families such that more than one residence is part of the structure. This is known as multifamily residence. Multifamily residences have a variety of building styles including low-rise and high-rise levels. The size and height of the building makes these more applicable to commercial building architectural and construction practices and codes. But there is a fundamental difference between a residence and a commercial space. A residence is where a person lives; that is, eats, sleeps, bathes, relaxes, and raises a family. It is a focus on the living, and the quality of living, that sets apart a residential space from a commercial space, and therefore a residential building from a commercial building.

For the most part, the individual living spaces of a multifamily residences act like and are treated like the living space of a single-family residence. There is much in common with the faculty and utility of all different types of living space with respect to shelter, comfort, health, privacy, and security. The common aspects of residences are discussed in this chapter.

While the living spaces of different types of residences are roughly similar, the buildings can be vastly different with respect to climate, structure, construction, and climate control.

It is important for residential buildings to provide the utility needs of the residents. At the same time, it is the desire of society to have buildings that minimize the amount of natural resources needed to operate. Homes are constructed using combinations of natural and engineered materials to provide structure, insulation, barriers, and protection from the elements. Other materials provide utility for the home in the form of lighting, plumbing, appliances for cooking and storing food, appliances for heating and cooling, and appliances for work and leisure activities. In a nutshell, for residences to provide their utility and function, they are consumers of resources. Energy and water are the most notable resources consumed by residential buildings after construction is complete. The residential structure also expels waste as either soiled liquid, solid waste, or exhaust fumes from combustion processes. There are many other forms of exhaust such as offgassing that must be dealt with in the residential buildings for the consideration of occupant health. Processes to resolve these unwelcome exhaust products require the use of additional energy or water resources.

ENERGY USE: RESIDENTIAL BUILDING SECTOR

The residential sector is responsible for 20% of all final energy consumption in the United States and accounted for almost 56% of the final energy used by all buildings in 2014. Similarly, residential buildings dominate the European building stock, representing about 25% of the final energy consumption in the European Union and 65% of the energy used by the buildings sector. Similar statistics are also observed in other parts of the world, such as China and Russia (Figure 17-1).

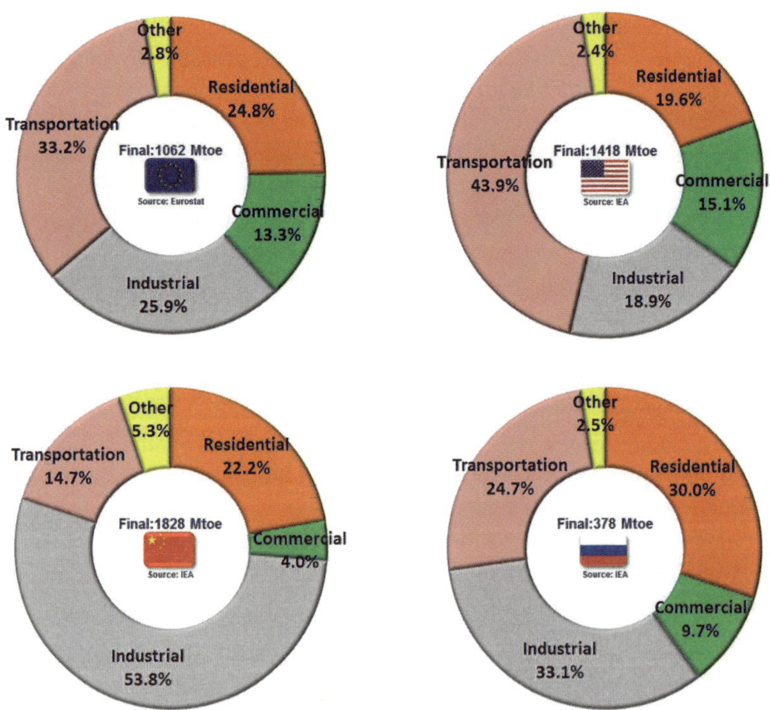

Figure 17-1 Final energy use breakdown in the United States, the European Union, China, and Russia. Mtoe: Million tonnes of oil equivalent (1 Mtoe = 39.683 MBtu = 41.868 GJ = 11,630 kWh).

From economic, environmental, and energy security perspectives, a sector that consumes one-fifth of all the energy used in most countries and more than half of all energy used by buildings around the world demands significant attention.

According to U.S. Census data, there are more than 135 million dwellings in the United States; similar European data indicate 216 million dwelling units in Europe. The projection is that by 2030 this number will grow to about 141 million in the United States and 241 million in the European Union. Given that number is increasing, efficiency must increase as well. More than 74% of all existing homes in the United States were constructed before 1989 and the widespread adoption of model energy codes governing their construction. Similarly, about 68% of the European residential buildings were constructed before the 1980s when energy building regulations were very limited. By almost any measure, most of these homes are likely underinsulated, have poorly performing fenestration, have significant envelope air leakage, need upgrades to all HVAC&R components and delivery systems, and contain outdated and inefficient lighting systems when compared to today's basic energy code minimums. In addition, these homes should be treated as systems that provide good indoor environmental quality for people. These needs define significant opportunity for energy, carbon, and peak power savings within the residential sector.

STANDARDS

Standard 90.2—Energy Efficient Building Design

ASHRAE Standard 90.2 (ASHRAE 2007) provides minimum requirements for energy-efficient design and construction of low-rise residential buildings (single- to multifamily of three stories or fewer above grade, and modular houses). The standard applies to the building envelope, heating equipment and systems, air-conditioning equipment and systems, domestic water heating equipment and systems, and provisions for overall building design alternatives and trade-offs. ASHRAE's Standard 90.2 was used as the basis for energy efficiency in residential construction in Kuwait (known as Kuwait 90.2) and was modified to account for variations between local building practices, climatic conditions, and culture.

At the time of this writing, ASHRAE and IES are revising Standard 90.2 with a goal of making it 50% more efficient than the 2006 International Energy Conservation Code (IECC). The first public review draft was released at the end of 2016. The new version is intended to deliver an accurate, flexible, performance-based tool to enable user creativity in meeting the performance objectives, including detailed rules governing the energy modeling and analyses.

The rule set is based on ANSI/RESNET/ICC 301 with specific exceptions and adjustments for building size. As a performance-based standard, the revised Standard 90.2 will focus on whole-building energy performance. It will combine several new building envelope, HVAC, lighting, and equipment technologies that are available in the market to meet the overall building performance target. Certified performance of insulation, fenestration and envelope air sealing are prioritized, while testing and verifying the envelope air leakage is mandatory. HVAC and water heating system performance is essential to achieving the performance target. Plumbing system design, insulation levels, and controls are prioritized, while requirements for HVAC system design, installation, commissioning, and verification are integral elements. Cost-effective advances in lighting technology—from lamps to control systems—can also help deliver even greater levels of lighting energy savings than current minimum code, providing revised modeling rules for quantifying residential lighting energy, credits for the use of vacancy sensors, dimmers and other control devices and revised lighting allowances for interior, exterior, garage and other residential lighting. Finally, the standard recognizes the important role of renewable energy and on-site power systems to help achieve the building performance targets. It emphasizes load minimization and HVAC performance strategies first so that any on-site power systems used can have maximum impact toward the overall building performance goals.

Standard 62.2—Ventilation and IAQ

ASHRAE Standard 62.2 (ASHRAE 2016) defines the minimum requirements for mechanical and natural ventilation systems and the building envelope intended to provide acceptable indoor air quality in low-rise residential buildings. There are three primary sets of requirements that involve whole-building ventilation, local demand-controlled exhaust, and source control. There are also a number of secondary requirements that focus on properties of specific items needed to achieve the main objectives of the standard.

Additional explanatory and educational material on how to achieve better IAQ in low-rise residential buildings is available in ASHRAE Guideline 24 (ASHRAE 2015). It provides explanatory and educational material on how to achieve good IAQ that may go beyond minimum code and air distribution requirements in Standard 62.2: for example, carbon monoxide (CO) alarms, better air filtration, and unvented combustion appliances. It also provides information relevant to ventilation and IAQ concerning envelope and system design, material selection, commissioning and installation, and operation and maintenance.

Standard 152—Thermal Distribution Systems

ASHRAE Standard 152 (ASHRAE 2014a) quantifies the delivery efficiency of duct systems, based on factors including location, leakage, and insulation of ductwork. It applies to single-family detached and attached residences with independent thermal systems. The standard provides estimates of the efficiency of thermal distribution systems that may be used in energy consumption or system capacity estimates. Computer simulation tools can be tested according to Standard

140 (ASHRAE 2014b) that also includes certification of residential energy performance analysis tools.

The standard also provides thermal distribution system efficiencies for both heating and cooling systems. Thermal distribution system efficiency is calculated for seasonal conditions (for energy consumption) or design conditions (for system sizing). This results in a total of four outputs from the method of test. This standard does not address the effectiveness of the tested system to provide comfort in the conditioned space or to deliver the designed or required airflow to individual rooms within the conditioned space.

A spreadsheet tool developed by Lawrence Berkeley National Laboratory (LBNL) and modified by the National Renewable Energy Laboratory (NREL), assists with the calculation of seasonal distribution system efficiency. A web link reference is given in the references list of this chapter. This calculation is required by the House Simulation Protocols (Wilson et al. 2014) when the simulation tool being used does not allow detailed duct modeling.

International Energy Conservation Code—IECC

The International Energy Conservation Code (IECC) is a model code regulation addressing the design of energy-efficient building envelopes and installation of energy-efficient mechanical, lighting, and power systems through requirements emphasizing performance (IECC 2015). The code contains separate provisions for low-rise residential buildings and for commercial buildings. The IECC-Residential Provisions establish minimum regulations for energy-efficiency using prescriptive and performance-related provisions, founded on broad-based principles that make possible the use of new materials and new energy-efficient designs. The 2015 edition included a new option for residential buildings using the energy rating index (ERI) path, in addition to the prescriptive and performance paths of previous versions. The target ERI score can be met through a wide range of performance options to demonstrate compliance. The ERI performance path needs to also meet the other mandatory code requirements of the IECC, follow hot-water piping provisions, and comply with the minimum insulation and window envelope performance requirements.

Residential Green Building Standard—ICC/ASHRAE 700

This voluntary standard was first published in 2009 and serves as an important resource for designers, contractors, developers, and policy makers, and as the industry benchmark for new construction, additions, and alterations of residential projects. The third edition of the National Green Building Standard was developed by the International Code Council (ICC), ASHRAE, and the National Association of Home Builders (NAHB) of the United States. The ICC/ASHRAE 700-2015 (ICC/ASHRAE 2015) covers sustainable sites, energy efficiency, water efficiency, materials and resource use, indoor environmental quality, operations and maintenance, and building owner education. The NGBS is the only residential green building rating system approved by the American National Standards Institute (ANSI).

The standard contains few minimum criteria but allows the builder or developer great flexibility in selecting green building practices. Projects receive points in each subject area for reaching certain performance or construction goals. There are four green certification levels for homes—bronze, silver, gold, and emerald. Land developments can earn one, two, three, or four stars. Projects require inspection by third-party verifiers at the rough-in stage and upon completion. NAHB provides accreditation for the verifiers as well as the final certification. The NGBS serves as the basis for several United States federal, state and local green building programs, and more than 93,000 U.S. homes have been certified as of 2016.

European Legislation

The main European legislation directly affecting the residential sector are the Energy Performance of Buildings Directive (EPBD 31/2010/EC), the Directive on Energy Efficiency (EED 2012/27/EU), the Directive on Energy Labelling (ELD 2010/30/EU) and the Directive on Eco-Design (EDD 2009/125/EC). All relevant European Union legislation can be accessed at the EUR-Lex web link reference that is given in the references list of this chapter.

Under the EPBD, the main provisions mandate that: (a) energy performance certificates (EPC) be included in all advertisements for the sale or rent; (b) all new buildings must be nearly zero energy buildings (nZEB) by December 31, 2020 (public buildings by December 32, 2018); and (c) minimum energy performance requirements be set for all new buildings, major building renovations, and for replacing or retrofitting building elements (e.g., heating and cooling systems, roofs, and walls). The nZEB requirements are defined at national level (BPIE 2015). For new residential buildings, most jurisdictions aim to have an annual source energy use less than 50 kWh/m^2 (15.8 kBtu/ft^2).

Under EED, there are specific energy savings requirements on existing European buildings, requiring EU member states to establish national plans for renovating their existing building stock. Most of the national building renovation strategies focus on residential buildings. The specific national strategies are part of the National Energy Efficiency Action Plans (NEEAP). They provide an overview of the country's national building stock, identify key policies that the country intends to use to stimulate renovations, and provide an estimate of the expected energy savings that will result from renovations.

Under the ELD and EED directives, energy efficiency regulations focus on specific commercial products that are also used in the residential sector (Bertoldi et al. 2016). Energy labels encourage consumers to purchase more energy-efficient products, which also leads to lower household energy bills. They refer to energy-using products (EuP), which are products that use, generate, transfer or measure energy.

Examples include boilers, air-conditioners, water heaters, water storage, stoves, freezers, pumps, motors, refrigerators and cooking appliances, consumer electronics, and lighting (more information can be found under DG Energy and DG Growth in the references). These labels also refer as to other energy-related products (ErP) that do not use energy (e.g., windows, thermal insulation, showerheads) but have an impact on the energy performance of buildings. All products should meet minimum requirements for energy efficiency or be banned from use. The legislation also covers 'packages' of space heating and water heating products (e.g., an air-to-water heat pump, temperature controller, and solar thermal system) or heating boilers coupled with renewables. Noncondensing boilers have been phased out of the market as of the end of 2015. The impact of ecodesign, energy labeling, ENERGY STAR, and potentially new forthcoming actions is estimated at 103 Mtoe (million tons oil equivalent, see definition in Figure 17-1) savings in 2020, of which ~52% come from the residential sector (Kemma et al. 2016). Several countries around the world such as the United States, Australia, Brazil, China, and Japan are implementing similar actions, contributing to a converging worldwide regulatory environment.

PERFORMANCE

Testing and Quality Assurance

As national residential building codes have evolved over the past decade, the testing of certain aspects of home construction has become increasingly important. The International Energy Conservation Code (IECC) for residences has specified since 2009 that envelope leakage be maintained below specific levels. In 2012, the IECC implemented requirements that residential air distribution systems (ducts and air handlers) meet or exceed specific thresholds of air leakage. Additionally, ASHRAE Standard 62.2-2016 (ASHRAE 2016) effectively requires the testing of envelope leakage in determining the mechanical ventilation requirements for acceptable indoor air quality in homes. These codes and standards specifications for residential envelope and duct leakage have led to increasing testing in residential construction. Although research-level ASTM standards for performing these types of tests exist, there has been a national effort to provide less time consuming and less expensive methods of satisfying these recent codes and standards requirements. In 2016, ANSI/RESNET/ICC Standard 380-2016 (ANSI/RESNET 2016) was promulgated to focus on the testing of residential envelope leakage, air distribution system leakage, and ventilation fan flow to meet the market demand for more cost-effective test methods.

There have also been significant national efforts to improve installation quality in residential construction. For example, The Air Conditioning Contractors of America (ACCA), an industry association, has developed ANSI/ACCA 5 QI-2015 to specify the quality installation and verification protocols for residential HVAC systems. The Residential Energy Services Network (RESNET), to ensure that manufacturer's installation requirements are followed, requires its certified practitioners to grade the installation quality of the insulation. These efforts are reinforced and bolstered by federal programs such as the Environmental Protection Agency's ENERGY STAR New Homes program and the Department of Energy's Zero Energy Ready Homes program, which require high-quality field installations.

Performance Ratings

Performance ratings for homes are becoming increasingly important in the residential construction marketplace. In the United States, the most widely used voluntary residential performance rating is the Home Energy Rating System (HERS) Index determined in accordance with the provisions of ANSI/RESNET/ICC Standard 301-2014. The purpose of the HERS Index is to provide home owners and stakeholders in the home construction, finance and real estate markets a measure of the energy performance of homes. As of August 2016, approximately 1.9 million U.S. homes have been inspected and received a HERS rating. During 2016, more than 206,000 homes received the HERS rating. (RESNET 2017).

The HERS Index was incorporated into the 2015 International Energy Conservation Code (IECC) as an alternative code compliance mechanism called the Energy Rating Index (ERI). One major distinction between the HERS Index (or the ERI) and prescriptive IECC code compliance is that the HERS Index considers all of the energy uses in the home rather than just the code-regulated uses of heating, cooling, hot water, and lighting energy.

A certified HERS Rater performs a home inspection and a series of diagnostic tests (e.g., blower door test, duct leakage test, combustion analysis) to determine, for example, the effectiveness of envelope thermal insulation, the amount and location of air leaks in the building envelope and HVAC distribution ducts, and existing or potential combustion safety issues. An accredited rating software is then used to calculate a rating score on the HERS Index.

The U.S. Department of Energy has also created a performance rating called the Home Energy Score (HES). This performance rating is not a relative rating that compares the rated home to a reference home but rather is based solely on the absolute source energy use of the rated home. The HES rating scale is from 1 to 10, with 10 being the best rating obtainable and 1 being the worst. The home is rated on this scale based on its projected source energy use as compared with the absolute range of home energy use for all homes within the geographic location of the home. The HES was created primarily for use in existing homes as a simplified mechanism to encourage home energy retrofits. As of October 2017, more than 74,000 HES evaluations have been completed (U.S. Department of Energy (DOE), Better Buildings).

Thirty-five U.S. jurisdictions have laws or executive actions mandating energy performance rating. Most of them cover various types of buildings, including commercial, multifamily and single-family homes. Several other efforts are in place throughout the world. Information on worldwide practices for building energy performance ratings and disclosures and examples of labels and certificates are available from the Institute for Market Transformation website (reference at the end of this chapter).

The European Union introduced mandatory energy labeling requirements for its member states in 2006 under the provisions of the European Directive (EPBD 91/2002/EC) on the energy performance of buildings. All new and existing buildings that undergo major refurbishment should meet minimum energy performance requirements and an energy performance certificate (EPC) is issued for all buildings or building units (e.g., apartment) that are rented out or sold, and they are valid for 10 years. Most of the European member states enforced the process after 2007 primarily issuing EPCs for new residential buildings. Certification of existing buildings and commercial buildings often followed at a later stage. Frontrunners such as Denmark, Germany, The Netherlands, and the United Kingdom have had similar schemes for new residential buildings since the 1980s (e.g., the Standard Assessment Procedure [SAP] in the UK).

Almost all EU member states have implemented an asset rating for new and small existing buildings, and some use operational rating for large complex commercial and public buildings. There are national differences in the qualifications, training, and/or accreditation of experts that perform an energy audit, the calculation tools used for the analysis, and the level of detail of the EPC's content. The EPC always includes a rating with an energy label commonly using a letter scale (e.g., A or A+ for high energy performance to G for low energy performance) and specific recommendations for improving the energy performance existing buildings (Arcipowska et al. 2014). Under the EPBD recast (31/2010/EC) there is more emphasis on the EPC rating disclosure to prospective tenants or buyers at time of lease or sale and the mandate to list the rating during property advertisement in public media. The new challenge is the mandate towards nearly zero energy buildings for all new buildings as of January 2021.

Several voluntary schemes are also available in Europe. A notable effort is the voluntary Active House label for new and renovated homes supported by the Active House Alliance. It aims to create healthier, more comfortable lives for occupants while also delivering environmental and climate benefits. Active House offers a holistic approach to building design and performance on the basis of three main principles: energy (e.g., demand, supply, primary energy performance) comfort (e.g., indoor thermal environment, air quality, daylight), and environment (e.g., environmental load, water consumption, sustainable construction), and their subparameters.

Another popular voluntary energy efficiency standard is Passive House (Passivhaus) for new buildings and for home renovations through the EnerPHit Retrofit Plan. It is especially popular in German-speaking European countries and Scandinavia, standardizing the design and construction of low energy buildings, including homes. Emphasis is placed on super insulation mostly suitable for cold climates (e.g., typical U-factors in the range of 0.10 to 0.15 $W/m^2·K$ [0.018 to 0.026 $Btu/h·ft^2·°F$]), elimination of thermal bridges, low-e glazings (U-factor less than 0.80 $W/m^2·K$ [0.141 $Btu/h·ft^2·°F$]), along with efficient mechanical ventilation (e.g., at least 75% heat recovery).

Similar concepts and labels are also introduced in France (i.e., Bâtiment Basse Consommation—BBC Effinergie label, or low energy building) and Switzerland (e.g., Minergie), which are most popular for residential buildings.

In the UK, the Code for Sustainable Homes (CSH) is an environmental assessment method for rating and certifying the performance of new homes. It was developed along the lines of the well-recognized and used environmental assessment and rating system of Building Research Establishment Environmental Assessment Method (BREEAM®) introduced in the 1990s. CSH addresses wide-ranging environmental and sustainability issues and enables developers, designers and building managers to demonstrate the environmental credentials of their buildings to clients, planners and other parties. The assessment involves the evaluation of the performance of a building in terms of specification, design, construction, and use. The rating is derived from a standard set of performance measures ranging from management processes to ecology that are set against various benchmarks.

Recently, a new rating system (Estidama Pearl Rating System) was introduced in the Emirate of Abu Dhabi that follows a point-based system along the lines of LEED and BREEAM but tailored to the Middle East region. The Pearl Villa Rating System (PVRS) aims to promote the construction of sustainable villas (e.g., large houses containing one dwelling for use by a single household or multifamily buildings with less than three stories) and improve the quality of life. The PVRS encourages water, energy, and waste minimization and local material use and aims to improve supply chains for sustainable and recycled materials and products. The PVRS respects the traditional fareej residential design and supports historical priorities of solar shading, outdoor thermal comfort, and internal privacy.

ARCHITECTURE AND DESIGN

Economy plays a big part in design of homes. The residence needs to have appeal to be desirable for living, in the same way that it needs to provide comfort for living. Comfort requires balance between indoor thermal conditions—mostly temperature and air quality—to maintain not just adequate healthful living, but thriving, and unnoticeable freshness of the air that does not require extra effort on the part of the occupant to ensure healthy living. This creates a natural economy for the building from time of construction, during its time of use and necessary upkeep, and in maintaining the inherent value of the property for eventual resale throughout the life of the dwelling. The goal of the architect is to tie all of this together to design appeal into the function in a way that best uses materials and resources to serve the living function of the dwelling, as well as manage its waste output to minimize the impact on the environment.

Several factors influence the appeal of a residential structure, such as location (urban, suburban or rural), and many others that focus on the look of the structure. In a similar way, the geographic location has a major influence on the design with respect to function of the home: the operation of and maintaining the indoor environment. The living space by definition is not empty because of inside living activities and therefore requires mechanical systems to manage the indoor temperature, humidity, and freshness. Freshness of the air is typically the opposite of thermal conditions in that usually the outdoor air has better healthful freshness than indoor air because of indoor pollutants constantly being added to the environment. Both passive and mechanical systems are crucial elements to the home comfort, and it is in these systems where the challenge of minimizing the use of natural resources becomes the art of green building.

New Energy Consuming Systems

A recent trend in residential buildings is an increase in use of electrically driven vehicles that is now integrating the transportation function into the overall residential energy usage economy. This trend connects the electrical vehicle to the residential building for refueling, and thus adds to the building overall energy demand.

Materials Use

Generally speaking, residential building construction in North America is different from commercial buildings by the use of timber (wood) products, using both milled dimensional lumber and engineering composite materials, versus the primarily steel-based construction practices used in commercial buildings. Concrete in residential construction is mostly limited to foundations and slab flooring, and as mortar for brick sheathing. As such the principles of construction for stability in the structure and for tight enclosures apply to residential buildings as well as they do to commercial buildings.

Building construction practices differ in other parts of the world. For example, central and southern European houses are dominated by concrete and masonry construction. The use of materials like stones, bricks, and concrete increase the building's thermal mass (usually contained in walls, floors, roofs) that absorbs, stores, and then releases significant amounts of heat at a later time. More information is available on this topic in Chapter 6. Thermal mass could have a positive effect on the indoor conditions during summer and winter. It regulates the magnitude of indoor temperature swings, reduces peak cooling loads in summer and acts as a thermal barrier in winter. Traditional architecture, especially in southern European countries, has successfully demonstrated these positive effects for passive, naturally ventilated houses. This type of construction practices combined with reduced infiltration high performance windows result in airtight envelope assemblies.

Thermal Performance

The thermal performance of a building can be defined by the amount of energy it needs for heating and cooling to maintain the desired comfort level of the occupants. There are many factors to consider in the thermal performance, including occupancy in the number of people and their activity level, which relates to the internal heat gain in the space. Other factors are the air barrier or envelope; insulation of the envelope including walls, floor, and ceiling; windows in terms of wall surface area; and placement with respect to the sun. There are other factors but these mentioned are the most significant to the home energy performance.

Thermal Air Barrier. Good thermal design practices require a tight barrier against both conductive losses as well as convective losses. In smaller structures such as single family dwellings, the volume of the home is small enough that the dominant thermal control is to reduce the convective or air exchange losses (losses in this sense could include either heat lost or gained to/from the environment). This is accomplished by use of an air barrier. The barrier, typically plastic sheeting or built into the construction board, must enclose the entire structure in a continuously connected or joined fashion. It is necessary that the barrier is designed into the building architecture before construction begins, including the foundation. Improper installation is a common cause of poor air barrier performance.

Measurement of the tightness of the thermal envelope is done by means of a pressure and flow evaluation, such as the blower door test. This test measures the volume of air that passes through a fan used to pressurize (or depressurize) the space (home) to a given pressure differential, typically 50 Pa (0.2 in. w.c.). In the United States, the measured value would be expressed in cubic feet per minute (cfm) or CFM50, and is converted to an air change rate per hour (ACH50) by dividing the measured value by the volume of the space or house and converting minutes to hours. The ACH50 value is normalized to the building volume, and can be used for comparing buildings of different sizes for compliance to green home standards.

Thermal Insulation. The air change rate defines the convective thermal performance and is a dynamic variable, whereas the conductive thermal performance is a static variant addressed by insulation. There are two major considerations for insulation: the resistance values (R-value) of the insulating layer(s) in walls, floors, and ceilings and reducing losses through bridging. The overall thermal transmittance of the wall assembly, called the U-factor, combines the R-values of the individual components of the wall assembly. Thermal bridging is essentially a thermal bypass to the surrounding insulation layers of the structure. Generally, thermal bridging occurs around structural members of the building. The bridging effect is reduced by using layers of materials. In this regard, using thicker dimensional lumber (for example using 2 × 6 in. instead of 2 × 4 in. studs) for framing increases the available insulation level. It is possible to widen the spacing between the studs to reduce the material usage and thermal bridging effect.

Thermal Performance of Windows. Windows provide a different challenge for managing conductive heat gain or loss to the space. Windows are necessary for daylighting, natural ventilation, and the desirable living function of the home. Windows must be designed into the structure to use the natural daylight and winter solar heat gain during daylight hours. Well-designed shading of the windows is necessary to limit the amount of solar gain during summer months. Aside from daylighting and natural ventilation during temperate seasons, windows designed on the north side of the house in the northern hemisphere are mostly used for exterior view, daylighting, and design appeal.

Similar to wall structures, window thermal transmittance (U-factor) of fenestration takes into account heat transfer from conduction and thermal bridging, convection, and longwave infrared radiation in the window design. The solar heat gain coefficient (SHGC) for a window is a ratio of how much incident solar radiant heat is admitted through the glass to the building interior. Ranging from 0 to 1, a lower number equates to lower heat gain by the window. In regions where the cooling load is dominant, it is naturally desirable to minimize solar heat gain. In regions where heating is the dominant load, it becomes desirable to use windows as a way of offsetting the heat demand during the cold seasons. That brings us back to the shading discussion because, even in higher heating load areas, there is need to reduce summer heat gains through use of shading design techniques. In addition, a prudent use of a robust load calculation tool will help to identify the optimal window performance factors (U-factor and SHGC) and avoid costly retrofits or lower effective energy performance.

A third criteria for window performance is air leakage, and this corresponds to the discussion on infiltration in the thermal air barrier section above.

Ground Source Thermal Performance. Basement, cellar, or slab constructions are made of high-mass masonry products for strength and stability against the forces of the solid soil and hydraulic pressures from the earth. For best performance, insulation and the air barrier must be continuous around the masonry construction as well. The air barrier serves also as a barrier against unwanted soil gas migration as well as prevention of moisture migration through the masonry. Design of the architecture demands careful details of how the barrier continues into the wall, as well as details of where and how continuation barriers are connected. Thermal conduction is minimized by using layers of thermal insulation board. Caution should be exercised to ensure moisture resistance and prevent mold-related deterioration by selecting materials that do not absorb moisture, and have good thermal insulation properties (high R-value).

To the homeowner, there is value in comfort and in reduced energy loads by sealing and insulating masonry walls, especially during the cold season. This building technique reduces cold temperature discomfort for the occupants by improving the interior surface temperatures, especially on walking surfaces. When the occupant feels more comfortable, there is less tendency to overcompensate by adjusting the thermostat to a higher setting. Using the barrier to reduce moisture migration through masonry surfaces reduces need for dehumidification. Using both a thermal barrier (insulation) and air barrier (plastic also serves as a water barrier) improves both comfort and moisture management will thus reduce the energy needs of the home. This is the ultimate goal of thermal performance and energy conservation.

The air barrier tightness introduces other needs related to air freshness for the occupants. Factors affecting the thermal performance of the residence must take into account the following issues:

- Infiltration.
- Moisture—introduced by occupant metabolism, cooking, and cleaning (bathing), as well as by infiltration though walls and concrete.
- Solar—gains or losses are mostly through glazing surfaces (windows).
- Internal heat gains—this includes living activities of basic metabolism, exercise, lighting, cleaning, cooking, and use of electronics and other appliances. Heat gains from lighting are decreasing with the advancement of lighting technology, such as LED lighting, but heat gains from electronic equipment have risen over the past couple of decades.

Indoor Environmental Quality

Indoor environmental quality (IEQ) implies that there is a level of satisfaction to be achieved with respect for the needs of the occupants. The IEQ components are developed and discussed in Chapter 8. Residential buildings have similar needs for IEQ as commercial buildings; however, activities in the home may differ in some ways because the occupant may have more control of their environment than in commercial buildings. The economy of a residence varies by the occupant's ability to afford different improvement factors associated with applying higher levels of green concepts into the design and construction of the residence. In many jurisdictions, use of legislated building codes reinforces the use of lower-energy applications and higher levels of comfort based on a knowledgeable and health-focused framework of standards such as ASHRAE Standard 62.2 in applying indoor air quality.

Indoor Air Quality. Chapter 8 of this guide discusses indoor air quality (IAQ) and references other sources for thorough coverage of air quality issues. It applies to residential buildings similarly as to other types of buildings. However, in residential buildings, the level of comfort may be more acutely felt because there is a longer period of occupancy by the residents than with other types of buildings. Pollutants may have a higher concentration indoors than outdoors. To address IAQ is to understand and control common pollutants that can harm effects to the health of the occupants. Some practical guidance is illustrated in Figure 17-2. A good resource on best practices for design, construction, and commissioning for profes-

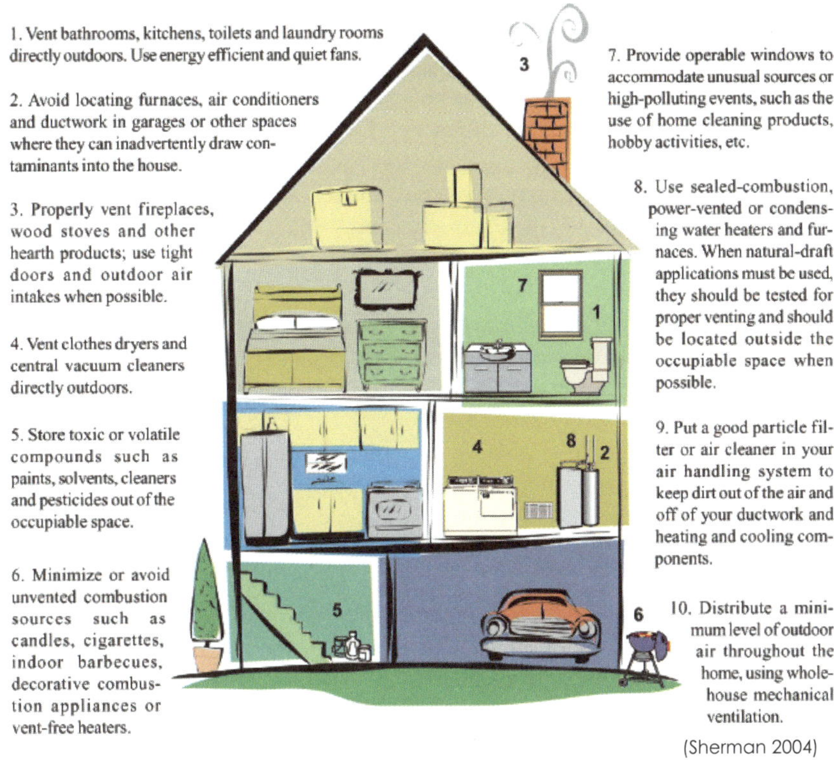

Figure 17-2 Top 10 ways to ensure good indoor air quality at home.

sionals concerned with IAQ in buildings is available in the ASHRAE *Indoor Air Quality Guide* (ASHRAE 2009).

The three main steps to control IAQ are as follows:

1. **Eliminate sources of indoor pollutants.** These can include cleaners, animals (pets), amid building materials made with offgassing chemicals, which includes furniture, carpet, and other textiles. Indoor air chemistry is complex because compounds can undergo chemical reactions in the presence of other reactive species and generate more harmful indoor compounds than those involved in the reaction. It is not always possible to eliminate pollutants entirely; for example, managing the offgassing by using sealed storage containers for cleaners is a way of achieving the nearly ideal condition. Natural pollutants such as allergens, dampness and mold, and radon should also be considered.

2. **Ensure proper space ventilation.** Ventilation is designed to minimize lingering effects of pollutants that cannot be eliminated from the space in step 1. For proper mechanical ventilation, it is necessary to manage moisture and thermal energy so there is no direct loss of energy from the space through the fresh air exchange. Energy recovery ventilators (ERVs) are appliances that facilitate exchange of polluted indoor air with fresh outdoor air. They are designed to transfer the enthalpy (heat and moisture) to the appropriate direction with high level of effectiveness. ERVs require an energy source to move air in two directions as well as any rotating enthalpy transfer media.

   Proper natural ventilation strategies, depending on outdoor conditions, should ensure the minimum outdoor air requirements. This may be a challenge in some areas. Exposure may be associated with location, socioeconomic factors, quality of housing, and other factors. For example, high-density urban cities may have residential areas close to traffic emissions. Socially deprived population groups tend to be more exposed to outdoor air pollutants (e.g., living in areas with high traffic) and indoor air pollutants (e.g., fuel poverty and inappropriate use of heating and cooking devices). Caution should be exercised when focusing on airtightness and lower ventilation rates that may increase chemical and biological pollutant concentrations.
3. **Filtration of the air.** Filtration requires energy to move air through a porous media that entraps the unwanted materials from the air. Particle filters or air cleaners need proper maintenance.

Illumination. Lighting is necessary for occupant satisfaction and perceived comfort. Where possible, daylighting should be used, provided the pros and cons as mentioned in Chapter 11 can be managed to provide the utility function and minimize possible discomforts and thermal gains to the home. Electrical lighting that requires planning so that wires and fixtures are managed to prevent sources of air leakage and in preventing thermal bridges where possible. Where it is not possible to add lighting without penetrating the thermal barriers, sealing must be made to prevent infiltration and load gains. Efficient design of lighting is covered in Chapter 11.

Acoustics. Depending on the setting of the home, acoustics can be a significant source of dissatisfaction to IEQ. With a majority of residences being in urban settings, noise issues from traffic, and neighbors are to be accounted for. The thermal barrier discussed in the Thermal Performance section will reduce migration of noise as well as air. Windows are especially important when considering noise reduction. The number of panes of class, and how well the windows are sealed will affect overall acoustic results. Other common factors for consideration include location of internal mechanical systems and appliances that have rotating or vibrating parts. Isolating these mechanical systems in isolated spaces will reduce noise penetration. Placement of external mechanical equipment such as compressor bearing condenser unit away from high-occupancy rooms, bedrooms, and close proximity to neighbors should be considered as well.

Designing and sizing ducts and registers to keep flow velocity less than 900 and 750 fpm (4.6 and 3.8 m/s), respectively, will have the best results.

FUTURE TRENDS

Zero Net Energy Homes

According to the U.S. Department of Energy (DOE) a zero energy building, or what is also referred to as a *net zero energy* or *zero net energy* building, is "an energy-efficient building where, on a source energy basis, the actual annual delivered energy is less than or equal to the on-site renewable exported energy." The first step is to increase efficiency (i.e., including efficient building construction, efficient systems and appliances, operations and maintenance, and change user behavior) and then address remaining needs with on-site renewable energy generation. Useful how-to guides and resources are available from U.S. DOE on best practices, lessons learned, and recommendations on strategies within typical construction budgets. See also the Digging Deeper sidebar for net zero energy research results.

Notable efforts are also underway in Europe. The Energy Performance of Buildings Directive (EPBD 31/2010/EC) provides a framework definition for nearly zero energy buildings (nZEBs) for all buildings, including residential. Accordingly, nZEBs are buildings that have a very high energy performance. The nearly zero or very low amount of energy required should be covered to a significant extent by energy from renewable sources, including energy from renewable sources produced on-site or nearby.

As of January 2021, all new residential buildings in the European Union will have to comply with the mandatory nZEB requirements. In some cases, like the Brussels Capital Region, the nZEB requirements were officially defined in 2011 and enforced as of 2015 and today nZEB requirements are mandatory for all new buildings. Some EU member states are also targeting existing homes that undergo major renovations to comply with nZEB requirements, although at less strict limits.

In most European countries, the nZEB definitions refer to maximum primary (source) EUIs as the main indicator, while in others (e.g., the United Kingdom, Norway, and Spain) carbon emissions are used as the main indicator or as a complementary indicator to primary energy use (BPIE 2015). Most jurisdictions aim to have an annual primary EUI not higher than 50 kWh/m^2·yr (15.8 kBtu/ft^2), although there are large national variations, such as less than 20 kWh/m^2 (6.3 kBtu/ft^2) in Denmark and 160 kWh/m^2 (50.7 kBtu/ft^2) in Austria (BPIE 2015). Different requirements are usually set for single-family houses and apartment buildings, and adjusted for different climate zones. Relevant information on nZEB European national definitions, best practices, indicators for energy performance, passive and active systems, and use of renewables are available at the ZEBRA2020 data tool (http://zebra2020.eu/tools/). Similar resources on energy performance indicators for residential buildings are available in the Building Performance Institute (BPIE) data hub and the EU Building Stock Observatory. The main construction and technical characteristics, as well as refurbishment trends in the residential building stocks of 20 European countries, and a tool for monitoring refurbishment activities are available from a recent European project (EPISCOPE, see reference at the end of this chapter). Regional and national studies in Europe highlight the trends and developments in energy upgrading of the European housing stock (Visscher et al. 2016).

Solar Heating

Solar thermal technologies provide heat that can be used for any low-temperature heat application in residential buildings. Solar energy is commonly used for domestic hot-water (DHW) heating. The technology is mature and has been commercially available in many countries for several decades. In typical DHW solar heating applications, water is heated directly in the collector or indirectly by a heat transfer fluid that is heated in the collector, passes through a heat exchanger, and transfers its heat to the DHW. The heat transfer fluid is transported by either natural or forced circulation. Natural circulation occurs by natural convection (thermosiphoning), whereas forced circulation uses pumps. Because of the intermittent nature of solar radiation, hot-water storage and an auxiliary heater must be installed to handle demand.

The trend in solar heating is toward systems that provide both DHW and space heating; these are termed solar combi systems. As a result of the higher demand expected, solar combi systems require significantly larger solar collector areas. Important issues to consider are building integration of the solar collectors and system design and installation (Bucker and Riffat 2015). Currently, solar combi systems represent about 25 of the solar thermal systems installed worldwide.

Thermosiphon solar DHW heating systems operate at low, self-regulated collector flow rates. Their performance is generally better in temperate climates than active systems (using a pump to circulate the fluid) operating at conventional flow rates and is comparable to pump systems operating at low collector flow rates. Thermosiphon systems dominate the market in residential and low-rise building applications in warm climates (e.g., South America, southern Europe [flat-plate collectors], China [evacuated tubes], and Australia). The collector fluid is circulated by natural convection, eliminating the need for the pump and controller of an active system. The hot-water storage tank is placed above the collector area.

Solar combi systems are usually combined with radiant floor heating (low water temperature operation) or other hydronic heat emitter systems. The systems need careful design to prevent overheating and handle stagnation problems during summer, because of the oversize of the system compared to the low thermal demand.

This problem can be handled by using a specific solar collector field configuration and connection with an expansion vessel, collector drainback, cooling devices in the collector loop, and a heat discharge loop. The solar collector area is between 12 to 20 m^2 (129 to 215 ft^2) for a single-family house (Balaras et al. 2010). The storage volume per unit gross solar collector area depends on design weather conditions, ranging from 26.8 L/m^2 (0.66 gal/ft^2) up to 116.7 L/m^2 (2.86 gal/ft^2). Depending on the size of the solar collector field, hot water storage, local weather conditions and loads, solar combi systems may cover 10% to 60% of the combined DHW and space heating demand.

For additional information on solar heating equipment see Chapter 37 of the 2016 *ASHRAE Handbook—HVAC Systems and Equipment*. Chapter 35 of the 2015 *ASHRAE Handbook—HVAC Applications* covers solar energy use, sizing, installation, operation, and maintenance issues. The Solar Heating and Cooling Programme (SHC) of the International Energy Agency (IEA) provides numerous practical information and guidance for all aspects of solar thermal energy (www.iea-shc.org). The U.S. Department of Energy has also detailed information on selecting, installing, and maintaining solar systems, along with local incentives and solar water heating federal tax credits that run through 2021 (https://energy.gov/energysaver/solar-water-heaters).

Digging Deeper

NIST NET ZERO ENERGY RESIDENTIAL TEST FACILITY: NET ZERO ENERGY AND BEYOND

Introduction. Goals for achieving net zero energy performance have been established in the United States and around the world (City of Melbourne 2014; EPBD 2010; IEA 2014). A net zero energy building (ZEB) is an energy-efficient building where, on a source energy basis, the actual annual delivered energy is less than or equal to the on-site renewable exported energy (DOE 2015).

The NZERTF. The Net Zero Energy Residential Test Facility (NZERTF) was constructed at the National Institute of Standards and Technology (NIST) in Gaithersburg, MD to support the development and adoption of cost-effective net zero energy (NZE) designs, technologies, and construction methods. The unoccupied and unfurnished two-story house has a basement and attic, and is similar in size and aesthetics to homes in the surrounding communities. Internal loads, energy and water usage of a virtual family of two adults and two children were simulated according to daily schedules (Omar and Bushby 2013).

Many technologies are used in the house to achieve the NZE goals including a 10.2 kW photovoltaic (PV) system, a high-efficiency air-to-air heat pump, a small-duct high-velocity (SDHV) heat pump, a solar hot-water system, and a

Digging Deeper

heat recovery ventilator (HRV). All floors of the house, including the attic, are within the conditioned space. A central heat pump system provides supply air to all floors except the attic. Passive air transfer grilles connect the basement to the first floor and attic to the second floor of the house. More information on the NZERTF design can be found in Pettit et al. (2014).

Build tight, ventilate right. To provide a high-quality indoor environment within the NZERTF, two fundamental principles were employed: "build tight, ventilate right" and contaminant source control. The first principle was pursued by constructing the building with a tight exterior envelope. While this approach is not new, particularly in northern Europe, the United States is still catching up with the latest airtightness construction practice (Chan et al. 2012).

The goal for the NZERTF was that it be extremely airtight through the use of a continuous air barrier system with a goal air change rate of less than 1.0 h^{-1} at 50 Pa (0.2 in. of water) as measured with a whole-house pressurization test. Blower door tests were performed at the NZERTF to confirm the envelope airtightness met the design targets the final result yielding a building envelope airtightness of 0.63 h^{-1} at 50 Pa (0.2 in. of water). This is tighter than the requirements in LEED and ENERGY STAR rating systems and slightly leakier than the Passivhaus requirement. The normalized leakage value for the house is 0.06, which is tighter than well over 99% of U.S. homes based on statistical analysis of the Lawrence Berkeley National Laboratory Residential Diagnostics Database (ResDB) by Chan et al. (2013).

A balanced and ducted HRV system supplies outdoor air to the first floor kitchen and second floor bedrooms, while drawing air for heat recovery from one bathroom located on the first floor and two on the second floor. To comply with the minimum ventilation requirements in ASHRAE Standard 62.2-2010 (ASHRAE 2010a), the HRV was sized to deliver 137 m^3 h^{-1} (80 cfm) of outdoor air. This rate did not include any infiltration credit.

Thermal Comfort. Dry-bulb temperature, globe temperature, and relative humidity were monitored in several rooms in the house to verify the heating and cooling system is providing a thermally acceptable environment relative to the criteria in ASHRAE Standard 55-2010 (ASHRAE 2010b).

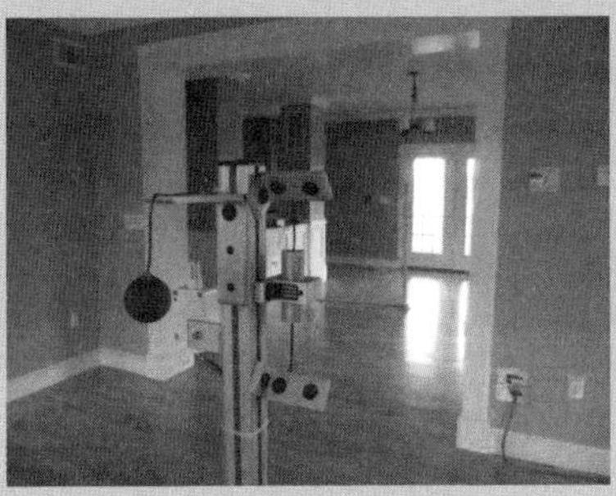

Net Zero Energy. A one-year demonstration period to demonstrate that the NZERTF could achieve net zero began July 1, 2013. An air-to-air heat pump system was used to provide the space heating/cooling. The house was operated as a single zone with constant thermostat set points of 23.8°C (74.8°F) and 21.1°C (70.0°F) during the cooling and heating sea-sons, respectively. Two solar thermal collectors were used in conjunction with a 303 L (80 gal) storage tank. For the 12 months (July 2013 through June 2014) the house produced 484 kWh more electrical energy than it consumed (Fanney et al. 2015). The solar thermal hot-water system provided 54% of the energy required to meet the domestic hot-water load over 12 months.

Ventilation Matters. Ventilation is not free, but it is worth it. The impact of the outdoor air delivered by the HRV at the NZERT was evaluated by Ng and Payne (2016). The annual heat pump energy required to condition outdoor air supplied by an HRV was 7% less than that required if outdoor air were supplied without heat recovery. Even though there is a penalty associated with the additional fan power required with an HRV, the cost of operating the HRV essentially paid for itself with the heat pump energy savings when compared with ventilation without heat recovery. In the NZERTF, when the HRV was turned off, the concentrations of chemicals increased at least sixfold and up to more than eightfold (Poppendieck et al. 2016).

Modeling the NZERTF. Results from the NZERTF home were used as in-puts into a simulation model to predict energy use and indoor air quality. Energy use for the first year of operation from July 2014 to July 2015, was 12,900 kWh (44 MMBtu), which was 5% less than simulated by model. (Fanney et al. 2015). The annual PV production was 13500 kWh (46 MMBtu), which was 6% less than the simulated PV production.

Digging Deeper

Air quality measurements were taken from the basement, living room, master bedroom, and attic and analyzed for formaldehyde and other pollutants. Detailed discussion on the real-time formaldehyde measurements can be found in the work Poppendieck et al. (2016) and of the validated coupled model in Ng et al. (2016).

The simulated NZERTF annual average formaldehyde concentration was lower than the formaldehyde concentration associated with a cancer risk of 1 in 10,000 and the California Office of Environmental Health Hazard Assessment, Chronic Reference Exposure Level (OEHHA cREL). At a 25% lower outdoor air ventilation rate, the ASHRAE 62.2-2010 rate, the simulated annual average concentration of formaldehyde increased 17% with an associated energy savings of only 4%.

Conclusion. The NIST NZERTF was able to demonstrate that, for a residential building, net zero could be achieved while meeting the needs and comfort of occupants. The NZERTF results provide valuable input to simulations that support and guide the understanding of how indoor air quality is coupled to the energy use of the building. The results show that, although some benchmarks of VOC emissions are achievable related to the basic construction materials of the house, more ideal conditions require still more improvement. This is especially the case when other sources of VOCs are introduced with furniture and occupant contributions that were not included in the NZERTF.

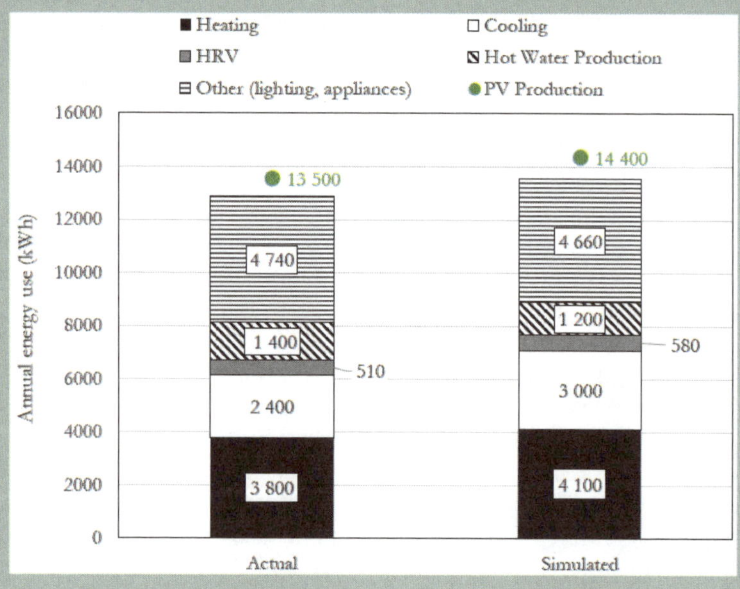

For more information and a list of all publications related to the NZERTF, please visit http://www.nist.gov/el/nzertf/.

References for this Facility

ASHRAE 2010a. Standard 62.2-2010: *Ventilation and Acceptable Indoor Air Quality in Low-Rise Residential Buildings*. Atlanta: ASHRAE.

ASHRAE 2010b. ANSI/ASHRAE Standard 55-2010: *Thermal Environmental Conditions for Human Occupancy*. Atlanta: ASHRAE.

Bernheim, A., P. White, and A. Hodgson 2014. High performance indoor air quality specification for net zero energy homes. NIST GCR 14-980. Gaithersburg, MD: National Institute of Standards and Technology.

Chan, W.R., J. Joh, and M.H. Sherman 2012. Analysis of Air Leakage Measurements from Residential Diagnostics Database. LBNL-5967E. Berkeley, CA, Lawrence Berkeley National Laboratory.

Chan, W.R., J. Joh, and M. H. Sherman 2013. Analysis of air leakage measurements of US houses. *Energy and Buildings* 66: 616–25.

City of Melbourne 2014. Zero Net Emissions by 2020. Melbourne, Australia: City of Melbourne.

DOE. 2012. 2011 *Buildings Energy Data Book*. Washington, D.C.: U.S. Department of Energy.

REFERENCES AND RESOURCES

Published

ANSI/RESNET/ICC. 2016. ANSI/RESNET/ICC 380-2016, *Standard for Testing Airtightness of Building Enclosures, Airtightness of Heating and Cooling Air Distribution Systems, and Airflow of Mechanical Ventilation Systems*. Oceanside, CA. Residential energy Services Network, Inc.

Arcipowska A., F. Anagnostopoulos, F. Mariottini, and S. Kunkel. 2014. Energy performance certificates across the EU. Brussels: Buildings Performance Institute Europe.

ASHRAE. 2007. ASHRAE Guideline 1.1, *HVAC&R Technical Requirements for the Commissioning Process*. Atlanta: ASHRAE.

ASHRAE. 2007. ASHRAE/IES Standard 90.2-2007, *Energy Efficient Design of Low-Rise Residential Buildings*. Atlanta: ASHRAE.

ASHRAE. 2009. *Indoor Air Quality Guide: Best Practices for Design, Construction and Commissioning*. Atlanta: ASHRAE.

ASHRAE. 2013. ANSI/ASHRAE Standard 202, *Commissioning Process for Buildings and Systems*. Atlanta: ASHRAE.

ASHRAE. 2014a. ANSI/ASHRAE Standard 152-2014, *Method of Test for Determining the Design and Seasonal Efficiencies of Residential Thermal Distribution Systems*. Atlanta: ASHRAE.

ASHRAE. 2014b. ANSI/ASHRAE Standard 140-2014, *Standard Method of Test for the Evaluation of Building Energy Analysis Computer Programs*. Atlanta: ASHRAE.

ASHRAE. 2014c. ASHRAE Guideline 14-2014, *Measurement of Energy and Demand Savings*. Atlanta: ASHRAE.

ASHRAE. 2015. Guideline 24-2015, *Ventilation and Indoor Air Quality in Low-Rise Residential Buildings*. Atlanta: ASHRAE.

ASHRAE. 2016. ANSI/ASHRAE Standard 62.2-2016, *Ventilation and Acceptable Indoor Air Quality in Low-Rise Residential Buildings*. Atlanta: ASHRAE.

Balaras, C.A., E. Dascalaki, P. Tsekouras, and A. Aidonis. 2010. High solar combi systems in Europe. *ASHRAE Transactions* 116(1):408–15.

Bertoldi, P., López-Lorente, J., Labanca, N. 2016. Energy Consumption and Energy Efficiency Trends in the EU-28. EUR 27972 EN. doi 10.2788/581574.

BPIE. 2015. *Nearly zero energy buildings definitions across Europe*. Brussels: Building Performance Institute.

Buker M.S. and S.B. Riffat. 2015. Building integrated solar thermal collectors – A review. *Renewable and Sustainable Energy Reviews* 51:327-346.

EC. 2016. *Guidelines accompanying Regulation (EU) No 1254/2014 with regard to the energy labelling of residential ventilation units and regulation (EU) No 1253/2014 with regard to ecodesign requirements for ventilation units*. Brussels: European Commission, DG Growth.

ICC. 2015. ICC/ASHRAE National Green Building Standard. eISBN-13: 978-0-86718-751-9.

IECC. 2015. International Energy Conservation Code. Washington, DC: International Code Council.

IRC. 2015. International Residential Code. Washington, DC: International Code Council.

Kemna, R., L. Wierda, and S. Aarts. 2016. *Ecodesign Impact Accounting*. Delft: Van Holsteijn en Kemna B.V. (VHK).

RESNET. 2014. ANSI/RESENT/ICC 301-2014. Standard for the Calculation and Labeling of the Energy Performance of Low-Rise Residential Buildings using an Energy Rating Index. Oceanside, CA: Residential Energy Services Network.

RESNET. 2017. Record number of homes HERS rated in 2016: Over 206,000 Oceanside, CA: RESNET. www.resnet.us/blog/record-number-of-homes-hers-rated-in-2016-over-206000/

Sherman, M. 2004. ASHRAE's New Residential Ventilation Standard. *ASHRAE Journal* 46(1): S149-156.

Visscher, H., E. Dascalaki, I. Sartori. 2016. Towards an energy efficient European housing stock: Monitoring, mapping and modelling retrofitting processes:

Special issue of Energy and Buildings. *Energy and Buildings* 132:1–154. http://dx.doi.org/10.1016/j.enbuild.2016.07.039.

Weiss, W. and F. Mauthner. 2016. Solar heat worldwide. IEA Solar Heat & Cooling Programme. Gleisdorf: International Energy Agency.

Wilson, E., C.E. Metger, S. Horowitz, and R. Hendron. 2014. *2014 Building America House Simulation Protocols*. Golden, CO: National Renewable Energy Laboratory (NREL).

Online

Abu Dhabi Urban Planning Council. Estidama Rating System.
https://www.upc.gov.ae/en/estidama.

Active House Alliance
www.activehouse.info.

Air Conditioning Contractors of America (ACCA)
www.acca.org/certification/residential-design.

American Council for an Energy-Efficient Economy (ACEEE)
http://smarterhouse.org/.

ASHRAE Indoor Air Quality Guide
http://iaq.ashrate.org.

ASHRAE Residential Buildings Resources
www.ashrae.org/resources--publications/bookstore/residential-buildings-resources.

Building Performance Institute
www.bpi.org.

Building Research Establishment, BRE Global Code for Sustainable Homes (CSH)
www.thecsh.co.uk.

Building Research Establishment Environmental Assessment Method (BREEAM) rating program
www.breeam.com/.

ENERGY STAR 6.0 for Windows, Doors and Skylights
https://www.energystar.gov/sites/default/files/Windows_Doors_and_Skylights_Program_Requirements%20v6.pdf.

ENERGY STAR New Homes
https://www.energystar.gov/newhomes

EPISCOPE-TABULA, European residential building typology and energy performance indicators for European housing stocks
www.episcope.eu.

EU Building Stock Observatory
https://ec.europa.eu/energy/en/eubuildings.

Eurovent, Europe's Industry Association for Indoor Climate (HVAC), Process Cooling, and Food Cold Chain Technologies
https://eurovent.eu/.

European Commission, DG Energy, Nearly zero-energy buildings
https://ec.europa.eu/energy/en/topics/energy-efficiency/buildings/nearly-zero-energy-buildings.
European Commission, DG Energy Products.
http://ec.europa.eu/energy/en/topics/energy-efficiency/energy-efficient-products.
European Commission, DG Growth.
http://ec.europa.eu/growth/industry/sustainability/ecodesign_en.
European Commission, Joint Research Centre, Energy Efficiency
http://iet.jrc.ec.europa.eu/energyefficiency/.
European Commission. Buildings Energy Efficiency.
https://ec.europa.eu/energy/en/topics/energy-efficiency/buildings.
EUR-Lex, Access to European Union Law
http://eur-lex.europa.eu.
Institute for Market Transformation. Building Rating.
www.buildingrating.org.
International Energy Agency (IEA), Solar Heating and Cooling Programme (SHC)
www.iea-shc.org.
National Association of Home Builders (NAHB)
www.nahb.org/.
Passive House Institute (PHI)
http://passiv.de.
Passive House Institute US (PHIUS)
www.phius.org/home-page
U.S. Department of Energy (DOE), Better Buildings
https://betterbuildingssolutioncenter.energy.gov/home-energy-score/
U.S. Department of Energy (DOE), Building America
http://energy.gov/eere/buildings/building-america-bringing-building-innovations-market
U.S. Department of Energy (DOE), Energy Saver, Solar water heaters
https://energy.gov/energysaver/solar-water-heaters
U.S. Department of Energy (DOE), Zero Energy Buildings
http://energy.gov/eere/buildings/zero-energy-buildings
U.S. Department of Energy (DOE), Zero Energy Ready Home
http://energy.gov/eere/buildings/zero-energy-ready-home
U.S. Department of Energy. ASHRAE Standard 152 Spreadsheet Tool.
http://energy.gov/eere/buildings/downloads/ashrae-standard-152-spreadsheet
Zebra 2020 Data Tool
www.zebra-monitoring.enerdata.eu/
http://zebra2020.eu/tools/

ASHRAE GreenTip #17-1

Mechanical Balanced Energy Recovery Ventilation (ERV)

GENERAL DESCRIPTION

The primary reason to ventilate a space is to provide for indoor air quality that does not harm the human beings living in and using the space. It has been noted that outdoor air is generally far less harmful to the occupants than indoor air. The American Lung Association lists the solutions to better indoor air quality (IAQ), in order of effectiveness as: (1) source control, (2) ventilation, and (3) filtration. Once you have done all that you can do with eliminating or lower indoor contaminants (1), the next best thing you can do is to ventilate the space with outdoor air. As building standards continue to improve the thermal performance by increasing airtightness of the building shell, the result is reduced natural air changes of the indoor space and worsening of the IAQ. It thus becomes necessary to add continuous mechanical ventilation to keep the IAQ healthy.

Any type of ventilation increases the energy cost of the building. There are many strategies to ventilate a building, but the most energy efficient and beneficial to IAQ is by ventilating using a balanced mechanical heat or energy recovery ventilator, ERV (energy recovery in most instances).

An ERV will continuously exhaust a set amount of air from the building while simultaneously bringing in the same amount of fresh air to the building. The two separate airstreams pass through an energy transferring mechanism within the ERV. This mechanism transfers the energy (heat and moisture) from the leaving air to the incoming air, or vice versa depending on the season, thus lowering the energy load being introduced into the building via the mechanical ventilation. Most ERVs also filter the incoming air to some degree.

When evaluating an ERV, there are four key metrics that should be compared: (1) sensible transfer efficiency, (2) latent transfer efficiency, (3) power used for moving the air, and (4) level of filtration. Other options include defrost strategy, control capabilities, annual service requirements, and upfront cost.

WHEN/WHERE IT'S APPLICABLE

All indoor spaces benefit from increased ventilation with regard to IAQ, with the assumption that the outdoor air is better than the indoor air. The biggest unanswered question is how much ventilation does the space need or, better stated, what ventilation rate will achieve good indoor air quality for the space. This ventilation rate is typically driven by code in the commercial sector. Residentially by states, it is not uniform. ASHRAE 62.2 provides a fallback of recommendations. Some states have adopted this as code, and some have no code. The underlying issue is that there is little scientific health-related experimental or other data that defines what good indoor air quality is, or at least no consensus. So, one should look to code if there is a commercial application, then to state/local building code if it is residential. Once you know what daily or hourly air change rate is needed, then one can start looking at the most efficient way to deliver it while also weighing in IAQ benefits and associated costs.

When looking at pros and cons of this technology, one must first understand what is being compared. The assumption herein is that you have decided a needed air change rate. Now to determine how to get it. Comparing natural infiltration, exhaust only, supply only, balanced, and, finally, the method of choice—mechanical balanced energy recovery ventilation.

PRO

- Lowest continued energy cost
- Highest level of IAQ, given outdoor conditions

CON

- Highest first cost
- Higher annual maintenance
- Air moving fans can add noise

KEY ELEMENTS OF COST

Equipment cost, installation cost, and continued operational cost all play a part in evaluating the delivery of a volume air change rate that will result in better IAQ.

First Cost

- Equipment cost H
- Installation cost H

Recurring Cost

- Device energy use S/H
- Overall energy impact to the building L
- Maintenance S/H
- Education of operator H

SOURCES OF FURTHER INFORMATION:

American Lung Association
http://www.lung.org/our-initiatives/healthy-air/indoor/at-home/ventilation-buildings-breathe.html.
U.S. Consumer Product Safety Commission
https://www.cpsc.gov/Safety-Education/Safety-Guides/Home/The-Inside-Story-A-Guide-to-Indoor-Air-Quality.
U.S. Environmental Protection Agency
https://www.epa.gov/indoor-air-quality-iaq/improving-indoor-air-quality.

CHAPTER EIGHTEEN

BUILDING-TYPE GREENTIPS

INTRODUCTION

The GreenTips contained in this book focus on a specific technology or design concept that could be considered for a green building. However, one can also compile a list of recommendations that relate to a specific type of building. Similar to the building-type-specific ASHRAE Advanced Energy Design Guides, this chapter contains a compilation of several building-type-specific set of GreenTips.

BUILDING-TYPE GREENTIPS

The intent of the GreenTips in this chapter is to give the design engineer a general idea of concepts that are common to certain building types. The focus is generally on HVAC-related design, although it should be apparent that there are a multitude of design choices that result in a truly green building.

ASHRAE
Building-Type
GreenTip #18-1

Performing Arts Spaces

GENERAL DESCRIPTION

Performing arts spaces include dance studios, black box theaters, recital halls, rehearsal halls, practice rooms, performance halls with stages and fixed seating, control rooms, back-house spaces, and support areas.

HIGH-PERFORMANCE STRATEGIES

Acoustics

1. Place less acoustically sensitive spaces around more sensitive spaces as acoustic buffers.
2. Work closely with the acoustic consultant, structural engineer, architect, and construction manager to integrate strategies that eliminate the distribution of vibration and equipment noise from the HVAC systems to the performance spaces.
3. Understand the different conditions for noise criteria.
4. Locate equipment as far away from acoustically sensitive spaces as practical. This includes rooftop air-handling units, which should not be installed above acoustically sensitive rooms.
5. The architect needs to provide enough space for the HVAC engineer to design duct systems to minimize noise transfer to acoustically sensitive spaces. Sound levels can be attenuated by the following: lowering air turbulence through proper duct design with lowered velocities, smooth bends, and preferred duct aspect ratios of 3:1 or lower; gradual diameter transitions; placing of terminal devices as far from elbows and takeoffs as possible; and designing appropriately long duct length leads to and from fans and bends. Absorptive liners can be added to attenuate high-frequency noise. These measures require space. Fan vibration transmission is reduced by flexible duct connectors, applying dampening lagging to ducts, low-pressure drop sound attenuators and silencers, or fiberglass lined plenums at

least ten times the duct cross-sectional area. For high-frequency noise, noise-cancellation technology is used.

6. Do not route piping systems through or above spaces that are acoustically sensitive.

Energy Considerations

1. Consider using demand-controlled ventilation for high-occupancy spaces.
2. Consider using heat recovery for spaces served by air-handling units (AHUs) with large amounts of outdoor air component. ANSI/ASHRAE/USGBC/IES Standard 189.1 Section 7 can serve as a guide for deciding to do energy recovery ventilation.
3. Consider strategies that allow the significant heat gain from the theatrical lighting equipment to stratify, rather than handle all of the equipment heat gain within the conditioned space zones in the building.
4. Because of the significant variation in the cooling load throughout the day, incorporating a thermal energy storage system into the central plant design will reduce the size of the chiller plant equipment, saving capital costs and energy and operational costs.
5. Consider the use of desiccant wheels to remove moisture to avoid the need for reheat in performance halls.

Occupant Comfort

1. Consider low-velocity underfloor air distribution (UFAD) strategies for large halls with fixed seating. High-velocity air entering under the feet of occupants causes discomfort due to convective heat losses, whereas low-velocity air should not provide such discomfort.
2. Consider separating stage-area air distribution from seating-area air distribution in halls.
3. Consider humidification control for all spaces where musicians and vocalists will practice, store musical instruments or sensitive equipment, and perform.

KEY ELEMENTS OF COST

1. If properly integrated, an underfloor air distribution system (UFAD) should not add significant capital costs to the project.
2. Heat recovery strategies should be assessed using life-cycle analyses. All components of the strategy must be taken into account, including the negative aspects such as adding fan static pressure and, therefore, using more fan energy when enthalpy wheel, desiccant wheel, or heat pipe strategies are considered.

SOURCES OF FURTHER INFORMATION

ASHRAE. 2017. *ASHRAE Handbook—Fundamentals.* Chapter 8, Sound and Vibration. Atlanta: ASHRAE.

ASHRAE. 2015. *ASHRAE Handbook—Applications.* Chapter 48, Noise and Vibration Control. Atlanta: ASHRAE.

ASHRAE. 2013. *UFAD Guide: Design, Construction and Operation of Underfloor Air Distribution Systems.* Atlanta: ASHRAE.

ASHRAE Building-Type GreenTip #18-2

Health Care Facilities

GENERAL DESCRIPTION

Healing patients is the primary mission of health care organizations, and the facilities where they practice or deliver those services must share and support that same mission. Health care facilities represent a wide range of space types that can consist of some doctors' offices that involve little to no special infrastructure, to research and specialty hospitals and diagnostic centers that involve an array of specialized medical equipment and supporting infrastructure components that cost into the tens of millions of dollars. Certainly, energy use and other sustainability features are significantly impacted by the type of health care facility.

Health care facilities and their systems must be flexible to accommodate remodeling and changes to the delivery of health care services, especially with anticipated changes to insurance and reimbursement. Changes are also driven by the advancement of medical technology and practices. Sustainable practices that are selected need to be capable of changing with the changes in the facility. The degree of flexibility is often an economic consideration. Another key health care consideration is system reliability and uptime. When sophisticated systems are evaluated for installation, care should be taken to ensure excellent equipment reliability and ease of maintenance.

The tips provided here are based on the common inpatient hospital facility considerations. Many outpatient facilities are similar to other office building types, and their green HVAC tips are likely to apply for some of these heath care facilities. Certainly, for some facilities, such as outpatient surgery facilities, it will be helpful to also review the office buildings (GreenTip #18-6).

Design teams must have extensive inpatient facility experience in addition to including green/sustainable experts. Many of the sustainable certification programs, guidebooks, and references promote an integrated design and delivery approach to incorporate sustainable features. By using an integrated design team of knowledgeable and experienced professionals who will carefully plan and execute sustainable options, the time, cost, and redesign will be minimized.

HIGH-PERFORMANCE STRATEGIES

The suggestions provided are focused on the uniqueness of inpatient health care facilities. They do not represent the vast array of sustainable ideas and tips that apply generally to all building types. In addition, the recommendations are more focused on those that impact HVAC. For recommendations and tips outside of HVAC, a few sources include U.S. Green Building Council's LEED for Healthcare, Green Guide for Healthcare, ANSI/ASHRAE Standard 189.3, Design, Construction, and Operation of Sustainable, High-Performance Health Care Facilities, and ANSI/ASHRAE/ASHE Standard 170-2013, Ventilation of Health Care Facilities. Another organization to investigate is Practice Greenhealth, and other publications to consider are GreenSource: The Magazine of Sustainable Design, Health Facilities Management, and HERD (Health Environments Research and Design Journal).

Safety and Infection Control

1. Consider high-efficiency particulate air (HEPA) filtration for all air-handling equipment serving the facility. Double HEPA filters may be used to compensate for lower airflow rates.
2. Consider air distribution strategies in operating rooms and trauma rooms that zone the spaces from most clean to least clean. Start with the cleanest zone being the operation/thermal plume location at the patient (the volume of air directly above the patient which is warmed due to the patient's body temperature), then the zone around the doctors, then the zone around the room, and finally, the zone outside the room.
3. Pressurize rooms consistent with American Institute of Architects (AIA) and/or ASHRAE guidelines (e.g., pressurized operating rooms, reduced pressure in rooms where patients are contagious, and installing adjacent vestibules to reduced pressure rooms to further control transmission of airborne pathogens).

4. Consider providing air exchange rates in excess of AIA guidelines in operating rooms, intensive care units, isolation rooms, trauma rooms, and patient rooms.
5. Provide redundancy of equipment for fail-safe operation and optimal full- and part-load energy-efficient operation.
6. Model intake/exhaust location strategies to ensure no reintroduction of exhaust into the building.
7. If adding atria containing plants, ensure that soils and plants are free of potential contaminates and include in the operations manuals the procedures to maintain as such.
8. Follow the steps outlined in ANSI/ASHRAE Standard 188-2015 *Legionellosis: Risk Management for Building Water Systems*. Take steps to prevent contamination from Legionella, especially for patients with compromised immune systems, to include the following.
   - Carefully design potable water systems to prevent Legionella growth in piping. Biofilms and scale can develop in piping where water may sit in dead legs or in piping of areas closed off for some reason. Limit dead legs to five pipe diameter lengths. Maintain hot water tanks at temperatures of at least 140°F (60°C). Legionella thrive and multiply when held at temperatures ranging from 95°F to 115°F (35°C to 46°C).
   - Maintain decorative fountains and cold water storage tank water temperatures below 68°F (20°C).
   - Use potable, noncirculated water at or below 68°F (20°C) for cold-water humidifiers.
   - Design piping systems to allow complete system flushing.
   - Specify point-of-use fittings to minimize bacteria growth and to regulate for safe, nonscalding shower temperatures.
   - Caution owners and operators of the importance of maintaining safe levels of water treatment chemicals for cooling towers, hot tubs, whirlpool baths, fountains, and swimming pools.

Energy Considerations

Hospitals typically have an Energy Utilization Index (EUI) about 2.5 times greater than that of office buildings. According to the World Health Organization, the use of electricity in the United

States adds more than $600 million in direct health costs and more than $500 billion in indirect costs. U.S. hospitals spend more than $500 billion on energy that typically represents 1% to 3% of their operating budget or around 15% of profits or net operating income (DOE 2009). Reducing energy costs can play a significant role in a hospital's organization profit or net operating income. Keep in mind, for a hospital operating on a 4% margin, it takes $25 of gross revenue to generate $1 of profit or net operating income. Every dollar of energy savings is a dollar added to the profit or net operating income. According to Target 100 and ASHRAE's Advanced Energy Design Guide for Large Hospitals, it is possible to reduce a new hospital's energy use by 50% or 60% over the minimum requirements to meet ANSI/ASHRAE/IESNA Standard 90.1-2004. Typically, the amount of energy savings comes down to an economic decision. Life-cycle economic cost analysis is recommended, including maintenance of the various components and systems. Often, the highest savings are achieved with systems that have fairly sophisticated equipment and more complicated operating control sequences. Appropriate assumptions must be applied to the cost of maintenance. Equally important, the design team needs to clearly outline to the building operators the expected maintenance requirements as part of the system selection process. This is an area commonly omitted from of the systems selection and economic analysis.

With the amount of energy savings potential, there are a significant number of options to consider. Energy modeling is recommended to help sort out the interplay of building envelope, lighting, HVAC, and service hot-water energy-saving options. Design teams and the health care operational staff should consider developing energy usage targets and goals at the onset of the project. Those will be useful in guiding the expectations and decision making.

For facilities to realize the potential of any design, three key steps are needed. First, ensure the systems are operating as intended. The design and construction team needs to functionally test and prove all systems are operating per the intent. Following occupancy, it should be an expectation that additional system shakedown will be needed. The actual operating conditions, both in terms of occupancy and facility operations as well as varying weather conditions, will uncover additional issues from that of the functional testing preoccupancy that will need to be addressed. Second, a real commitment to monitoring, benchmarking, and having a continuous improvement process in place is equally important to optimize the energy-saving potential. Third, which is really a subpart of the second idea, is a commitment to budget and fund proper maintenance over the life of the buildings and systems.

Two health-care-specific resources for energy-saving ideas for building envelope, lighting, HVAC, and service water heating are outlined for the various climate zones in the United States are part of the Advanced Energy Design Guides (AEDG) series:

- Advanced Energy Design Guide for Small Hospitals and Healthcare Facilities (30% better than ANSI/ASHRAE/IES Standard 90.1-1999).
- Advanced Energy Design Guide for Large Hospitals (50% better than ANSI/ASHRAE/IES Standard 90.1-2004). Large hospitals being defined as facilities over 100,000 ft^2 (9300 m^2).

The AEDGs provide case studies and how-to tips for implementing the various suggested strategies. Two other excellent studies that have been referenced already are from the work done through the University of Washington's Integrated Design Lab. See the report "Target100! Envisioning the High Performance Hospital: Implications for A New Low Energy, High Performance Prototype" (2010) and review the website Targeting 100! website (Reference listed at the end of this Green Tip). There are two critical observations from these reports that are consistent from the modeling results used in the two AEDG for health care facilities:

- Space heating (reheat) because of the mandatory air exchange requirements is the largest single energy use area.
- There is a significant amount of heat rejection from a variety of processes.

Energy recovery strategies that capture the heat which would otherwise be lost to use for space heating and water heating are great opportunities to explore. Recent modification to ASHRAE Standard 90.1 make the use of energy recovery more prevalent, and use the requirements of that Standard and Standard 189.1, Section 7 for direction on when energy recovery should be applied. The references provide a variety of ways that can be done. Other recommendations include using (1) variable-air-volume (VAV) systems in noncritical spaces working in conjunction with lighting occupancy sensors and (2) dedicated outdoor air systems (DOAS).

Occupant Comfort

Energy considerations are important, but an overall well-designed green health care facility will also include focus on other factors. These factors are addressed with the following recommendations for occupant comfort:

1. Acoustics of systems and spaces must be designed with patient comfort in mind.
2. Daylight and views should be provided while minimizing the HVAC load impact of these benefits.
3. Provide individual temperature control of patient rooms with the capability of adjustment by patient.
4. Building pressurization relationships/odor issues should be carefully mapped and addressed in the design and operation of the building.

KEY ELEMENTS OF COST

1. HEPA filtration costs are significant in both first cost and operating costs. The engineer should work closely with the infection control specialists at the health care facility to determine cost/benefit assessment of the filtration strategies.
2. Heat recovery strategies should be assessed using life-cycle analyses. All components of the strategy must be taken into account, including the negative aspects, such as adding fan static pressure and, therefore, using more fan energy when heat wheel or heat pipe strategies are considered. Again, use ASHRAE Standards 90.1 and 189.1 for design guidance (or code compliance).

SOURCES OF FURTHER INFORMATION

AHA. 2012. *Business Case Cost-Benefit Worksheet. Sustainability Roadmap for Hospitals*. Chicago, IL: The American Society for Healthcare Engineering, American Hospital Association. www.sustainabilityroadmap.org/strategies/businesscase.shtml.

AIA. 2006. *Guidelines for Design and Construction of Health Care Facilities.* Washington, D.C.: American Institute of Architects.

American Society of Healthcare Engineering. Green Guide for Health Care. www.gghc.org.

ASHRAE. 2012. *50% Advanced Energy Design Guide for Large Hospitals*. Atlanta: ASHRAE.

ASHRAE. 2011. *50% Advanced Energy Design Guide for Small to Medium Office Buildings*. Atlanta: ASHRAE.

ASHRAE. 2015. ANSI/ASHRAE/ASHE Standard 170, *Ventilation of Health Care Facilities*. Atlanta: ASHRAE.

ASHRAE. 2013. *HVAC Design Manual for Hospitals and Clinics*, 2nd ed. Atlanta: ASHRAE.

ASHRAE. 2015. *ASHRAE Handbook—HVAC Applications.* Chapter 8, Health Care Facilities. Atlanta: ASHRAE.

ASHRAE. 2015. Standard 188-2015, *Legionellosis: Risk Management with Building Water Systems*. Atlanta: ASHRAE.

ASHRAE. 2017. *ASHRAE Handbook—Fundamentals*. Chapter 10, Indoor Environmental Health, and Chapter 11, Air Contaminants. Atlanta: ASHRAE.

CDC. 2003. *Guideline for Environmental Infection Control in Health-Care Facilities*. 2003 Recommendations of CDC and the Healthcare Infection Control Practices Advisory Committee (HICPAC). www.cdc.gov/infectioncontrol/pdf/guidelines/environmental-guidelines.pdf

DOE. 2012. EERE Network News. New Hospital Energy alliance to Promote Clean Energy in Healthcare. www.energy.gov/articles/energy-department-s-hospital-energy-alliance-helps-partner-save-energy-and-money

George, R. 2012. (unpublished ASHRAE Chapter Presentation). Preventing legionella associated with building water systems. www.legionellaprevention.org

Health Care Without Harm and the Center for Maximum Potential Building Systems. 2004. Green Guide for Health Care™. http://www.gghc.org.

NFPA. 2015. NFPA 99: *Health Care Facilities Code*. Quincy, MA: National Fire Protection Agency.

Online Resources

Health Facilities Management
www.hfmmagazine.com/subscribe.

Health Environments Research and Design Journal
http://journals.sagepub.com/home/HER

Practice Greenhealth
https://practicegreenhealth.org/.

Target 100!
http://idlseattle.com/t100/.

The Center for Health Design
www.healthdesign.org.

WHO. 2009. WHO Guideline 2009–Natural Ventilation for Infection Control in Health-Care Settings. World Health Organization, Geneva, Switzerland.
http://www.who.int/water_sanitation_health/publications/natural_ventilation.pdf.

ASHRAE
Building-Type
GreenTip #18-3

Laboratory Facilities

GENERAL DESCRIPTION

Laboratory facilities are infrastructure intensive and include many different types of spaces. The HVAC systems for these different types of spaces must be designed to address the specific needs of the spaces being served. The first considerations should always be safety and system redundancy to ensure the sustainability of laboratory studies. Life-cycle cost analysis for different system options is critical in developing the right balance between first and operating costs.

HIGH-PERFORMANCE STRATEGIES

Safety

1. Design fume hoods and associated air distribution and controls to protect the users and the validity of the laboratory work.
2. Pressurize rooms to be consistent with the *ASHRAE Laboratory Design Guide*, 2nd edition (ASHRAE 2015a) and any other code-required standards. Use building pressurization mapping to develop air distribution, exchange rate, and control strategies.
3. Optimize air exchange rates to ensure occupant safety while minimizing energy usage.
4. Design storage and handling exhaust and ventilation systems for chemical, biological, and nuclear materials to protect against indoor pollution, outdoor pollution, and fire hazards.
5. Model intake/exhaust location strategies to ensure that lab exhaust air is not reintroduced back into the building's air-handling system.

Redundancy

1. Consider a centralized lab exhaust system with a redundant (n + 1) exhaust fan setup.

2. Redundant central chilled-water, steam, or hydronic heating, air-handling, and humidification systems should be designed for fail-safe operation and to optimize full-load and part-load efficiency of all equipment.

Energy Considerations

1. Recent modification to ASHRAE Standard 90.1 make the use of energy recovery more prevalent, and use the requirements of that Standard and Standard 189.1, Section 7 for direction on when energy recovery should be applied. Carefully design heat recovery for hazardous spaces.
2. Use VAV systems to minimize air exchange rates during unoccupied hours.
3. Consider the use of computational fluid dynamics (CFD) and other modeling software to establish safe air change rates for particular laboratory needs and follow up with monitoring to verify its effectiveness. Consider low-flow fume hoods with constant-volume controls where this concept can be properly applied.
4. Look for opportunities for sharing spaces to reduce conditioned space volume, (e.g., preparation areas and laboratory support spaces could be shared between labs). The number of laboratories can also be reduced if work schedules can be shifted and the same space can be used through the day and night.
5. Isolate and minimize the size of laboratory space for those processes requiring greater airflow for safety.

Occupant Comfort

1. Air systems should be designed to allow for a collaborative working environment. Acoustic criteria should be adhered to in order to maintain acceptable levels of noise control.
2. Daylight and views should be considered where lab work will not be adversely affected. When appropriate, glazed walls to other spaces or hallways will permit views without requiring the laboratory to be on an outside wall.

KEY ELEMENTS OF COST

1. Heat recovery strategies should be assessed using life-cycle analyses. All components of the strategy must be taken into account, including the negative aspects, such as adding fan static pressure and, therefore, using more fan energy when strategies using enthalpy wheels, desiccant wheels or heat pipes are considered. Again, use Standards 90.1 and 189.1 for design guidance.
2. Low-flow fume hoods should be evaluated considering the impact of reducing the sizes of air-handling, heating, cooling, and humidification systems.

SOURCES OF FURTHER INFORMATION

ASHRAE 2015a. A*SHRAE Laboratory Design Guide,* 2nd ed. Atlanta: ASHRAE.

ASHRAE. 2015b. *ASHRAE Handbook—HVAC Applications*. Chapter 16, Laboratories. Atlanta: ASHRAE.

Barnet, B. 2013. Energy Recovery in Lab Air-Handling Systems. *ASHRAE Journal* 55(9): 20–35.

Johnson, G.R. 2008. HVAC Design for Sustainable Lab. *ASHRAE Journal* 50(9):24–32.

Mathew, P., S. Greenberg, D. Sartor, P. Rumsey and J. Weale. 2008. Laboratory Performance: Metrics for Energy Efficiency. *ASHRAE Journal* 50(1):40-7.

NFPA. 2015. NFPA 45, *Standard on Fire Protection for Laboratories using Chemicals.* Quincy, MA: National Fire Protection Association.

Tschudi, W., Dale Sartor, Evan Mills and Tim Xu. 2002. H*igh-Performance Laboratories and Cleanrooms: A Technology Roadmap*. Berkeley, CA: Lawrence Berkeley National Laboratory.

Online

Labs 21, Environmental performance criteria http://labs21.lbl.gov/EPC/intro.htm.

ASHRAE Building-Type GreenTip #18-4

Student Residence Halls

GENERAL DESCRIPTION

Student residence halls are made up primarily of living spaces (e.g., bedrooms, living rooms, kitchen areas, common spaces, study spaces, and data/communications closets). Most of these buildings also have central laundry facilities, assembly/main lobby areas, and central meeting/study rooms. Some of these spaces also include classrooms, central kitchen and dining facilities, and other specialty facilities. Many of the strategies outlined below can also be applied to hotels and multiunit residential complexes, including urban luxury condominium developments.

HIGH-PERFORMANCE STRATEGIES

Energy Considerations

1. Recent modification to ASHRAE Standard 90.1 make the use of energy recovery more prevalent, and use the requirements of that Standard and Standard 189.1, Section 7 for direction on when energy recovery should be applied.
2. Use VAV systems with economizers or induction systems for public/common spaces and consider VAV also for sleeping areas.
3. Specify that the static pressure set point for the central supply fans be based on the most open VAV box position.
4. Specify demand-controlled ventilation for areas with variable use (i.e., study spaces, lounges, meeting rooms, etc.), following the requirements specified in ASHRAE Standards 90.1 and 189.1.
5. Use natural ventilation (operable windows) and hybrid natural ventilation strategies.
6. Use variable-speed drives for chilled-water pumps and HVAC fans, including fan-coil units.
7. Consider using a ground-source heat pump (GSHP) system.

8. Consider using skylights in corridors for daylighting. Also integrate electrical lighting with daylighting through controls.

Occupant Comfort

1. Systems should be designed to appropriately control noise in occupied spaces.
2. Daylight and views should be optimized, while minimizing load impact on the building.
3. Consider providing occupant control in all dorm rooms. However, provide overrides for the case when a window is open.

KEY ELEMENTS OF COST

1. While there is a premium to be paid in first cost for energy control measures (ECM), many utility companies have energy rebate programs that make this concept acceptable, even on projects with tight budgets.
2. Heat recovery strategies should be assessed using life-cycle analyses. All components of the strategy must be taken into account, including the negative aspects such as added fan pressure and, therefore, using more fan energy when enthalpy wheel, desiccant wheel, or heat pipe strategies are considered. Again, use Standards 90.1 and 189.1 for design guidance.
3. Hybrid natural ventilation strategies could be used, such as operable windows, properly designed vents, using the Venturi effect to optimize natural airflow through the building, and the shutdown of mechanical ventilation and cooling systems during ambient temperature ranges between 60°F and 80°F (16°C and 27°C). These measures will significantly reduce operating costs. The costs of the operable windows and vents will need to be weighed against the energy savings.

REFERENCES AND RESOURCES

Friedman, G. 2010. Energy-Saving Dorms. *ASHRAE Journal* 52(5): 20–4.

SOURCES OF FURTHER INFORMATION

ASHRAE. 2017. *ASHRAE Handbook—Fundamentals.* Chapter 16, Ventilation and Infiltration. Atlanta: ASHRAE.

BRESCU, BRE. 1999. *Natural Ventilation for Offices Guide* and CD-ROM. ÓBRE on behalf of the NatVent Consortium, Garston, Watford, UK.

CIBSE. 2005. *Natural Ventilation in Non-Domestic Buildings.* London: Chartered Institution of Building Services Engineers.

Kelty, C. 2017. 2017 ASHRAE Technology Award: Living While Learning. *ASHRAE Journal* 59(4):8.

ASHRAE Building-Type GreenTip #18-5

Athletic and Recreation Facilities

GENERAL DESCRIPTION

Athletic and recreational spaces include natatoriums, gymnasiums, cardio rooms, weight-training rooms, multipurpose rooms, courts, offices, and other support spaces.

HIGH-PERFORMANCE STRATEGIES

Energy Considerations

1. Follow the requirements of Standard 189.1 for the inclusion of demand-controlled ventilation for high-occupancy spaces.
2. Similarly, follow Standard 189.1 for the determination of when heat recovery for spaces served by AHUs with a significant amount of outdoor air component.
3. Consider strategies that allow the significant heat gain in high-volume spaces to stratify, rather than handling all of the heat gain within the conditioned space zones in the building.
4. Consider heat recovery/no-mechanical-cooling strategy for the pool area in moderate climates.
5. Incorporate occupied/unoccupied mode for large locker room and toilet room areas to set back the air exchange rate in these spaces during unoccupied hours and save fan energy.
6. Consider heating pool water with waste heat from pool dehumidification system.
7. Consider using a water-based geothermal heat pump system (GHPS).

Occupant Comfort

1. Consider CO_2 sensors or other mechanisms to determine the level of occupancy in all spaces that have infrequent, dense occupancy.

2. Consider high-occupancy and low-occupancy modes for air-handling equipment in gymnasiums using a manual switch and variable-frequency drives.
3. Consider hybrid natural ventilation strategies in areas that do not have humidity control issues (e.g., pools, training rooms).

KEY ELEMENTS OF COST

1. The pool strategy, item 6 in the Energy Considerations, should reduce first cost and operating costs.
2. Heat recovery strategies should be assessed using life-cycle analyses. All components of the strategy must be taken into account, including the negative aspects, such as adding fan static pressure and, therefore, using more fan energy when strategies using enthalpy wheels, desiccant wheels or heat pipes are considered. The location of outdoor air intakes and exhaust outlet points should be a factor in the original building design to minimize ducting length to get to the heat recovery units.
3. Demand-controlled ventilation adds minimal first cost and often provides paybacks in one to two years.

SOURCES OF FURTHER INFORMATION

ASHRAE. 2015. *ASHRAE Handbook—HVAC Applications*. Chapter 5, Places of Assembly, and Chapter 7, Educational Facilities. Atlanta: ASHRAE.

ASHRAE. 2016. *ASHRAE Handbook—HVAC Systems and Equipment*. Chapter 24, Desiccant Dehumidification and Pressure-Drying Equipment, and Chapter 25, Mechanical Dehumidifiers and Related Components, and Chapter 26 Air-to-Air Energy Recovery Equipment. Atlanta: ASHRAE.

Crowe, Derek. 2011. State-of-Art School. *ASHRAE Journal*. 53(5):36–42.

Lawrence, T.M. 2004. Demand-controlled ventilation and sustainability. *ASHRAE Journal* 46(12):117–21.

ASHRAE Building-Type GreenTip #18-6

Commercial Office Buildings

GENERAL DESCRIPTION

Commercial office buildings are made up primarily of office spaces, meeting rooms, and central core facilities such as toilet rooms, storage space, and utility rooms (including telephone and data). Some of these spaces also include central kitchen and dining facilities. The strategies outlined below can also be applied to most large-scale commercial office buildings.

HIGH-PERFORMANCE STRATEGIES

Energy Considerations

1. Consider a dedicated outdoor air system (DOAS) with total energy recovery (TER).
2. Use high-efficiency fan-coil units or chilled-beam/induction systems in conjunction with the DOAS concept.
3. Use natural ventilation and hybrid natural ventilation strategies.
4. Use energy conservation measures (ECM) for fan-coil units.
5. Use ground source heat pumps (GSHP) where feasible.
6. Consider using underfloor air distribution (UFAD).
7. Develop zone temperature adjustment strategies to allow for demand response programs during peak demand periods.
8. Include occupancy sensor control capability to shut off receptacle controlled plug loads when practical.
9. Incorporate energy-efficient lighting strategies that share integrated occupancy controls with the HVAC system.

Occupant Comfort

1. Systems should be designed to appropriately control noise in occupied spaces.
2. Daylight and views should be optimized, while minimizing heating and cooling load impact on the building.

KEY ELEMENTS OF COST

1. Although there is a premium to be paid in first cost for ECMs, many utility companies have energy rebate programs that make this concept acceptable, even on projects with tight budgets.
2. Energy recovery, UFAD, and DOAS strategies should be assessed using life-cycle analyses. Ventilation energy recovery should be evaluated based on the requirements of ASHRAE Standard 189.1 All components of the strategy must be taken into account, including the negative aspects, such as adding fan static pressure and, therefore, using more fan energy when heat wheel or heat pipe strategies are considered.
3. Hybrid natural ventilation strategies could be used such as using operable windows, properly designed vents, the Venturi effect to optimize natural airflow through the building, and the shutdown of mechanical ventilation and cooling systems during ambient temperature ranges between 60°F and 80°F (16°C and 27°C). This will significantly reduce operating costs. The costs of the operable windows and vents will need to be weighed against the energy savings.

REFERENCES AND RESOURCES

Mumma, S. A. 2011. DOAS misconceptions. *ASHRAE Journal* 53(8): 76–9.

SOURCES OF FURTHER INFORMATION

ASHRAE. 2011. *Advanced Energy Design Guide for Small to Medium Office Buildings: Achieving 50% Energy Savings*. Atlanta: ASHRAE.

ASHRAE. 2015. *ASHRAE Handbook—HVAC Applications*. Chapter 3, Commercial and Public Buildings. Atlanta: ASHRAE.

ASHRAE. 2017. *ASHRAE Handbook—Fundamentals*. Chapter 16, Ventilation and Infiltration. Atlanta: ASHRAE.

BRESCU, BRE. 1999. *Natural Ventilation for Offices Guide* and CD-ROM. ÓBRE on behalf of the NatVent Consortium, Garston, Watford, UK.

Dutton, S.M., D.Banks, S. Brunswick, W.J. Fisk. *Natural Ventilation in California Offices: Estimated Health Effects and Eco-*

nomic Consequences. Lawrence Berkeley National Laboratory. LBNL-6910E. October 2013. https://eta.lbl.gov/sites/all/files/publications/lbnl-6910e.pdf.

ASHRAE
Building-Type
GreenTip #18-7

K–12 School Buildings

GENERAL DESCRIPTION

K–12 school buildings are made up primarily of classrooms, gymnasiums, libraries, an auditorium, a central kitchen, and dining facilities. (Also see GreenTip #18-6: Commercial Office Buildings and GreenTip #18-5: Athletic and Recreation Buildings.)

HIGH-PERFORMANCE STRATEGIES

Energy Considerations

1. Follow Standard 189.1 for the determination of when heat recovery for spaces served by AHUs with a significant amount of outdoor air component.
2. Follow the guidance in ASHRAE's Advanced Energy Design Guide for K-12 School Buildings.
3. Use dedicated outdoor air systems (DOASs) with energy recovery combined with fan coils or chilled-beam/induction systems for spaces.
4. Use natural ventilation and hybrid natural ventilation strategies.
5. Use ECMs for fan-coil units.
6. Use GSHPs where feasible.

Occupant Comfort

1. Systems should be designed to appropriately control noise in occupied spaces.
2. Daylight and views should be optimized while minimizing load impact on the building.

KEY ELEMENTS OF COST

1. Although there is a premium to be paid in first cost for energy conservation measures (ECM), many utility companies have

energy rebate programs that make this concept acceptable, even on projects with tight budgets.

2. Energy recovery strategies should be assessed using life-cycle analyses. All components of the strategy must be taken into account, including the negative aspects, such as adding fan static pressure and, therefore, using more fan energy when heat wheel or heat pipe strategies are considered. The location of outdoor air intakes and exhaust outlet points should be a factor in the original building design to minimize ducting length to get to the heat recovery units.
3. Hybrid natural ventilation strategies could be used, such as using operable windows, properly designed vents, using the Venturi effect to optimize natural airflow through the building and the shutdown of mechanical ventilation and cooling systems during ambient temperature ranges between 60°F and 80°F (16°C and 27°C). These actions will significantly reduce operating costs. The costs of the operable windows and vents will need to be weighed against the energy savings.

SOURCES OF FURTHER INFORMATION

ASHRAE. 2011. *Advanced Energy Design Guide for K-12 School Buildings: Achieving 50% Energy Savings Toward a Net Zero Energy Building*. Atlanta: ASHRAE.

ASHRAE. 2015. *ASHRAE Handbook—HVAC Applications*. Chapter 7, Educational Facilities. Atlanta: ASHRAE.

ASHRAE. 2017. *ASHRAE Handbook—Fundamentals*. Chapter 16, Ventilation and Infiltration. Atlanta: ASHRAE.

BRESCU, BRE. 1999. *Natural Ventilation for Offices Guide* and CD-ROM. ÓBRE on behalf of the NatVent Consortium, Garston, Watford, UK.

Collaborative for High Performance Schools. Best Practices Manual. http://www.chps.net/dev/Drupal/node/288.

Mumma, S. A. 2011. DOAS misconceptions. *ASHRAE Journal* 53(8): 76–9.

Svensson C., and S.A. Aggerholm. 1998. Design tool for natural ventilation. ASHRAE IAQ 1998 Conference, October 24–27, New Orleans, LA.

ASHRAE Building-Type GreenTip #18-8

Data Centers

GENERAL DESCRIPTION

Data centers are energy-intensive buildings using 20 W/ft^2 (215 W/m^2) to greater than 1000 W/ft^2 (10 764 W/m^2). They support information technology (IT) spaces, electronic equipment, and electronic communications. These facilities are known by many different names: data centers, computer rooms, server rooms, telecom rooms, datacom, or information technology equipment (ITE) rooms. (The term ITE will be used in the remainder of this article to designate these spaces.) The HVAC systems for data centers can be as diverse as the data centers themselves and include various forms of air or liquid cooling. While the first considerations must be reliability of the infrastructure and IT systems to minimize the risks and losses associated with outages, it is also necessary to minimize energy consumption. To this end, many strategies of redundancy, operational practice, and service diversity are used, along with green practices such as outdoor air, free cooling, air containment, and automatic control. In the end, the purpose of a data center is always to provide a suitable environment (power, cooling, and security) for the electronic equipment in an energy-efficient, scalable, and reliable manner.

HIGH-PERFORMANCE STRATEGIES

Reliability and Scalability

1. Install only the necessary redundancy to meet the owner's reliability requirements; optimize redundancies.
2. Use zoning strategies for different types of ITE based on power densities and preferred cooling strategies for each type.
3. Rightsize the cooling and power infrastructure for the IT loads with sufficient stages to operate at best efficiency points as loads change.
4. Design scalable infrastructure that can change with the rapid changes in ITE and its software (data center occupants).

Energy Considerations

The first step is to review and follow ASHRAE Standard 90.4-2016, *Energy Standard for Data Centers*. Beyond that, consider the following.

1. Use air management best practices (e.g., separate hot/cold air-streams, maximize temperature differences across HVAC heat exchangers, eliminate air leaks, use preferred ITE equipment cooling classes, etc.) to achieve ITE inlet temperature uniformity as described in ASHRAE's third edition of *Thermal Guidelines for Data Processing Environments*.
2. Install monitoring and control equipment (building management systems) to measure efficiency and performance of HVAC as well as IT systems.
3. Install data center infrastructure management (DCIM) systems to measure and manage ITE and infrastructure systems more holistically.
4. Use cooling plant best practices (e.g., VFDs, optimize chiller/condenser supply temperatures, chilled water vs. DX, etc.).
5. Consider liquid cooling when heat loads or total cost of ownership (TCO) requirements demonstrate energy savings.
6. Avoid high-energy-consuming humidity controls such as reheat. Control humidity using dew point sensing. Consider alternatives such as desiccant dehumidification and adiabatic humidifying.
7. Use air or water economizers where TCO and risks are acceptable.
8. Improve power distribution efficiencies (e.g., evaluate transformer efficiency, reduce number of power conversions, use higher voltages, consider DC power, minimize power run distances, optimize redundancy strategy).
9. Capture and reuse waste heat where possible.
10. Determine optimal operating temperature and humidity ranges using the steps defined in ASHRAE's third edition of Thermal Guidelines for Data Processing Environments (The recommended range should be used as the starting point if no optimization has been performed.)

Electronic Equipment and Environmental Requirements

1. Implement server virtualization where applicable.
2. Collaborate with IT department on ITE purchases. When prudent, select the appropriate higher ASHRAE Classes of hardware (A2, A3, and A4) to permit higher operating temperatures, balance TCO, reliability, locale, and other owner objectives. The reader is referred to the third edition (2012) of Thermal Guidelines for Data Processing Environments.
3. Enable power management features on ITE.
4. Specify high-efficiency power supplies and eliminate unnecessary power conversions. Consider DC-powered hardware if available.
5. Remove “comatose” ITE hardware.
6. Perform technology refreshes to increase computer performance per watt.
7. Optimize IT equipment supply air conditions per Thermal Guidelines for Data Processing Environments.
8. Control HVAC systems based on entering air conditions at the ITE.
9. Determine appropriate gaseous and particulate contamination control measures for the data center location.

KEY ELEMENTS OF COST

1. Infrastructure redundancy comes with a significant capital cost as well as energy-efficiency penalties. The engineer should work to optimize the required redundancy without exceeding the owner's business requirements and risk models.
2. TCO is an important concept for data centers that requires careful evaluation of IT hardware classes, deployment and technology refresh strategies, system redundancies, energy efficiency, cooling transport media (air, water, refrigerant), scalability, etc.

SOURCES OF FURTHER INFORMATION

ASHRAE. 2005. *Design Considerations for IT Equipment Centers*. ASHRAE Datacom series Book No. 3. Atlanta: ASHRAE.

ASHRAE. 2008. *High Density Data Centers, Case Studies and Best Practices*. ASHRAE Datacom Series Book No. 7. Atlanta: ASHRAE.

ASHRAE. 2009. *Thermal Guidelines for Data Processing Environments*, 3rd ed. ASHRAE Datacom Series Book No. 1. Atlanta: ASHRAE.

ASHRAE. 2009. *Best Practices for Datacom Facility Energy Efficiency*, 2nd ed. ASHRAE Datacom Series, Book No. 6. Atlanta: ASHRAE.

ASHRAE. 2012. ANSI/ASHRAE Standard-127R, *Method of Testing for Rating Computer and Data Processing Room Unitary Air Conditioner*. Atlanta: ASHRAE.

ASHRAE. 2012. P*ower Trends and Cooling Applications*, 2nd ed. ASHRAE Datacom Series Book No. 2. Atlanta: ASHRAE.

ASHRAE. 2012. *Particulate and Gaseous Contamination in Datacom Environments*. ASHRAE Datacom Series Book No. 8. Atlanta: ASHRAE.

ASHRAE. 2015. *ASHRAE Handbook—HVAC Applications*. Chapter 19, Data Centers and Telecommunication Facilities. Atlanta: ASHRAE.

ASHRAE. 2016. ANSI/ASHRAE Standard 90.4-2016. *Energy Standard for Data Centers*. Atlanta: ASHRAE.

OTHER REFERENCES AND RESOURCES

Sources for climate data and sustainable design strategies based on climate include, but not limited to the following plus historical and current publications by authors such as Victor Olgyay and Murray Milne (UCLA):

Published

ASHRAE. 2012. *Thermal Guidelines for Data Processing Environments*. Atlanta: ASHRAE.

ASHRAE. 2015. *ASHRAE Handbook—HVAC Applications*. Chapter 44, Building Envelopes (includes "Quick Design Guide for High-Performance Building Envelopes"). Atlanta: ASHRAE.

ASHRAE. 2017. *ASHRAE Handbook—Fundamentals*. Chapter 14, Climatic Design Information. Atlanta: ASHRAE.

ASHRAE. 2017. ANSI/ASHRAE/USGBC/IES Standard 189.1-2017, *Standard for the Design of High-Performance Green Buildings Except Low-Rise Residential Buildings*. Atlanta: ASHRAE. (Will be incorporated into the 2018 version of the *International Green Construction Code* [IgCC].)

Lechner, N. 2001. *Heating Cooling and Lighting*. New York: John Wiley.

Said, S.A.M., H.M. Kadry, and B.I. Ismail. 1996. Climate conditions for Saudi Arabia. *ASHRAE Transactions* 102(1):37–44.

Wimmers, G., and the Light House Sustainable Building Center. "Passive Design Toolkit for Homes." 2009. City of Vancouver, British Columbia. http://vancouver.ca/files/cov/passive-home-design.pdf.

Online

ASHRAE Advanced Energy Design Guides
www.ashrae.org/standards-research--technology/advanced-energy-design-guides.

Climate Zone, for international climate data
http://climate-zone.com/.

Energy Efficiency & Renewable Energy (EERE) DOE Buildings Performance Database
https://energy.gov/eere/buildings/building-performance-database.

Energy Efficiency & Renewable Energy (EERE) (international Weather Data compatible with EnergyPlus simulation software format). https://energyplus.net/weather.

Intergovernmental Panel on Climate Change (IPCC)
www.ipcc.ch.

National Oceanic and Atmospheric Administration (NOAA) National Climatic Data Center
www.ncdc.noaa.gov/.

World Meteorological Organization (WMO) (international climate data in several languages)
http://worldweather.wmo.int/.
Ask Nature
www.asknature.org.

Software

ASHRAE. 2017. Weather Data Viewer, Version 6.0 2017. Atlanta: ASHRAE.

CHAPTER NINETEEN

EXISTING BUILDINGS

INTRODUCTION

The growth rate of commercial and institutional floor area in the United States is presently 1.3% per year (U.S. Energy Information Administration [a]). The corresponding figure for residential floor area is 1.7% per year (U.S. Energy Information Administration [b]). Because the lifetimes of buildings are measured in decades, this means that it will take many decades before the majority of the building area in this country has been constructed to meet current relatively ambitious environmental codes and standards. This in turn suggests that the greatest opportunity to reduce the overall environmental impacts of buildings is to renovate existing buildings. Extending the life of an existing building through a renovation project means that the development of a new site can be avoided. A renovation project requires fewer materials than a new construction project, and provides an opportunity to make the building significantly more energy- and water-efficient. Finally, a renovation project makes it possible to deliver improved indoor environmental quality (IEQ) to building occupants.

This chapter addresses the energy performance of existing buildings, audits covering various aspects of sustainability and IEQ, how to improve building sustainability and IEQ, and finally, several different approaches to renovating existing buildings to enhance their performance in these areas.

ENERGY PERFORMANCE OF EXISTING BUILDINGS

Existing commercial and residential buildings are responsible for more than 40% of global primary energy use and approximately one-third of global greenhouse gas emissions. These figures apply to both developed and developing countries (U.N. Environment Programme). In the United States, commercial buildings are responsible for about one-third of these impacts, while residential buildings are responsible for the remaining two-thirds (U.S. Energy Information Administration [a] and [b]). Clearly, improving the performance of both existing commercial and residential buildings are key strategies for reducing world energy use and greenhouse gas emissions.

It may seem intuitive that newer buildings should be more energy efficient than older buildings. While this is true for residential buildings, it is not generally true for commercial buildings, at least in the United States. The data in Table 19-1 are taken from the 2009 Residential Energy Consumption Survey (EIA 2009) and the 2012 Commercial Building Energy Consumption Survey (EIA 2012).

Note that the average EUI of commercial buildings constructed between 1960 and 1979 is higher than those for commercial buildings constructed either before or after that time period. While significant opportunities often exist to improve the energy performance of buildings constructed in any time period, it would appear that commercial buildings built between 1960 and 1979 may be particularly good candidates for energy improvement projects.

BUILDING AUDITS

Building audits are typically performed to assign one of the labels or certifications described in Chapter 2, to meet a regulatory requirement, or to determine how to improve the performance of the building. These goals are often overlapping. Building audits range from the relatively simple (e.g., an audit to obtain an ENERGY STAR score) to the complex (e.g., an audit to obtain LEED certification). This section addresses energy audits, water audits, IEQ audits, and audits addressing other sustainability categories.

Energy Audits

Energy audits will perhaps be the most familiar type of building audit to readers of this book. Energy audits have been conducted at least since the mid-1970s, after the oil embargo focused attention on the amount of energy wasted by modern residential and nonresidential buildings alike. ASHRAE has been at the forefront of developing formal energy audit procedures, which are presented in detail in ASHRAE's *Proce-*

Table 19-1: Average Building Energy Use Intensity (EUI) by Year Constructed, Thousands of Btus per ft^2 (kWh/m^2)

Building Type	1959 or Before	1960 to 1979	1980 to 2012*
Residential	52.1 (164.3)	48.6 (153.3)	40.1 (126.5)
Commercial	67.3 (212.2)	90.2 (284.5)	80.4 (253.5)

*Residential building energy use intensity is for 1980–2009

dures for Commercial Building Energy Audits (ASHRAE 2011). This book identifies four energy audit levels of effort: the Preliminary Energy Use Analysis, the Level 1 Walk-Through Survey, the Level 2 Energy Survey and Analysis, and the Level 3 Detailed Analysis of Capital Intensive Modifications. Note that the boundaries between these levels of effort are somewhat fluid.

The Preliminary Energy Use Analysis includes gathering and analyzing energy bills for at least a one-year period, developing an EUI and an "Energy Cost Index" (ECI) based on gross conditioned floor area, and comparing these to those for buildings having similar characteristics. ENERGY STAR Portfolio Manager is a good source for such benchmark data (U.S. Environmental Protection Agency). Doing this allows the auditor to determine target energy use and cost indices, and thereby estimate the approximate savings that could be obtained by making building upgrades. ASHRAE Standard 100: Energy Efficiency in Existing Buildings provides a detailed table of target EUIs by building activity and climate zone (ASHRAE 2015).

The Level 1 Walk-Through Survey includes a relatively brief survey of the building focused on identifying a list of potential energy efficiency measures (EEMs), also often referred to as energy conservation measures (ECMs). These can range from no-cost measures (such as setting thermostats back or up at night) to capital intensive measures (such as replacing boilers or chillers). Approximate installation costs and energy cost savings may or may not be provided as part of this effort.

The Level 2 Energy Survey and Analysis includes a more in-depth building survey, as well as more accurate installation cost and cost savings estimates. Establishing cost savings requires a good understanding of baseline operating conditions, proposed operating conditions, utility rates, and interactions among ECMs. Cost savings may be calculated using either spreadsheet models or—particularly when there are significant interactions between ECMs—whole-building simulation models. The results of the Level 2 survey should allow the building owner or manager to determine which measures are worth developing further for potential implementation.

Finally, the Level 3 Detailed Analysis of Capital Intensive Modifications includes additional data-gathering as well as more rigorous engineering and economic analysis than is performed in a Level 2 survey. Sometimes referred to as an Investment Grade Audit (IGA), this level of effort is intended to be sufficient to allow the building owner or manager to make major capital investment decisions.

Water Audits

Water audits can be similar to energy audits, and are often included as part of an energy audit. Historically, the low cost of building water use compared to the cost of building energy use has resulted in auditors paying less attention to water than to energy. However, as the availability of water becomes more constrained in many geographical areas and as municipalities increase rates in order to repair aging water and sewer infrastructure, more emphasis is being placed on auditing buildings' water use and finding ways to reduce the same.

ASHRAE's *Performance Measurement Protocols for Commercial Buildings: Best Practices Guide* (ASHRAE 2012) sets out three water audit or "performance measurement procedure" levels of effort: Basic Evaluation, Diagnostic Measurement, and Advanced Analysis. The Basic Evaluation includes gathering and analyzing water bills for at least a one-year period, determining usage per unit area of the building and per unit area of irrigated landscape, and comparing these indices for those of buildings having similar characteristics. ENERGY STAR Portfolio Manager is a good source for such benchmark data. Doing this allows the auditor to determine the potential for savings. The Basic Evaluation also includes a simple walk-through survey to identify water use issues. On the basis of these data, the auditor should be able to develop recommendations for reducing water use.

The Diagnostic Measurement and Advanced Analysis are progressively more detailed efforts to be used if the recommendations implemented after performing the Basic Evaluation do not result in the desired level of savings. The Diagnostic Measurement procedure adds determining the number of people of each gender occupying the building and then calculating the daily water flow for each fixture type (water closets, urinals, and sinks). The Advanced Analysis procedure adds submetering of specialized uses, such as cooling tower makeup water, kitchens, subdivided landscape areas, and processes. The submeter data can then be used to develop a detailed building water use model. Completing one or both of these two procedures can reveal additional opportunities for reducing water used in the building and on the irrigated landscape.

IEQ Audits

Performance Measurement Protocols for Commercial Buildings: Best Practices Guide (ASHRAE 2012) also sets out three IEQ audit levels of effort. The IEQ categories addressed are thermal comfort, indoor air quality (IAQ), lighting, and acoustics. As is the case with water audits, the three levels of effort are Basic Evaluation, Diagnostic Measurement, and Advanced Analysis. The Basic Evaluation is a two-part process consisting of an occupant survey and field observations. The Performance Measurement Protocol recommends two online, automated surveys: one developed by the Center for the Built Environment at the University of California at Berkeley, the other by the Usable Buildings Trust based in the U.K. Both of these surveys have large databases of accumulated results against which occupant responses can be compared. Field observations should be conducted after completion of the occupant survey. The Protocol contains checklists and spreadsheets to help ensure that all relevant data is captured for the four IEQ categories considered. On the basis of the Basic Evaluation, the auditor should be able to recommend operational changes for improving building IEQ.

The Diagnostic Measurement and Advanced Analysis are progressively more detailed efforts to be used if further testing or design changes are required to fully address building IEQ issues. The Diagnostic Measurement procedure adds physical measurements (e.g., temperatures, infiltration airflows, outdoor airflows, illuminances, and sound levels) plus additional analysis based on these measurements. The Advanced Analysis procedure adds calling in specialists in the four IEQ categories as needed to perform additional testing and develop designs for building modifications. Completing one or both of these two procedures can establish a path toward resolving any IEQ issues that cannot be resolved through use of the Basic Evaluation procedure alone.

Audits Addressing Other Sustainability Categories

Guidance for auditing sustainability categories other than energy efficiency, water efficiency, or IEQ is offered by U.S. Green Building Council (USGBC 2014a) and BRE Global (BRE). In the USGBC publication, that guidance is often—but not always—embedded in the descriptions of individual LEED prerequisites and credits. For example, to obtain the "Light Pollution Reduction" credit in the "Sustainable Sites" category, the assessment team may take perimeter illuminance measurements with building exterior and site lights on and off to demonstrate that the former condition results in illuminance levels no more than 20% higher than the latter condition. BRE consistently provides assessment criteria and evidence that must be submitted to receive credit in each sustainability category. To obtain credit for minimizing "Night time light pollution," for example, the assessment team may submit (a) photos confirming the external luminaires are designed to restrict upward light and light spill, (b) confirmation that light is switched off after a set time, and/or (c) light survey results and recommendations.

Clearly, building audits can encompass a very wide range of activity. The building owner's requirements should always be elicited and documented at the outset. Those requirements should then drive the scope and depth of the audit process.

IMPROVING BUILDING SUSTAINABILITY AND IEQ

The ultimate goal of performing a building audit is often—though not always—to improve building performance (i.e., to reduce its operating costs, lessen its impact on the environment and/or to improve IEQ for building occupants). As part of its "Energy Savings Performance Contract" (ESPC), the U.S. Department of Energy has developed a comprehensive list of 19 technology categories (TCs) and associated ECMs. Despite the nominal focus on energy conservation and efficiency, these categories also include water conservation and efficiency and IEQ improvements. These categories and examples of associated building upgrades are shown in Table 19-2.

Many additional examples of building upgrades addressing energy use, water use, and IEQ are provided in Informative Annex E of ASHRAE Standard 100 (ASHRAE 2015).

U.S. Green Building Council (USGBC 2014a) provides ideas for improving building and site sustainability in areas not addressed by the U.S. Department of

Table 19-2: Department of Energy Technology Categories and Examples of Upgrades

TC	TC Description	Examples
1	Boiler plant improvements	Install oxygen trim controls to improve combustion efficiency. Install economizer in boiler flue to improve fuel utilization efficiency. Replace existing boilers with more efficient boilers to improve fuel utilization efficiency.
2	Chiller plant improvements	Reset chilled-water temperature during periods of low loads to improve chiller plant efficiency. Replace constant-volume primary pumping with variable primary or primary-variable secondary pumping to improve chiller plant efficiency. Replace existing chillers with heat recovery chillers or more efficient chillers.
3	Building automation systems (BASs)/energy management control systems (EMCSs)	Upgrade from pneumatic control to direct digital control. Integrate HVAC and electrical equipment into existing BAS/EMCS. Install a new BAS/EMCS to serve the entire facility.
4	Heating, ventilating, and air conditioning (HVAC) systems (excepting chillers and boilers)	Convert constant-volume air distribution system to variable-volume air distribution to reduce energy use. Replace existing packaged direct expansion (DX) units with more efficient packaged DX units or variable-refrigerant-flow (VRF) units. Install energy recovery wheels or heat pipes on dedicated outdoor air systems.

Table 19-2: Department of Energy Technology Categories and Examples of Upgrades *(Continued)*

TC	TC Description	Examples
5	Lighting improvements	Retrofit existing lighting fixtures with more efficient lamps and ballasts/drivers to reduce energy use and to improve light levels and/or color rendering. Install occupancy sensors to switch lights off when space is not in use. Provide daylighting via skylights or windows to reduce energy use and provide a connection to the exterior environment.
6	Building envelope modifications	Install or replace weather stripping and caulk window and door frames to reduce energy use and to improve thermal comfort. Replace windows to reduce energy use and to reduce glare. Install solar shading to reduce energy use and to increase potential for using daylighting.
7	Chilled water, hot water, and steam distribution systems	Repair and/or replace steam traps to reduce energy use. Insulate piping to reduce energy use. Replace electric domestic hot water (DHW) heaters with natural gas DHW heaters to reduce source energy use and energy costs.
8	Electric motors and drives	Replace larger existing AC motors with more efficient motors. Replace smaller existing AC motors with electrically commutated motors to reduce energy use and/or improve controllability. Install variable-speed drives on motors to reduce energy use and/or improve controllability.

Table 19-2: Department of Energy Technology Categories and Examples of Upgrades *(Continued)*

TC	TC Description	Examples
9	Refrigeration	Reset refrigerator or freezer unit temperatures to optimize food quality and reduce energy use. Consolidate multiple refrigerator or freezer units into a smaller number of units to reduce energy use. Replace refrigeration units with more efficient refrigeration units.
10	Distributed generation	Install cogeneration system to provide space or process heating, cooling, and electric power. Install steam microturbine in place of a large pressure reducing valve to provide electric power. Install fuel cells to provide electric power.
11	Renewable energy systems	Install solar hot-water system to provide domestic hot water or space heating. Install solar photovoltaic system to provide electric power. Replace HVAC units with ground-coupled heat pump system to reduce energy use.
12	Energy/utility distribution systems	Install power factor correction equipment to reduce power factor charges. Replace existing transformers with more efficient transformers. Install natural gas distribution system to enable fuel switching.
13	Water and sewer conservation Systems	Replace existing water closets and urinals with low-flow plumbing fixtures. Install weather-based irrigation system controls to reduce water use. Install a condensate drain piping system to reclaim water from air-handling units for reuse.

Table 19-2: Department of Energy Technology Categories and Examples of Upgrades *(Continued)*

TC	TC Description	Examples
14	Electrical peak shaving/load shifting	Program BAS/EMCS to shed loads in response to signals from electric utility or to maintain building demand below a preset maximum value. Install chilled-water or ice thermal storage system to shift chiller operation to off-peak hours. Install absorption, natural gas engine-driven, or steam turbine-driven chillers to reduce peak electric loads.
15	Energy cost reduction through rate adjustments	Switch to more favorable electric rate schedule. Consolidate electric meters to reduce monthly fixed charges. Purchase natural gas from third party supplier to obtain a more favorable rate.
16	Energy related process improvements	Implement recycling program for scrap materials or used packaging to reduce energy use and costs. Implement manufacturing improvements to reduce energy use. Implement water treatment process improvements to reduce energy use.
17	Commissioning	Review and reprogram BAS/EMCS to match equipment operating hours and temperature set points to the building owner's requirements. Adjust, repair, or replace economizer controls to make effective use of air-side free cooling. Rebalance airflows to reduce energy use and help ensure ventilation standards are met.

Table 19-2: Department of Energy Technology Categories and Examples of Upgrades *(Continued)*

TC	TC Description	Examples
18	Advanced metering systems	Install individual utility meters for each building on a campus to improve monitoring capability. Install submeters for large pieces of equipment or systems to improve monitoring capability. Install advanced metering infrastructure to automatically read utility meters.
19	Appliance/plug load reductions	Install client software to adjust computer power settings when computers are not in use. Install occupancy-sensor-based controllers to reduce energy use by vending machines when space is not occupied. Replace existing laundry equipment with more efficient laundry equipment.

Energy Technology Categories or ASHRAE Standard 100, as well as for areas that are covered by those categories but that are unlikely to result in energy or water savings. Table 19-3 provides examples of upgrades suggested in the LEED rating system for operations and maintenance in the categories of Location and Transportation, Sustainable Sites, and Materials and Resources. Note that many of these examples consist of establishing policies and practices to achieve the desired goals, rather than installing new equipment or systems.

Special care must be taken when implementing energy-related upgrades to a historic building, defined as a structure that is listed or that is eligible for listing in a governmentally recognized registry of historic buildings or places. ASHRAE's proposed Guideline 34P, *Energy Guideline for Historical Buildings*, provides advice on how to approach this type of project. Perhaps the most important single piece of advice offered to the design team is to research and identify all applicable requirements imposed by the governing authorities at the outset of the project. The Energy Guideline provides detailed guidance on envelope modifications and upgrades, HVAC system improvements and additions, and lighting system improvements and additions. In addition to improving the energy performance of the building, it is also always important (1) to preserve the character of the historic building, and (2) to ensure that any changes made are reversible, as the service life of the upgrades will likely be less than the preservation life expectancy of the historic building.

Table 19-3: Three LEED Sustainability Categories and Examples of Upgrades

Sustainability Category	Examples
Location and Transportation	Set up a carpool matching website to promote carpooling. Provide preferential parking for employees who participate in a rideshare program. Provide subsidies to encourage employees to use public transit.
Sustainable Sites	Plant native or adapted vegetation to provide habitat and promote biodiversity. Install a system to capture and treat stormwater flowing off of impervious surfaces to reduce run-off. Clean high reflectance roofs and pavement periodically to reduce the heat island effect. Shield exterior lighting fixtures to reduce light pollution. Use manual or electric powered equipment (e.g., lawnmowers) for site maintenance to reduce pollution.
Materials and Resources	Establish a purchasing policy for ongoing consumables that sets requirements for post-consumer recycled content. Establish a purchasing policy for furniture and furnishings that requires wood products to be certified by the Forest Stewardship Council. Establish a waste reduction and recycling program that reuses, recycles, or composts specified percentages of ongoing and durable goods waste.

DELIVERY MECHANISMS FOR UPGRADES TO EXISTING BUILDINGS

There are three general approaches that may be taken to implementing upgrades and new policies (henceforth included under the term upgrades) for existing buildings. Those are: self-implementation by the building owner, design of upgrades by an engineering firm and implementation by a contractor or contractors, and design and implementation by an energy services company (ESCO). As is the case with audit levels of effort, the boundaries between these three approaches are somewhat fluid.

If a building owner has sufficient capability in-house, they may elect to conduct the building audit, design the upgrades desired, and implement the upgrades without using any consultants or contractors. Advantages to this approach include maximization of control over the project and—depending on labor rates—minimization of costs. Disadvantages include the diversion of labor resources to an effort that may not be part of the building owner's core mission and the inability to take advantage of any specialized expertise or skills that consultants and contractors may be able to offer.

More commonly, the building owner will engage with consultants to audit the building and design the upgrades desired. Those consultants may include engineers, architects, and others with specialized knowledge. The owner will then engage with a general contractor or individual trade contractors to implement the upgrades. This approach does not suffer from the two disadvantages listed above for the self-implementation approach. However, depending again on labor rates, this approach may cost more overall than self-implementation. An additional disadvantage to this approach may be that if the consultant(s) and general contractor do not provide adequate oversight and coordination of trade contractors, the upgrades may not perform as well as originally estimated. This is because energy conservation measures may interact with each other and with building systems in complex ways that cross the boundaries of particular trades and consultant expertise. If the upgrades perform suboptimally, projected energy savings and other benefits will not be fully realized.

Another common approach is for the building owner to engage with an energy services company (ESCO) to undertake all steps: auditing the building, designing the upgrades, and installing the upgrades. In essence, the ESCO acts as the general contractor for the entire process. The ESCO will typically hire consultants with specialized knowledge and contractors with specialized skills as needed. Most ESCOs will also guarantee a percentage of the energy and other utility cost savings anticipated to result from the building upgrades. Such a guarantee makes it easier for the building owner to obtain a loan from a lending institution if he or she is either unable or does not wish to pay the entire project cost upfront. This approach, known generally as "energy performance contracting," also does not suffer from the two disadvantages listed for the self-implementation approach. It may or may not cost more than the other two approaches, depending on labor rates and ESCO markups. Additionally, energy performance contracting is more likely to avoid the problem of under-performing upgrades than the other two approaches, particularly if a savings guarantee is included. A disadvantage to this approach is that the primary focus of ESCOs—as the name suggests—is on energy savings. If the building owner is mainly interested in making improvements that go beyond energy savings, the ESCO may not be the best entity to manage the overall upgrade process.

However, a final potential advantage to an ESCO-managed process is that it may be possible to use the guaranteed utility cost savings to pay for other upgrades that do not result in cost savings. If, to use an example provided in the previous section, a building owner wishes to increase site sustainability by planting native or adapted vegetation, it may be possible to use excess energy savings to pay for that improvement. Of course, it may be possible to implement this measure using the other two approaches as well, although the cost savings guarantee offered by the ESCO would make this easier to structure from a financial standpoint.

Governments and utility companies offer a variety of policies and incentives to support the implementation of energy upgrades to existing buildings. Some examples include: grants, loans, tax credits, rebate payments, and net metering of renewable energy installations. For projects in the United States, the Database of State Incentives for Renewables & Efficiency (DSIRE) is an excellent online resource for identifying these programs by ZIP code.

If a loan is needed to finance the building upgrade project, the lending institution will likely require that the utility cost savings be measured and verified in order to minimize its risk. The building owner may wish to measure and verify savings even if there is no loan involved, to confirm that estimated savings are being achieved. Two widely recognized measurement and verification (M&V) standards are the International Performance Measurement and Verification Protocol published by the Efficiency Valuation Organization and M&V Guidelines: Measurement and Verification for Performance-Based Contracts published by the U.S. Department of Energy. More information concerning measurement and verification is provided in Chapter 16.

REFERENCES AND RESOURCES

Published

ASHRAE. 2011. *Procedures for Commercial Building Energy Audits*, 2nd ed. Atlanta: ASHRAE.

ASHRAE. 2012. *Performance Measurement Protocols for Commercial Buildings: Best Practices Guide*. Atlanta: ASHRAE.

ASHRAE. 2015. ANSI/ASHRAE/IES Standard 100-2015, *Energy Efficiency in Existing Buildings*. Atlanta: ASHRAE.

FEMP. 2016. M&V Guidelines: Measurement and Verification for Performance Contracts, Version 4.0. Washington, D.C.: U.S. Department of Energy. https://energy.gov/sites/prod/files/2016/01/f28/mv_guide_4_0.pdf.

Sustainable Building and Climate Initiative. 2009. Building and Climate Change: Summary for Decision-Makers, https://europa.eu/capacity4dev/unep/document/buildings-and-climate-change-summary-decision-makers. Nairobi: United Nations Environment Programme.

USGBC. 2014a. LEED v4 for Building Operations and Maintenance. Washington, D.C.: U.S. Green Building Council.

USGBC. 2014b. LEED v4 User Guide. Washington, D.C.: U.S. Green Building Council.

Online

BRE, BREEAM In-Use Technical Manual, www.breeam.com/filelibrary/Technical%20Manuals/SD221_BREEAM-In-Use-International-Technical-Manual_V0.pdf

DSIRE, Database of State Incentives for Renewables & Efficiency www.dsireusa.org/.

Efficiency Valuation Organization, International Performance Measurement and Verification Protocol Library https://evo-world.org/en/products-services-mainmenu-en/protocols/ipmvp

EIA. 2009. *Residential Energy Consumption Survey*. Washington, D.C.: U.S. Department of Energy. www.eia.gov/consumption/residential/index.php.

EIA. 2012. *Commercial Buildings Energy Consumption Survey* (CBECS). Washington, D.C.: U.S. Department of Energy. www.eia.gov/consumption/commercial/.

European Commission, BUILD UP www.buildup.eu/.

Maldonado, Eduardo (Editor), Implementing the Energy Performance of Buildings Directive (EPBD): Featuring Country Reports 2012, www.epbd-ca.org/Medias/Pdf/CA3-BOOK-2012-ebook-201310.pdf.

U.S. Department of Energy (a), Department of Energy, Indefinite Delivery Indefinite Quantity, Multiple Award, Energy Savings Performance Contract energy.gov/sites/prod/files/2013/10/f3/generic_idiq_espc_contract.pdf

U.S. Environmental Protection Agency, ENERGY STAR Portfolio Manager www.energystar.gov/buildings/facility-owners-and-managers/existing-buildings/use-portfolio-manager.

CHAPTER TWENTY

EMERGING TRENDS AND EPILOGUE

INTRODUCTION

As we progress further into the twenty-first century, the goal of buildings using much less energy while providing all the shelter and comfort we need is attainable generally now only on a limited scale. Several emerging trends are noted and discussed in this chapter.

Energy (electricity) resources and generation will become more distributed as opposed to centralized. Whether it is in the form of collection of renewables, combined heat and power (CHP) or energy storage, there are major benefits to having the resources near the loads. Instead of energy flowing one direction from grid to buildings, the movement will be in both directions, and the exchange of information between the grid and the consumers will be the basis for a smart grid. To coordinate these bidirectional flows, buildings will need to be "smarter." The Internet of Things (IoT) that inhabit buildings need to work synergistically so that resources are used and optimized. We need buildings to be designed with resiliency as a high priority with our ever-changing climate and vulnerability to outside changes. To assure that buildings continue to perform optimally throughout their lives, they will need to be monitored and tracked with a credible universal metric.

The chapter discusses a few of the key emerging trends that are seen for the coming years in the sustainable built environment, and is divided up into the following sections:

- Distributed resources and energy consumption
- Smart buildings and the Internet of Things
- Virtual and augmented reality
- Resiliency
- Building labeling and reporting
- New construction techniques

DISTRIBUTED (DISTRICT) ENERGY SYSTEMS

Heating and/or cooling buildings using distributed (or district) energy (DE) systems can affect overall energy consumption in various ways, from modest increases or decreases to very dramatic decreases in fuel and energy consumption. The overall amount of energy consumed within the boundaries of an efficiently operated DE service area can be significantly reduced compared to the aggregate of a baseline whereby each building has its own dedicated boilers and chillers. Energy reduction within the buildings will be offset by the energy consumed in the central DE plants and in distributing the thermal energy from central plants to the customer buildings. If the central DE plant uses similar technology (e.g., gas boilers and electric chillers) as otherwise used in the individual buildings, there may be little or no net reduction in energy use. However, the larger (and generally more efficient) DE plant equipment more than offsets extra energy consumed in the distribution of the thermal energy to the buildings for at least a slight net reduction in overall energy consumption. If the centralized DE system includes the generation of electricity along with the cogeneration of combined heating and power (CHP) or combined cooling, heating, and power (CCHP), then the resulting net energy efficiency gains can be quite substantial. DE systems are more readily able to economically use natural renewable thermal energy, such as geothermal heat for district heating systems or cold, deep-water sources (e.g., lakes or oceans) for district cooling systems.

Central energy plants associated with DE systems generally have higher levels of operational efficiency and reliability compared to the alternative of multiple, dispersed, smaller boilers and chiller plants within individual buildings. This is because larger DE plants can more easily justify sophisticated design, automated optimized control systems, more attentive maintenance programs, and more highly trained and focused operations and maintenance personnel. Central energy plants are also better able to match the system load with central equipment versus part-load, especially at part-load throughout the district network.

Reductions in overall fuel and energy consumption, for non-CHP systems, are mainly achieved through the ability of DE plants to more readily and more economically use alternative technologies than is the case for individual buildings. These technologies include but are not limited to dual-fuel boilers; alternative fuel boilers (including renewable fuels such as low-energy content landfill gas, municipal solid waste, and wood waste); high-efficiency boilers; high-efficiency chillers; alternative energy-efficient refrigerants (e.g., ammonia); nonelectric chiller plants (e.g., absorption chillers, engine-driven chillers, or turbine-driven chillers); hybrid chiller plants (with various combinations of electric and nonelectric chillers); and energy-efficient series or series-parallel chiller configurations for high ΔT systems;

Finally, with the increased use of renewable energy resources on either side of the electric meter, the ability to store energy at the point of use, will become increasing important. The storage of heat, cooling and electricity, in active or passive systems, will help make the energy collected by renewable dispatchable—and dramatically more valuable—to the property owner and to the electric grid operator.

SMART BUILDINGS AND THE INTERNET OF THINGS

One recent trend in developing smart buildings and their interactions with a smart grid is the coming of wireless communications between control and monitoring devices. Another related trend is the rapid growth in the Internet of Things or Internet of Energy Things (IoeT).

The concept of the IoT is that all objects, such as a light in a room, are connected to the Internet directly or indirectly via a network of sensors reporting their state, such as the temperature, humidity, and occupancy of a room (see Figure 13-1). In the case of lighting, the status can be detected and directly controlled remotely. A room's status can be detected and controlled remotely by controlling the state of other connected devices, such as a HVAC unit.

The IoT is premised on the notion that more data about the current state of the objects within some domain, such as a building, and the factors that can affect them, such as weather, will enable the domain to be managed more precisely in line with organizational goals. Thus, a smart building can be more energy efficient because it will dynamically adjust to internal and external environmental changes by processing real-time digital data streams to make smart decisions about issues such as internal thermal comfort while at the same time accounting for external factors, such as real-time electricity prices.

A smart building cannot exist in isolation from the environment, and thus it is reliant on a set of complementary external data streams to use its smarts. It needs access to digital data streams describing, for example, forthcoming weather conditions, occupancy demands, and electricity prices. The challenge for smart building operators will be to intelligently merge internal and external data streams to fully exploit the IoT capabilities embedded within their buildings.

VIRTUAL AND AUGMENTED REALITY

Virtual reality (VR) is a technology that immerses the person in a computer-generated environment that is viewed and experienced in lieu of the real world. Augmented reality (AR) is an overlay technology that enhances a person's view of the world. The user is still part of and aware of their surroundings, just with additional information or views provided. The use of VR or AR is just now (as of this writing in 2017) coming on-line in some limited applications. However, there is a vast potential for both VR and AR to become part of the mainstream in the daily functions of building design, construction, and operations. For example, AR could be used to help visualize construction documents on the construction site. A field technician could view the maintenance history data on a piece of equipment, or see graphics outlining how to perform a maintenance task.

These new methods for communication could also help improve coordination between collaborators across the room or across the globe during the design phase of a project. VR techniques could allow an architect to walk their clients and other design team members through a building in three dimensions before it is even built in an experience much more telling than some of the fly-through visualization done with current design software packages. The whole concept of how design reviews are done is likely to change as these technologies mature and reach the mainstream market.

There is also a possibility to combine some features of VR and AR into an experience that some call mixed-reality (MR). For example, MR can project holographic images onto the real world experience that can be manipulated in real time.

RESILIENCY

Though many people think that the major impacts of destabilizing our climate are decades away, professionals in the insurance industry have the data to show we are already feeling some impacts. The 2016 World Economic Forum concluded that the “failure of climate change mitigation and adaptation” was the number one risk for the next 10 years in terms of potential societal impact (World Economic Forum, 2016). Claims for extreme events are increasing approximately 40% per decade and one national insurance group has reduced homeowner’s damage policy limits in Florida from $1.2 million to $400,000 with a goal of getting that down to $100,000 (Mills, no date). Creating structures and communities that can not only survive extremes of wind, flooding, temperature but be places of sanctuary and shelter for populations greater than the buildings were designed for. Resilient design, as defined by the Resilient Design Institute, is “the intentional design of buildings, landscapes, communities, and regions in response to vulnerabilities to disaster and disruption of normal life”. In 2005, Hurricane Katrina devastated parts of New Orleans and led to a humanitarian crisis with the follow-up housing of nearly 16,000 people who did not or were not able to evacuate. Following Hurricane Sandy, the mayor of New York summoned 200 experts to study what could be done and what should be done to make the city more resilient and safe, in each building type and its specific population. For instance, whereas commercial building resiliency would be a business decision for the owner, multifamily residential must provide for essential needs such as safety, drinking water, habitable temperatures, and functioning stairs and elevators. Climate zone, community type, topography, and building type all have bearing on the prioritizing of the threats.

The concept of resiliency differs from sustainability, but many feel that resiliency is complementary to sustainability. They are not mutually exclusive, and some aspects of sustainability can contribute to the overall resiliency of a building.

BUILDING LABELING AND REPORTING

Information on a building's energy use is the critical first step in making the necessary changes and choices that reduce energy use and costs. The ASHRAE Building Energy Quotient (Building EQ) program provides an easily understood scale to convey a building's energy use in comparison to similar buildings, occupancy types, and climate zone, while also providing building owners with building-specific information that can be used to improve building energy performance. A detailed discussion on Building EQ is given in Chapter 2 of this guide.

This program is similar to a program initiated in 2002 by European member countries called the Energy Performance of Buildings Directive (EPBD). Its BUILD UP web portal has been created to help facilitate the exchange of information on building energy efficiency. One key part of the EPBD was the requirement for energy performance certificates to be issued when any building (commercial or residential) is constructed, sold, or rented to a new tenant. A detailed discussion on this program was also given in Chapter 2.

Approximately 10 years after the initiation of the European energy performance certificate program, cities in the U.S. began have building energy performance documented as well. The number of cities requiring energy reporting continues to grow with several large cities such as New York, Boston, Chicago, and Atlanta now requiring some form of energy performance reporting.

NEW CONSTRUCTION TECHNIQUES

Two other new trends are also transforming how we design and construct buildings. For one, there are now buildings being designed and built for high-rise buildings being built with wood frame construction. Wood has a very high strength to weight ratio and is lighter weight than steel or concrete (thus lighter loading to support). Engineered wood products such as glue-laminated timber for structural framing members make this possible.

Another recent development is the application of 3D printing technologies to building construction. For example, a 3D printed office building with a 250 m^2 (2700 ft^2) footprint was completed in Dubai 2016 (Reuters 2016). A large 3D printer using extruded cement layers completed the basic framing structure in 17 days. A five-story apartment building was also completed using 3D printing concepts in China in 2015 (Stampler 2015).

REFERENCES AND RESOURCES

Mills, E. n.d. Responding to climate change—The insurance industry perspective. Berkeley, CA: Lawrence Berkeley National Laboratory. http://evanmills.lbl.gov/pubs/pdf/climate-action-insurance.pdf.

Sukhtian, L., S. Aboudi, and L Ireland. 2016. Dubai says opens world's first functioning 3D-printed office. *Reuters*. http://www.reuters.com/article/us

-emirates-printing-3d/dubai-says-opens-worlds-first-functioning-3d-printed-office-idUSKCN0YF1OB

Stampler, L. 2015. A Chinese company 3D-printed this five-story apartment building. *Time*. http://time.com/3674557/3d-printed-apartment-building-winsun/.

Online

ASHRAE Building Energy Quotient
www.buildingenergyquotient.org/.

European Commission, BUILD UP, The European Portal for Energy Efficiency in Buildings.
www.buildup.eu/en.

Foster, S., C. Weems, R. Matyasovski, and T. Palefski. n.d. Resiliency: From Surviving to Thriving. *High Performance Buildings*
www.hpbmagazine.org/Topics-Opinions/Resiliency-From-Surviving-to-Thriving/.

World Economic Forum 2016
www.weforum.org/agenda/2016/01/what-are-the-top-global-risks-for-2016/.

References and Resources

These listings do not necessarily imply endorsement or agreement by ASHRAE or the authors with the information contained in the documents.

BOOKS AND SOFTWARE PROGRAMS

ACGIH. 1999. *Bioaerosols: Assessment and Control*. J. Nacher, ed. Cincinnati, Ohio: American Conference of Governmental Industrial Hygienists. www.acgih.org.

AGCC. 1994. *Applications Engineering Manual for Direct-Fired Absorption*. Washington, DC: American Gas Cooling Center.

RSMeans. 2010. *Green building: Project planning and cost estimating*, third ed. Hoboken, NJ: John Wiley & Sons.

AIA. 1996. *Environmental Resource Guide*. Edited by Joseph Demkin. New York: John Wiley & Sons.

AIA. 2006. *Guidelines for Design and Construction of Health Care Facilities*. Washington, D.C.: American Institute of Architects.

Alvarez, S., E. Dascalaki, G. Guarracino, E. Maldonado, S. Sciuto, and L. Vandaele. 1998. *Natural Ventilation of Buildings—A Design Handbook*. F. Allard, ed. London: James & James Science Publishers Ltd.

Argiriou, A., A. Dimoudi, C.A. Balaras, D. Mantas, E. Dascalaki, and I. Tselepidaki. 1996. *Passive Cooling of Buildings*. M. Santamouris and A. Asimakopoulos, eds. London: James & James Science Publishers Ltd.

ARI. 1997. *ARI STANDARD 275-97, Standard for Application of Sound Rating Levels of Outdoor Unitary Equipment*. Arlington, VA: Air-Conditioning and Refrigeration Institute.

Arthus-Bertrand, Y. 2001. *Earth from Above: 365 Days*. New York: Harry N. Abrams.

Arthus-Bertrand, Y. 2002. *Earth From Above*. New York: Harry N. Abrams.

ASHRAE. 1975. ASHRAE Standard 90-75, *Energy Conservation in New Building Design*. Atlanta: ASHRAE.

ASHRAE. 1988. *Active Solar Heating Systems Design Manual*. Atlanta: ASHRAE.

ASHRAE. 1995. *Commercial/Institutional Ground-Source Heat Pump Engineering Manual*. Atlanta: ASHRAE.

ASHRAE. 1996. ASHRAE Guideline 1-1996, *The HVAC Commissioning Process*. Atlanta: ASHRAE.

ASHRAE. 2000. Interpretation IC-62-1999-30 (August) of *ANSI/ASHRAE Standard 62-1999, Ventilation for Acceptable Indoor Air Quality*. Atlanta: ASHRAE.

ASHRAE. 2004. *Advanced Energy Design Guide for Small Office Buildings*. Atlanta: ASHRAE.

ASHRAE. 2006. *Advanced Energy Design Guide for Small Retail Buildings*. Atlanta: ASHRAE.

ASHRAE. 2007. ASHRAE Guideline 1.1, *HVAC&R Technical Requirements for the Commissioning Process*. Atlanta: ASHRAE.

ASHRAE. 2009. *The ASHRAE Guide for Buildings in Hot and Humid Climates,* 2nd edition. Atlanta: ASHRAE.

ASHRAE. 2009. *Indoor Air Quality Guide: Best Practices for Design, Construction and Commissioning*. Atlanta: ASHRAE.

ASHRAE. 2013. ASHRAE Guideline 0-2013, *The Commissioning Process*. Atlanta: ASHRAE.

ASHRAE. 2013. *HVAC Design Manual for Hospitals and Clinics*, 2nd ed. Atlanta: ASHRAE.

ASHRAE. 2014. ASHRAE Guideline 14-2014, *Measurement of Energy and Demand Savings*. Atlanta: ASHRAE.

ASHRAE. 2014. *ASHRAE Handbook—Refrigeration*. Atlanta: ASHRAE.

ASHRAE. 2015. ASHRAE Guideline 13-2015, *Specifying Direct Digital Control Systems*. Atlanta: ASHRAE.

ASHRAE. 2015. *ASHRAE Handbook—HVAC Applications*. Atlanta: ASHRAE.

ASHRAE. 2015. ANSI/ASHRAE/IESNA Standard 100-2015, *Energy Conservation in Existing Buildings*. Atlanta: ASHRAE.

ASHRAE. 2015. *Fundamentals of Water System Design,* 2nd edition. Atlanta: ASHRAE.

ASHRAE. 2016. *ASHRAE Handbook—HVAC Systems and Equipment*. Atlanta: ASHRAE.

ASHRAE. 2016. ANSI/ASHRAE Standard 62.1-2016, *Ventilation for Acceptable Indoor Air Quality*. Atlanta: ASHRAE.

ASHRAE. 2016. ANSI/ASHRAE Standard 62.2-2016, *Ventilation and Acceptable Indoor Air Quality in Low-Rise Residential Buildings*. Atlanta: ASHRAE.

ASHRAE. 2017. *ASHRAE Handbook—Fundamentals*. Atlanta: ASHRAE.

ASHRAE. 2016. ANSI/ASHRAE/IESNA Standard 90.1-2016, *Energy Standard for Buildings Except Low-Rise Residential Buildings*. Atlanta: ASHRAE.

ASHRAE. 2017. ANSI/ASHRAE Standard 52.2-2017, *Method of Testing General Ventilation Air-Cleaning Devices for Removal Efficiency by Particle Size.* Atlanta: ASHRAE.

ASHRAE. 2017. ANSI/ASHRAE Standard 55-2017, *Thermal Environmental Conditions for Human Occupancy*. Atlanta: ASHRAE.

ASHRAE. 2017. ANSI/ASHRAE/USGBC/IES Standard 189.1-2017, *Standard for the Design of High-Performance Green Buildings Except Low-Rise Residential Buildings*. Atlanta: ASHRAE. (Will be incorporated into the 2018 version of the *International Green Construction Code* [IgCC].)

ASPE. 1998. *Domestic Water Heating Design Manual*. Chicago: American Society of Plumbing Engineers.

ASTM. 1998. *ASTM D 6245-1998: Standard Guide for Using Indoor Carbon Dioxide Concentrations to Evaluate Indoor Air Quality and Ventilation*. Philadelphia: ASTM International.

Baker, N. 1995. *Light and Shade: Optimising Daylight Design*. European Directory of sustainable and energy efficient building. London: James & James Science Publishers Ltd.

Balcomb, J.D. 1984. *Passive Solar Heating Analysis: A Design Manual*. Atlanta: ASHRAE.

Bauman, F.S., and A. Daly. 2003. *Underfloor Air Distribution Design Guide*. Atlanta: ASHRAE.

Beckman, W., S.A. Klein, and J.A. Duffie. 1977. *Solar Heating Design*. New York: John Wiley & Sons, Inc.

BRESCU, BRE. 1999. *Natural Ventilation for Offices Guide and CD-ROM*. ÓBRE on behalf of the NatVent Consortium, Garston, Watford, UK, March.

CEC. 2001. *Nonresidential Alternative Calculations Methods Manual*. Sacramento, CA: California Energy Commission.

CEC. 2002. *Part II: Measure Analysis and Life-Cycle Cost 2005, California Building Energy Standards*, P400-02-012. Sacramento, CA: California Energy Commission.

Crosbie, M.J., ed. 1998. *The Passive Solar Design and Construction Handbook*. Steven Winter Associates. New York: John Wiley & Sons, Inc.

Del Porto, D., and C. Steinfeld. 1999. *The Composting Toilet System Book*. New Bedford, MA: The Center for Ecological Pollution Prevention.

Dorgan, C., and J.S. Elleson. 1993. *Design Guide for Cool Thermal Storage*. Atlanta: ASHRAE.

Elleson, J.S. 1996. *Successful Cool Storage Projects: From Planning to Operation*. Atlanta: ASHRAE.

EPA. 2010. *Guide to Purchasing Green Power*. Washington, D.C.: U.S. Environmental Protection Agency. www.epa.gov/grnpower/buygreenpower/guide.htm.

Esty, D.C., and A.S. Winston. 2006. *Green to Gold: How Smart Companies Use Environmental Strategy to Innovate, Create Value, and Build Competitive Advantage*. New Haven, CT: Yale University Press.

Evans, B. 1997. Daylighting design. *In Time-Saver Standards for Architectural Design Data*. New York: McGraw-Hill.

FEMP. 2008. *M&V Option A Detailed Guidelines.* Washington, DC: U.S. Department of Energy, Federal Energy Management Program.

Givoni, B. 1994. *Passive and Low Energy Cooling of Buildings*. New York: Van Nostrand Reinhold.

Gottfried, D. (Continuous updating). *Sustainable Building Technical Manual, Green Building Design, Construction, and Operations*. Washington, DC: U.S. Green Building Council. www.usgbc.org.

Hawkin, P., A. Lovins, and L.H. Lovins. 1999. *Natural Capitalism*. Little Brown.

Heerwagen, J.H., J.C. Montgomery, W.C. Weimer, and J.G. Heubach. 1995. Assessing the human and organizational impacts of green buildings. Pacific Northwest Laboratory, Richland, WA.

Henning, H-M., ed. 2004. *Solar Assisted Air-Conditioning in Buildings—A Handbook For Planners*. Wien: Springer-Verlag.

Houghton, J.T., G.J. Jenkins, and J.J. Ephraums, eds. 1990. *Climate Change: The IPCC Scientific Assessment.* Intergovernmental Panel on Climate Change. New York: Cambridge UP.

Howell, J., R. Bannerot, and G. Vliet. 1982. *Solar-Thermal Energy Systems: Analysis and Design.* New York: McGraw-Hill.

IECC. 2015. *International Energy Conservation Code*. Available from www.iccsafe.org.

IES. 2011. *Lighting Handbook.* New York: Illuminating Engineering Society of North America.

IES. 1998. *RP-33-99, Lighting for Exterior Environments; RP-20-98, Lighting for Parking Facilities*. New York: Illuminating Engineering Society of North America. www.iesna.org.

IESNA. 2002. Proposed revisions to ASHRAE/IESNA Standard 90.1-2001. Illuminating Engineering Society of North America, Energy Management Committee.

Kavanaugh, S.P., and K. Rafferty. 1997. *Ground-Source Heat Pumps: Design of Geothermal Systems for Commercial and Institutional Buildings*. Atlanta: ASHRAE.

Kinsley, M. 1997. *Economic Renewal Guide*. Colorado: Rocky Mountain Institute.

Klein, S., and W. Beckman. 2001. *F-Chart Software.* The F-Chart Software Company., Madison, WI. www.fchart.com.

Kowalski, W.J. 2006. *Aerobiological Engineering Handbook: Airborne Disease and Control Technologies.* New York: McGraw-Hill.

Kreith, F., and J. Kreider. 1989. *Principles of Solar Engineering*, 2d ed. Wasington, DC: Hemisphere Pub.

Ludwig, A. 1997. *Builder's Greywater Guide and Create an Oasis with Greywater.* Oasis Design.

Macher, J., ed. 1999. *Bioaerosols: Assessment and Control.* Cincinatti, Ohio: American Conference of Governmental Industrial Hygenists.

McIntosh, I.B.D., C.B. Dorgan, and C.E. Dorgan. 2002. *ASHRAE Laboratory Design Guide.* Atlanta: ASHRAE.

Mendler, S.F., and W. Odell. 2000. *HOK Guidebook to Sustainable Design*. New York: John Wiley & Sons, Inc.

NADCA. *General Specifications for the Cleaning of Commercial Heating, Ventilating and Air Conditioning Systems*. Washington, DC: National Air Duct Cleaners Association.

Nattrass, B., and M. Altomare. *The Natural Step for Business; Wealth, Ecology and the Evolutionary Corporation*. Gabriola Island, British Columbia: New Society Publishers.

NBI. 1998. *Gas Engine Driven Chillers Guideline*. Fair Oaks, CA: New Buildings Institute and Southern California Gas Company. www.newbuildings.org /downloads/guidelines/GasEngine.pdf.

NBI. 2001. *Advanced Lighting Guidelines*. Fair Oaks, CA: New Buildings Institute. www.newbuildings.org/lighting.htm.

NFPA. 2004. *NFPA 45, Standard on Fire Protection for Laboratories using Chemicals*. Quincy, MA: National Fire Protection Association.

NFPA. 2005. *NFPA 99, Standard for Health Care Facilities*. Quincy, MA: National Fire Protection Association.

NRC. *Photovoltaic Systems Design Manual*. Ottawa, Ontario, Canada: Natural Resources Canada, Office of Coordination and Technical Information.

NRC. RETScreen (Renewable Energy Analysis Software). Natural Resources Canada, Energy Diversification Research Laboratory, Ottawa, Ontario, Canada. www.retscreen.net.

NSF. 2000. *NSF/ANSI 5-2000e: Water Heaters, Hot Water Supply Boilers, and Heat Recovery Equipment*. Ann Arbor, MI: National Sanitation Foundation.

PG&E. 2002. *Codes and Standards Enhancement Report, Code Change Proposal for Cooling Towers.* California: Pacific Gas & Electric.

Savitz, A.W., and K. Weber. 2006. *The Triple Bottom Line*. San Fransisco: John Wiley & Sons, Inc.

Schaffer, M. 2005. *Practical Guide to Noise and Vibration Control for HVAC Systems,* 2nd edition. Atlanta: ASHRAE.

SMACNA. 2005. *HVAC Duct Construction Standards,* 3d ed. Chapter 1. Chantilly, VA: Sheet Metal and Air Conditioning Contractors National Association, Inc.

SMACNA. 2007. *IAQ Guidelines for Occupied Buildings Under Construction*. Chantilly, VA: Sheet Metal and Air Conditioning National Association, Inc.

Spitler, J. 2009. *Load Calculation Applications Manual*. Atlanta: ASHRAE.

Taylor, S., P. Dupont, M. Hydeman, B. Jones, and T. Hartman. 1999. *The CoolTools Chilled Water Plant Design and Performance Specification Guide*. San Francisco, CA: PG&E Pacific Energy Center.

Trane. 1994. *Water-Source Heat Pump System Design*. Trane Applications Engineering Manual, SYS-AM-7. Lacrosse, WI: Trane Company.

Trane Co. 1999. *Trane Applications Engineering Manual, Absorption Chiller System Design*, SYS-AM-13. Lacrosse, WI: Trane Company.

Trane. 2000. *Multiple-Chiller-System Design and Control Manual.* Lacrosse, WI: Trane Company.

Trane. 2005. *Trane Installation, Operation, and Maintenance Manual: Series R® Air-Cooled Rotary Liquid chillers*, RTAA-SVX01A-EN. Lacrosse, WI: Trane Company. www.trane.com/Commercial/Uploads/Pdf/1060/RTAA-SVX01A-EN_09012005.pdf.

UG. 2010. *Wise Energy Guide*. Chatham, ON: Union Gas. www.uniongas.com/residential/energyconservation/education/wiseEnergyGuide/WEG_Booklet_Web_Final.pdf/.

USGBC. 2001. *LEED Reference Guide, Version 2.0.* Washington, DC: U.S. Green Building Council.

USGBC. 2005. *LEED Reference Guide, Version 2.2.* Washington, DC: U.S. Green Building Council.

USGBC. 2009. *LEED 2009 for New Construction and Major Renovations.* Washington, D.C.: U.S. Green Building Council.

USGBC. 2009. *LEED 2009 Green Building Operations and Maintenance.* Washington, DC: U.S. Green Building Council.

Watson, D., and G. Buchanan. 1993. *Designing Healthy Buildings.* Washington, DC: American Institute of Architects.

Watson, R.D., and K.S. Chapman. 2002. *Radiant Heating and Cooling Handbook.* New York: McGraw-Hill.

Watsun. 1999. *WATSUN-PV—A computer program for simulation of solar photovoltaic systems, User's Manual and Program Documentation,* Version 6.1. Watsun Simulation Laboratory, University of Waterloo, Ontario, Canada.

Wolverton, B.C. 1996. *How to Grow Fresh Air: 50 Plants That Purify Your Home of Office.* Baltimore, MD: Penguin Books.

PERIODICALS AND REPORTS

Adamson, R. 2002. Mariah Heat Plus Power™ Packaged CHP, applications and economics. Second Annual DOE/CETC/CANDRA Workshop on Microturbine Applications, January 17–18, University of Maryland, College Park, MD.

Afify, R. 2008. Designing VRF systems. *ASHRAE Journal* 50(6):52–55.

Alliance for Sustainable Built Environments. Existing buildings hold the key. www.awarenessintoaction.com/whitepapers/how-existing-buildings-high-performing-green-leed-certified.html.

Andersson, L.O., K.G. Bernander, E. Isfält, and A.H. Rosenfeld. 1979. Storage of heat and coolth in hollow-core concrete slabs. Swedish experience and application to large, American style buildings. Second International Conference on Energy Use and Management, Lawrence Berkeley National Laboratory, LBL-8913.

ASHE. 2002. Green healthcare design guidance statement. American Society of Healthcare Engineering, Chicago.

Aynur, T., Y. Hwang, and R. Radermacher. 2008. Experimental evaluation of the ventilation effect on the performance of a VRV system in cooling mode—Part I: Experimental evaluation. *HVAC&R Research* 14(4):615–30.

Aynur, T., Y. Hwang, and R. Radermacher. 2008. Simulation evaluation of the ventilation effect on the performance of a VRV system in cooling mode—Part II: Simulation evaluation. *HVAC&R Research* 14(5):783–95.

Bahnfleth, W.P., and W.S. Joyce. 1994. Energy use in a district cooling system with stratified chilled water storage. *ASHRAE Transactions* 100(1):1767–78.

Balaras, C.A. 1995. The role of thermal mass on the cooling load of buildings. An overview of computational methods. *Energy and Buildings* 24(1):1–10.

Balaras, C.A., E. Dascalaki, P. Tsekouras, and A. Aidonis. 2010. High solar combi systems in Europe. *ASHRAE Transactions* 116(1):408–15.

Balaras, C.A., H.-M. Henning, E. Wiemken, G. Grossman, E. Podesser, and C.A. Infante Ferreira. 2006. Solar cooling—An overview of european applications and design guidelines. *ASHRAE Journal* 48(6):14–22.

Bauman, F., and T. Webster. 2001. Outlook for underfloor air distribution. *ASHRAE Journal* 43(6):18–25.

Beebe, J.M. 1959. Stability of disseminated aerosols of *Pastuerella tularensis* subjected to simulated solar radiations at various humidities. *Journal of Bacteriology* 78:18–24.

Bisbee, D. 2003. Pulse-power water treatment systems for cooling towers. Energy Efficiency & Customer Research & Development, Sacramento Municipal Utility District.

Braun, J.E. 1990. Reducing energy costs and peak electrical demand through optimal control of building thermal storage. *ASHRAE Transactions* 96(2):876–88.

CBE. 2007. Summary report: Control strategies for mixed-mode buildings. Berkeley, CA: Center for Built Environment. http://www.cbe.berkeley.edu/research/pdf_files/SR_MixedModeControls2007.pdf.

CEC. 1996. Source energy and environmental impacts of thermal energy storage. Completed by Tabors, Caramanis & Assoc. for the California Energy Commission.

CEC. 2004. Monitoring the energy-use effects of cool roofs on California commercial buildings, CEC PIER study report. California Energy Commission.

CEC. 2005. Title 24 2005 Reports and Proceedings. California Energy Commission. www.energy.ca.gov.

Cendón, S.F. 2009. *New and Cool: Variable Refrigerant Flow Systems*. AIArchitect.

CETC. 1996a. Technical report on Bentall Corporation Crestwood 8 C-2000 Building. CANMET Energy Technology Centre, Natural Resources Canada, Ottawa, Ontario, Canada.

CETC. 1996b. Technical report on Green on the Grand C-2000 Building. CANMET Energy Technology Centre, Natural Resources Canada, Ottawa, Ontario, Canada.

CFR. 1995. Energy conservation voluntary performance standards for commercial and multi-family high rise residential buildings; mandatory for new federal buildings. Code of Federal Regulations 10 CFR 435, Office of the Federal Register, National Archives and Records Administration, Washington, DC.

Chapman, K.S. ASHRAE RP-907, Final report, Development of design factors for the combination of radiant and convective in-space heating and cooling systems. Atlanta: ASHRAE.

Chapman, K.S. 2002. Development of radiation transfer equation that encompasses shading and obstacles. *ASHRAE Transactions* 108(2):997–1004.

Chapman, K.S., J.M. DeGreef, and R.D. Watson. 1997. Thermal comfort analysis using BCAP for retrofitting a radiantly heated residence. *ASHRAE Transactions* 103(1):959–65.

Chapman, K.S., J.E. Howell, and R.D. Watson. 2001. Radiant panel surface temperature over a range of ambient temperatures. *ASHRAE Transactions* 107(1):383–89.

Chapman, K.S., J. Rutler, and R.D. Watson. 2000. Impact of heating systems and wall surface temperatures on room operative temperature fields. *ASHRAE Transactions* 106(1):506–14.

Cho, Y.I., S. Lee, and W. Kim. 2003. Physical water treatment for the mitigation of mineral fouling in cooling-tower water applications. *ASHRAE Transactions* 109(1):346–57.

Coad, W.J. 1999. Conditioning ventilation air for improved performance and air quality. *Heating/Piping/Air Conditioning*, September.

Corbett, T. *Green Buildings and Risk*. www.aepronet.org/ge/no43.html.

Darlington, A., M. Chan, D. Malloch, C. Pilger, and M.A. Dixon. 2000. The biofiltration of indoor air: Implications for air quality. *Indoor Air 2000* 10(1):39–46.

Darlington, A.B., J.F. Dat, and M.A. Dixon. 2001. The biofiltration of indoor air: Air flux and temperature influences the removal of toluene, ethylbenzene, and xylene. *Environ Sci Technol* 35(1):240–46.

Darlington, A., M.A. Dixon, and C. Pilger. 1998. The use of biofilters to improve indoor air quality: The removal of toluene, TCE, and formaldehyde. *Life Support Biosph Sci* 5(1):63–69.

DeGreef, J.M., and K.S. Chapman. 1998. Simplified thermal comfort evaluation of MRT gradients and power consumption predicted with the BCAP methodology. *ASHRAE Transactions* 104(1):1090–97.

Deru, M., and P. Torcellini. 2007. *Source Energy and Emissions Factors for Energy Use in Buidlings.* Technical Report NREL/TP_550-38617, National Renewable Energy Laboratory, Golden, CO.

DOE. *The International Performance Measurement and Performance Protocol* (IPMPP). U.S. Department of Energy, Office of Energy Efficiency and Renewable Energy, Washington, DC.

DOE. 2002. *Energy Tip Sheet #1*, May. Washington, DC: U.S. Department of Energy, Office of Industrial Technologies, Energy Efficiency, and Renewable Energy.

DOE. 2005. Energy impact of commercial building controls and performance diagnostics: Market characterization, energy impact of building faults and energy savings potential. Washington, D.C.: U.S. Department of Energy.

DOE. 2006. Solar buildings. Transpired air collectors ventilation preheating. DOE/GO-102001-1288. Washington, DC: U.S. Department of Energy.

Duffy, G. 1992. Thermal storage shifts to saving energy. *Eng. Sys.* July/August: 32–38.

EDR. 2009. *Chilled Water Plant Design Guide*. Energy Design Resources. www.energydesignresources.com/Resources/Publications/DesignGuidelines/tabid/73/articleType/ArticleView/articleId/422/Default.aspx.

Ehrt, D., M. Carl, T. Kittel, M. Muller, and W. Seeber. 1994. High performance glass for the deep ultraviolet range. *Journal of Non-Crystalline Solids* 177:405–19.

EIA. Table 10.8, Photovoltaic cell and module shipments by type, trade, and prices, 1982–2008. U.S. Energy Information Administration. www.eia.doe.gov/aer/txt/stb1008.xls.

El-Adhami, W., S. Daly, and P.R. Stewart. 1994. Biochemical studies on the lethal effects of solar and artificial ultraviolet radiation on *Staphylococcus aureus*. *Arch Microbiol* 161:82–87.

Ellis, M.W., and M.B. Gunes. 2002. Status of fuel cell systems for combined heat and power applications in buildings. *ASHRAE Transactions* 108(1):1032–44.

EPA. *Storm water management for construction activities*. EPA document No. EPA-8320R-92-005. Washington, D.C.: U.S. Environmental Protection Agency.

EPA. Toxicity Characteristic Leaching Procedure (TCLP) Test, 40 CFR Part 260. U.S. Environmental Protection Agency, Washington, DC.

EPA. Universal Waste Rule, 40 CFR Part 260. U.S. Environmental Protection Agency, Washington, DC.

EPA. 2010. *How to Conserve Water and Use It Effectively*. U.S. Environmental Protection Agency, Washington, DC. www.epa.gov/owow/nps/chap3.html.

EPIA. 2010. 2014 global market outlook for photovoltaics until 2014. European Photovoltaic Industry Association. www.epia.org/fileadmin/EPIA_docs/public/Global_Market_Outlook_for_Photovoltaics_until_2014.pdf.

Epoch Times. 2009. Water efficiency a priority for green buildings. www.theepochtimes.com/n2/content/view/17751.

Erickson, D.C., and M. Rane. 1994. Advanced absorption cycle: Vapor exchange GAX. ASME International Absorption Heat Pump Conference, January 19–21, New Orleans, LA.

ESTIF. 2010 Solar thermal markets in Europe—Trends and market statistics 2009. www.estif.org/fileadmin/estif/content/market_data/downloads/2009%20solar_thermal_markets.pdf.

EU. 2008. Climate change: Commission welcomes final adoption of Europe's climate and energy package. Europa press release IP/08/1998, European Union, Brussels.

European Commission. 2002. Energy Performance of Buildings, Directive 2002/91/EC of the European Parliament and of the Council. *Official Journal of the European Communities*, Brussels.

European Commission. 2010. Energy Performance of Buildings, Directive 2010/31/EU of the European Parliament and of the Council. *Official Journal of the European Communities*, Brussels.

Fagan, D. 2001. A comparison of storage-type and instantaneous heaters for commercial use. *Heating/Piping/Air Conditioning Engineering*, April.

Fernandez, R.O. 1996. Lethal effect induced in *Pseudomonas aeruginosa* exposed to ultraviolet-A radiation. *Photochem & Photobiol* 64(2):334–39.

Fiorino, D.P. 1994. Energy conservation with stratified chilled water storage. *ASHRAE Transactions* 100(1):1754–66.

Fiorino, D.P. 2000. Six conservation and efficiency measures reducing steam costs. *ASHRAE Journal* 42(2):31–39.

Galuska, E.J. 1994. Thermal storage system reduces costs of manufacturing facility (Technology Award case study). *ASHRAE Journal*, March.

Gansler, R.A., D.T. Teindil, and T.B. Jekel. 2001. Simulation of source energy utilization and emissions for HVAC systems. *ASHRAE Transactions* 107(1):39–51.

Gleason, G.W. 1951. Energy—Choose it wisely today for safety tomorrow. *ASH&VE Transactions* 57.

Goetzler, W. 2007. Variable refrigerant flow systems. *ASHRAE Journal* 46(4): 24–31.

Goss, J.O., L. Hyman, and J. Corbett. 1996. Integrated heating, cooling and thermal energy storage with heat pump provides economic and environmental solutions at California State University, Fullerton. *Proceedings of the EPRI International Conference on Sustainable Thermal Energy Storage, Minneapolis, MN,* pp. 163–67.

Grondzik, W.T. 2001. The (mechanical) engineer's role in sustainable design: Indoor environmental quality issues in sustainable design. HTML presentation available at www.nettally.com/gzik/ieq/ieq-sust_files/frame.htm.

Guz, K. 2005. Condensate water recovery. *ASHRAE Journal* 47(6):54–56.

GWEC. 2010. Global wind 2009 report. Global Wind Energy Council, Brussels, Belgium. www.gwec.net/index.php?id=167.

Hayter, S., P. Torcellini, and R. Judkoff. 1999. Optimizing building and HVAC systems. *ASHRAE Journal* 41(12):46–49.

Heerwagen, J. 2001. Do green buildings enhance the well being of workers? www.edcmag.com.

Hilten, R.N. 2005. An analysis of the energetics and stormwater mediation potential of greenroofs. Master's thesis, University of Georgia, Athens, GA.

HPAC. 2004. Innovative grocery store seeks LEED certification. *HPAC Engineering* 27:31.

ICC. 2010. *International green construction code*, Public version 1.0. International Code Council, Washington, DC.

IUVA. 2005. General guideline for UVGI air and surface disinfection systems. IUVA-G01A-2005 International Ultraviolet Association, Ayr, Ontario, Canada, www.iuva.org.

Jarnagin, R.E., R.M. Colker, D. Nail and H. Davies. 2009. ASHRAE Building EQ. *ASHRAE Journal* 51(12):18–21.

Jones, B., and K.S. Chapman. ASHRAE RP-657, Final report, Simplified method to factor mean radiant temperature (MRT) into building and HVAC system design. Atlanta: ASHRAE.

Kainlauri, E.O., and M.P. Vilmain. 1993. Atrium design criteria resulting from comparative studies of atriums with different orientation and complex interfacing of environmental systems. *ASHRAE Transactions* 99(1):1061–69.

Keeney, K.R., and J.E. Braun. 1997. Application of building precooling to reduce peak cooling requirements. *ASHRAE Transactions* 103(1):463–69.

Kintner-Meyer, M., and A.F. Emery. 1995. Optimal control of an HVAC system using cold storage and building thermal capacitance. *Energy and Buildings* 23:19–31.

Kosik, W.J. 2001. Design strategies for hybrid ventilation. *ASHRAE Journal* 43(10):18–24.

Larsson, N., ed. 1996. *C-2000 Program Requirements*. Natural Resources Canada, Ottawa, Ontario, Canada.

Lawrence, T.M. 2004. Demand-controlled ventilation and sustainability. *ASHRAE Journal* 46(12):117–21.

Lawrence, T. 2008. Selecting CO_2 criteria for space monitoring. *ASHRAE Journal* 50(12):18–27.

Lawrence, T., J. Perry, and P. Dempsey. 2010. Capturing condensate by retrofitting AHUs. *ASHRAE Journal* 52(1):48–54.

Lawson, S.H. 1988. Computer facility keeps cool with ice storage. *HPAC* 60(88):35–44.

Leaman, A., and B. Bordass. 1993. Building design, complexity and manageability. *Facilities*, Vol. 11.

LeMar, P. 2002. Integrated energy systems (IES) for buildings: A market assessment. Final report by Resource Dynamics Corporation for Oak Ridge National Laboratory, Contract No. DE-AC05-00OR22725.

Mathaudhu, S.S. 1999. Energy conservation showcase. *ASHRAE Journal* 41(4):44–46.

Mauthner, F. and W. Weiss. 2013. Solar heat worldwide: Markets and contribution to the energy supply 2011. Edition 2013. IEA Solar Heating and Cooling Programme, May 2013. International Energy Agency, Paris. http://www.iea-shc.org/solar-heat-worldwide.

McDonough, W. 1992. *The Hannover principles: Design for sustainability*. Presentation, Earth Summit, Brazil.

McKurdy, G., S.J. Harrison, and R. Cooke. Preliminary evaluation of cylindrical skylights. 23rd Annual Conference of the Solar Energy Society of Canada Inc., Vancouver, British Columbia.

Morris, W. 2003. The ABCs of DOAS. *ASHRAE Journal* 45(5).

Mumma, S.A. 2001. Designing dedicated outdoor air systems. *ASHRAE Journal* 43(5):28–31.

NREL. 2010. NREL highlights utility green power leaders. NREL press release NR-1910. Golden, CO: National Renewable Energy Laboratory.

O'Neal, E.J. 1996. Thermal storage system achieves operating and first-cost savings (Technology Award case study). *ASHRAE Journal* 38(4).

Palmer, J.M., and K.S. Chapman. 2000. Direct calculation of mean radiant temperature using radiant intensities. *ASHRAE Transactions* 106(1):477–86.

Patnaik, V. 2004. Experimental verification of an absorption chiller for BCHP applications. *ASHRAE Transactions* 110(1):503–507.

Pearson, F. 2003. ICR 06 Plenary Washington. *ASHRAE Journal* 45(10).

Rautiala, S., S. Haatainen, H. Kallunki, L. Kujanpaa, S. Laitinen, A. Miihkinen, M. Reiman, and M. Seuri. 1999. Do plants in office have any effect on indoor air microorganisms? *Indoor Air 99: Proceedings of the 8th International Conference on Indoor Air Quality and Climate, Edinburgh, Scotland*, pp. 704–709.

Reindl, D.T., R.A. Gansler, and T.B. Jekel. 2001. Simulation of source energy utilization and emissions for HVAC systems. *ASHRAE Transactions* 107(1):39–51.

Reindl, D.T., D.E. Knebel, and R.A. Gansler. 1995. Characterising the marginal basis source energy and emissions associated with comfort cooling systems. *ASHRAE Transactions* 101(1):1353–63.

Rosenbaum, M. 2002. A green building on campus. *ASHRAE Journal* 44(1):41–44.

Rosfjord, T., T. Wagner, and B. Knight. 2004. UTC microturbine CHP product development and launch. Presented at the 2004 DOE/CETC Annual Workshop on Microturbine Applications, January 20–22, Marina del Rey, CA.

Ruud, M.D., J.W. Mitchell, and S.A. Klein. 1990. Use of building thermal mass to offset cooling loads. *ASHRAE Transactions* 96(2):820–29.

Scheatzle, David. ASHRAE RP-1140 Final report, Establishing a base-line data set for the evaluation of hybrid HVAC systems. Atlanta: ASHRAE.

Schell, M., and D. Int-Hout. 2001. Demand control ventilation using CO_2. *ASHRAE Journal* 43(2):18–24.

Stack, A., J. Goulding, and J.O. Lewis. Shading systems: Solar shading for the European climates. DG TREN, Brussels, Belgium. http://erg.ucd.ie/down_thermie.html.

Svensson C., and S.A. Aggerholm. 1998. Design tool for natural ventilation. ASHRAE IAQ 1998 Conference, October 24–27, New Orleans, LA.

Taylor, S. 2002. Primary-only vs. primary-secondary variable flow chilled water systems. *ASHRAE Journal* 44(2):25–29.

Taylor, S.T. 2005. LEED and Standard 62.1. *ASHRAE Journal Sustainability Supplement*, September.

Taylor, S., and J. Stein. 2002. Balancing variable flow hydronic systems. *ASHRAE Journal* 44(10):17–24.

Torcellini, P.A., N. Long, and R. Judkoff. 2004. Consumptive water use for U.S. power production. *ASHRAE Transactions* 110(1):96–100.

Torcellini, P., S. Pless, M. Deru, B. Griffith, N. Long, and R. Judkoff. 2006. Lessons learned from case studies of six high-performance buildings. Technical report, NREL/TP-550-37542. National Renewable Energy Laboratory, Golden, CO.

Torcellini, P., N. Long, S. Pless, and R. Judkoff. 2005. Evaluation of the low-energy design and energy performance of the Zion National Park Visitor Center, NREL Report No. TP-550-34607. National Renewable Energy Laboratory, Golden, CO. www.eere.energy.gov/buildings/highperformance/research_reports.html.

Trane. 2000. Energy conscious design ideas—Air-to-air energy recovery. *Engineers Newsletter* 29(5).

USGBC. LEED-NC v 2.2 Rating System. U.S. Green Building Council, Washington, DC.

USGBC. 2009. LEED Version 3. U.S. Green Building Council, Washington, DC.

von Kempski, D. 2003. Air and well being—A way to more profitability. *Proceedings of Healthy Buildings 2003, Singapore*, pp. 248–354.

von Kempski, D. 2009. Going beyond green—green perception A paradigm change in IAQ for the well-being and health of the occupants. *Proceedings of Healthy Buildings 2009, Syracuse, New York,* paper 138.

von Paumgartten, P. *Existing Buildings Hold the Key.* Alliance for Sustainable Built Environments. www.awarenessintoaction.com/whitepapers/how-existing-buildings-high-performing-green-leed-certified.html (28 Sep. 2010).

Waterfall, P.H. 1998. *Harvesting Rainwater for Landscape Use.* http://ag.arizona.edu/pubs/water/az1052/.

Watson, R.D., K.S. Chapman, and L. Wiggington. 2001. Impact of dual utility selection on 305 m^2 (1000 ft^2) residences. *ASHRAE Transactions* 107(1):365–70.

Watson, R.D., K.S. Chapman, and J.M. DeGreef. 1998. Case study: Seven-system analysis of thermal comfort and energy use for a fast-acting radiant heating system. *ASHRAE Transactions* 104(1).

Waye, K.P., J. Bengtsson, R. Rylander, F. Hucklebridge, P. Evans, and A. Clow. 2002. Low frequency noise enhances cortisol among noise sensitive subjects during work performance. *Life Science* 70(7):745–58.

Webber, M.E. 2008. *Energy versus Water: Solving Both Crises Together.* Scientific American Special Editions. www.sehn.org/tccenergyversuswater.html.

Zagreus, L., C. Huizenga, E. Arens, and D. Lehrer. 2004. Listening to the occupants: A Web-based indoor environmental quality survey. *Indoor Air 2004* 14(suppl 8):65–74.

WEBSITES

Advanced Buildings Technologies and Practices
www.advancedbuildings.org.

AGORES—A Global Overview of Renewable Energy Sources (the official European Commission renewable energy information centre and knowledge gateway, with a global overview of RES)
www.agores.org.

Air-Conditioning, Heating and Refrigeration Institute
www.ahrinet.org/.

Alliance for Sustainable Built Environments
www.awarenessintoaction.com/whitepapers/how-existing-buildings-high-performing-green-leed-certified.html.

Alliance to Save Energy
http://ase.org .

American Council for an Energy-Efficient Economy
www.aceee.org.

American Gas Association
www.aga.org.

The American Institute of Architects
www.aia.org.

American Society of Healthcare Engineering
www.ashe.org.

American Society of Healthcare Engineering, *Green Guide for Health Care*
www.gghc.org.
American Society of Plumbing Engineers
http://aspe.org.
American Solar Energy Society
www.ases.org.
American Wind Energy Association
www.awea.org.
Architecture 2030 Challenge
www.architecture2030.org/.
Architectural Energy Corporation
www.archenergy.com.
Armstrong Intelligent Systems Solutions
www.armstrong-intl.com.
ASHRAE
www.ashrae.org.
ASHRAE Building eQ
www.buildingeq.com.
ASHRAE Engineering for Sustainability
www.engineeringforsustainability.org/.
ASHRAE Standard Project Committee 191
http://spc191.ashraepcs.org/.
ASHRAE's Sustainability Roadmap
http://images.ashrae.biz/renovation/documents/sust_roadmap.pdf.
ASHRAE TC 8.3, recently sponsored programs and presentations
http://tc83.ashraetcs.org/programs.html.
Athena Institute
www.athenasmi.ca.
BetterBricks
www.betterbricks.com
Building Energy Efficiency Research Project at the Department of Architecture, The University of Hong Kong
www.arch.hku.hk/research/beer/.
Building for Environmental and Economic Sustainability (BEES)
www.bfrl.nist.gov/oae/software/bees.html.
BuildingGreen
www.greenbuildingadvisor.com.
Building Research Establishment Environmental Assessment Method (BREEAM®) rating program
www.breeam.org.
California Debt & Investment Advisory Commission document
www.treasurer.ca.gov/cdiac/publications/p3.pdf.

California Energy Commission
www.energy.ca.gov.
California Energy Commission, Renewable Energy Program
www.energy.ca.gov/renewables/.
The Canadian Renewable Energy Network
www.canren.gc.ca/.
The Center for Health Design
www.healthdesign.org.
Center for the Built Environment, Occupant Indoor Environmental Quality Survey
www.cbesurvey.org.
Center of Excellence for Sustainable Development
www.sustainable.doe.gov.
Center of Excellence for Sustainable Development, Smart Communities Network
www.smartcommunities.ncat.org.
Centre for Analysis and Dissemination of Demonstrated Energy Technologies
www.caddet-re.org.
Collaborative for High Performance Schools
www.chps.net.
Cool Roof Rating Council
www.coolroofs.org.
DDC Online
www.ddc-online.org.
Dena
www.dena.de/en/.
Earth Energy Society of Canada
www.earthenergy.ca/.
ecoENERGY for Renewable Power
www.ecoaction.gc.ca/ecoenergy-ecoenergie/power-electricite/index-eng.cfm.
EN4M Energy in Commercial Buildings (software tool)
www.eere.energy.gov/buildings/tools_directory/software.cfm/ID=299/.
Energy Design Resources
www.energydesignresources.com/publication/gd/.
ENERGY STAR®
www.energystar.gov.
Energy Trust of Oregon
www.energytrust.org.
Europe's Energy Portal
www.energy.eu/#renewable.
European Biomass Association
www.aebiom.org.
European Biomass Industry Association
www.eubia.org/about_biomass.0.html.

Euporean Commission, BUILD UP program
www.buildup.eu/.
European Commission, Concerted Action Energy Performance of Buildings Directive
www.epbd-ca.org.
European Commission (DG TREN), SARA Project
www.sara-project.net.
European Photovoltaic Industry Association
www.epia.org.
European Renewable Energy Centres Agency
www.eurec.be.
European Solar Thermal Industry
http://solarheateurope.eu/welcome-to-solar-heat-europe/.
European Wind Energy Association
https://windeurope.org/.
EUROSOLAR—The European Association for Renewable Energies e.V.
www.eurosolar.org.
F-Chart Software
www.fchart.com.
The Federal Energy Management Program Continuous Commissioning (SM) Guidebook
www1.eere.energy.gov/femp/program/om_guidebook.html.
Geoexchange
www.geoexchange.org.
Global Wind Energy Council
www.gwec.net.
Green Building Advisor
www.greenbuildingadvisor.com.
Green Energy In Europe
www.ecofys.com/.
Green Globes
www.greenglobes.com.
Green Roof Industry Information Clearinghouse and Database
www.greenroofs.com.
GreenSpec® Product Guide
www.buildinggreen.com/menus.
Green Venture
www.greenventure.on.ca.
Hannover Principles
http://www.mcdonough.com/wp-content/uploads/2013/03/Hannover-Principles-1992.pdf.

Harvard University Office for Sustainability, Green Building Resource, Design Phase Guide
www.green.harvard.edu/theresource/new-construction/design-phase/bidding-construction/www.aepronet.org/ge/no43.html.

Heat Pump Centre
www.heatpumpcentre.org.

Heschong Mahone Group
www.h-m-g.com/projects/daylighting/projects-PIER.htm.

Illuminating Engineering Society of North America
www.iesna.org.

Intergovernmental Panel on Climate Change
www.ipcc.ch.

International Code Council
www.iccsafe.org.

International Energy Agency, Bioenergy
www.ieabioenergy.com.

International Energy Agency, Photovoltaic Power Systems Programme
www.iea-pvps.org.

International Energy Agency, Solar Heating And Cooling Programme
www.iea-shc.org.

International Energy Agency, Task 38
www.iea-shc.org/task38.

International Hydropower Association
www.hydropower.org.

International Hydropower Association Fact Sheet
www.hydropower.org/downloads/F1_the_contribution_of_hydropower.pdf.

International Living Building Institute, The Living Building Challenge
http://living-future.org/lbc.

International Organization for Standardization family of 14000 standards, www.iso.org/iso-14001-environmental-management.html.

International Performance Measurement and Verification Protocol
www.evo-world.org.

International Solar Energy Society
www.ises.org.

Irrigation Association
www.irrigation.org.

Labs 21, Environmental Performance Criteria
www.labs21century.gov.

Lawrence Berkeley National Laboratories
www.lbl.gov/.

Lawrence Berkeley National Laboratories, Environmental Energy Technologies Division
http://eetd.lbl.gov/.
Lawrence Berkeley National Laboratory, Radiance
http://radsite.lbl.gov/radiance/HOME.html.
Massachusetts State Building Code
www.mass.gov/bbrs/code.htm.
Minnesota Sustainable Design Guide
www.sustainabledesignguide.umn.edu.
Pierce, D.G. 2003. Public-private partnerships: Demystifying the process. Mondaq:
http://www.mondaq.com/canada/article.asp?articleid=23737.
M&V Guidelines for Federal Energy Projects, Version 3.0.
http://mnv.lbl.gov/keyMnVDocs/femp.
National Building Controls Information Program
www.buildingcontrols.org.
National Hydropower Association
www.hydro.org.
National Oceanic and Atmospheric Administration
www.noaa.gov.
National Renewable Energy Laboratory
www.nrel.gov.
National Renewable Energy Laboratory, Building-Integrated PV
www.nrel.gov/pv/building_integrated_pv.html.
National Renewable Energy Laboratory, Buildings Research
www.nrel.gov/buildings/.
National Renewable Energy Laboratory, Solar Energy Research Facility
www.nrel.gov/pv/facilities_serf.html.
Natural Resources Canada, Commercial Buildings Incentive Program
http://nrcan.gc.ca/evaluation/reprap/2001/cbip-pebc-eng.php.
Natural Resources Canada, EE4 Commercial Buildings Incentive Program
http://canmetenergy-canmetenergie.nrcan-rncan.gc.ca/eng/software_tools/ee4.html.
Natural Resources Canada, Office of Energy Efficiency
http://oee.nrcan.gc.ca/commercial/newbuildings.cfm?attr=20.
Natural Resources Canada, RETScreen
(software for renewable energy analysis)
www.retscreen.net.
The Natural Step
www.naturalstep.org.
New Buildings Institute
www.newbuildings.org.

New York's Battery Park City Authority
www.batteryparkcity.org/page/index_battery.html.
New York City Department of Design and Construction, Sustainable Design
www.nyc.gov/html/ddc/html/design/sustainable_home.shtml.
National Oceanic and Atmospheric Administration
www.noaa.gov.
North Carolina Cooperative Extension Service, Water Quality and Waste Management
www.bae.ncsu.edu/programs/extension/publicat/wqwm.
North Carolina Solar Center, DSIRE
www.dsireusa.org/.
Oak Ridge National Laboratory, Bioenergy Feedstock Information Network
http://bioenergy.ornl.gov/.
Oikos: Green Building Source
www.oikos.com.
Option B: Methods by Technology
www.energyautomation.com/pdfs/ipmvp-vol2.pdf.
Photovoltaic Resource Site
www.pvpower.com.
QuikWater, High Efficiency Direct Contact Water Heaters
www.quikwater.com.
RealWinWin, Inc.
www.realwinwin.com.
Regional Climate Data
www.wrcc.dri.edu/rcc.html.
Renewable Energy Deployment Initiative
www.nrcan.gc.ca/redi.
Rising Sun Enterprises
www.rselight.com.
Rocky Mountain Institute
www.rmi.org.
Savings by Design
www.savingsbydesign.com.
School of Photovoltaic and Renewable Energy Engineering,
University of New South Wales
www.pv.unsw.edu.au.
Sheet Metal and Air Conditioning Contractors' National Association
www.smacna.org.
SkyCalc
www.energydesignresources.com/resource/129/.
Solar Energy Industries Association
www.seia.org.

Spirx Sarco Design of Fluid Systems, Steam Learning Module
www.spiraxsarco.com/learn/modules.asp.
Sustainable Building Challenge
www.iisbe.org/.
Sustainable Building Guidelines
www.ciwmb.ca.gov/GreenBuilding/Design/Guidelines.htm.
Sustainable Buildings Industry Council
www.sbicouncil.org.
Sustainable Communities Network
www.sustainable.org.
Sustainable Sources
www.greenbuilder.com.
Sustainable Sources Bookstore
bookstore.greenbuilder.com/index.books.
Texas Manual on Rainwater Harvesting
www.twdb.state.tx.us/publications/reports
/rainwaterharvestingmanual_3rdedition.pdf.
"Tips for Daylighting with Windows"
http://windows.lbl.gov/pub/designguide/designguide.html.
Tool for the Reduction and Assessment of Chemical and Other Environmental Impacts (TRACI)
www.epa.gov/nrmrl/std/sab/traci/.
Trane Company
www.trane.com.
Tri-State Generation and Transmission Association, Inc.
http://www.tristategt.org/.
Usable Buildings Trust
www.usablebuildings.co.uk.
U.S. Department of Energy, Energy Efficiency and Renewable Energy
www.eere.energy.gov/.
U.S. Department of Energy, Energy Efficiency and Renewable Energy, Biomass program
www1.eere.energy.gov/biomass.
U.S. Department of Energy, Energy Efficiency and Renewable Energy, Biomass program, Resources for Consumers Web page
www1.eere.energy.gov/biomass/for_consumers.html.
U.S. Department of Energy, Energy Efficiency and Renewable Energy, Energy Savers
www.energysavers.gov/.
U.S. Department of Energy, Energy Efficiency and Renewable Energy, Zero Energy Buildings
http://zeb.buildinggreen.com.

U.S. Department of Energy, Federal Energy Management Program
www.eere.energy.gov/femp.

U.S. Department of Energy, Federal Energy Management Program Guidelines
www1.eere.energy.gov/femp/pdfs/mv_guidelines.pdf.

U.S. Department of Energy, High Performance Buildings Institute
www1.eere.energy.gov/buildings/commercial_initiative/.

U.S. Energy Information Administration,
Commercial Building Energy Consumption Survey
www.eia.doe.gov/emeu/cbecs/.

U.S. Environmental Protection Agency
www.epa.gov.

U.S. Environmental Protection Agency, Building Energy Analysis Tool, eQuest
www.energydesignresources.com/resource/130/.

U.S. Environmental Protection Agency, Green Chemistry
www.epa.gov/greenchemistry/.

U.S. Environmental Protection Agency, Water Sense Program
http://epa.gov/watersense/.

U.S. Green Building Council
www.usgbc.org.

U.S. Green Building Council, Leadership in Energy and Environmental Design,
Gold building, Hewlett Foundation
www.usgbc.org/Docs/Certified_Projects/Cert_Reg67.pdf.

U.S. Green Building Council, Leadership in Energy and Environmental Design,
Green Building Rating System™
www.usgbc.org/DisplayPage.aspx?CategoryID=19.

U.S. Green Building Council, Regional Priority Credit Listing
www.usgbc.org/rpc.

The Whole Building Design Guide
www.wbdg.org.

World Meteorological Organization, World Radiation Data Centre
http://wrdc-mgo.nrel.gov/.

The World's Water
www.worldwater.org.

Terms, Definitions, and Acronyms

ΔT	=	change or change in temperature T
AC	=	air conditioning
ACGIH	=	American Conference of Governmental Industrial Hygienists
AHU	=	air-handling unit
AIA	=	American Institute of Architects
ANSI	=	American National Standards Institute
BAS	=	building automation system
BEES	=	Building for Environmental and Economic Sustainability
BF	=	ballast factor
BIM	=	building information modeling
bioclimatic chart	=	a psychrometric chart with a plot of the conditions for primarily passive strategies
BOMA	=	Building Owners and Managers Association
BREEAM®	=	Building Research Establishment Environmental Assessment Method
brownfield	=	real estate property that is, or potentially is, contaminated
Btu	=	British thermal unit
C	=	centigrade (temperature scale)
CAN	=	Canada
CAV	=	constant air volume
CBE	=	Center for the Built Environment
CCHP	=	combined cooling, heating, and power
CDT	=	cold deck temperature
CFC	=	chlorofluorocarbon
cfm	=	cubic feet per minute
CFR	=	current facility requirements
charette	=	collaborative and interdisciplinary effort to solve a design problem within a limited time
CHP	=	combined heating and power
CHW	=	chilled water
condenser	=	device to dissipate (get rid of) excess energy in A/C systems
COP	=	coefficient of performance
CxA	=	commissioning authority
daylighting	=	lighting (of a building) using daylight directly or indirectly from the sun
D/B	=	design/build
dB(A)	=	A-weighting decibels

D/B/B	=	design/bid/build
DDC	=	direct digital control
DE	=	district energy
DG	=	distributed generation
DHW	=	domestic hot water
DOAS	=	dedicated outdoor air system
DOE	=	U.S. Department of Energy
DSF	=	double skin facade
DX	=	direct expansion
E	=	ventilation effectiveness
EA	=	energy and atmosphere
EDC	=	environmental design consultant
EDG	=	engine-driven generator
ENERGY STAR®	=	a government-backed program/rating system that helps consumers achieve superior energy efficiency
energy source	=	on-site energy in the form in which it arrives at or occurs on a site (e.g., electricity, gas, oil, or coal)
energy resource	=	raw energy that (1) is extracted from the earth (wellhead or mine mouth), (2) is used in the generation of the energy source delivered to a building site (e.g., coal used to generate electricty), or (3) occurs naturally and is available at a site (e.g., solar, wind, or geothermal energy)
enthalpy	=	the thermodynamic property of a system resulting from the combination of observable properties (per unit mass) thereof: namely, the sum of internal energy and flow work; flow work is the product of volume and specific mass (i.e., energy transmitted into or out of a system or transmitted across a system boundary)
entropy	=	a measure of the molecular disorder of a system, such that the more mixed a system is, the greater its entropy, and the more orderly or unmixed a system is, the lower its entropy
E&O	=	errors and omissions
EPA	=	U.S. Environmental Protection Agency
EPBD	=	Energy Performance of Buildings Directive
EPC	=	energy performance certificate
EU	=	European Union
F	=	Fahrenheit (temperature scale)
F-chart	=	method of calculating solar fraction
fenestration	=	window treatment
GHG	=	greenhouse gas
gpf	=	gallons per fixture
gpm	=	gallons per minute
GSHP	=	ground-source heat pump
guideline	=	within ASHRAE, a document similar to a standard, but with less strict rules on developing a consensus and not written in code language
HCFC	=	hydrochlorofluorocarbons
HEPA	=	high-efficiency particulate air
HID	=	high-intensity discharge
HVAC&R	=	heating, ventilating, air-conditioning, and refrigerating
hybrid ventilation	=	combination of natural and mechanical outdoor air ventilation
hydronic	=	pertaining to liquid flow
IAQ	=	indoor air quality

IDP	=	integrated design process
IEA	=	International Energy Agency
IEC	=	indirect evaporative cooling
IEQ	=	indoor environmental quality
IES	=	Illuminating Engineering Society of North America
insolation	=	entry into a building of solar energy
IPCC	=	Intergovernmental Panel on Climate Change
IPMVP	=	International Performance Measurement and Verification Protocol
K	=	Kelvin or absolute (temperature scale)
kW	=	kilowatt
kWh	=	kilowatt-hour
kW_R	=	refrigeration cooling capacity in kW
latent load	=	thermal load due strictly to effects of moisture
LCA	=	life-cycle assessment
LCCA	=	life-cycle cost analysis
LCEA	=	life-cycle environmental assessment
LD	=	liquid desiccant
LED	=	light-emitting diode
LEED	=	Leadership in Energy and Environmental Design
leeward	=	the downwind side—or side the wind blows away from
L/f	=	liters per fixture
low-E	=	low emissivity
LPD	=	lighting power density
LP	=	liquefied petroleum
L/s	=	liters per second (airflow and water flow)
media	=	energy forms distributed within a building, usually air, water, or electricity
MERV	=	minimum efficiency reporting value
mhp	=	motor horsepower
M&V	=	measurement and verification
NADCA	=	National Air Duct Cleaning Association
NC	=	noise criteria
NO_x	=	oxides of nitrogen
NOAA	=	National Oceanic Atmospheric Administration
nonrenewables	=	energy resources that have definite, although sometimes unknown, quantity limitations
NR	=	natural refrigeration
NREL	=	National Renewable Energy Laboratory
NZEB	=	net zero energy building
OC	=	on center
O&M	=	operations and maintenance
OPR	=	Owner's Project Requirements
OSHA	=	Occupational Safety and Health Administration
P3	=	public-private partnership
parametric analysis	=	in situations where multiple parameters affect an outcome, an analysis that determines the magnitude of one or more parameter's impact alone on that outcome
plug loads	=	loads (electrical or thermal) from equipment plugged into electrical outlets
POTW	=	publicly owned treatment works
precooling	=	cooling done prior to the time major cooling loads are anticipated
PV	=	photovoltaic

R (as in R-19)	=	thermal heat transfer resistance
RC	=	room criteria
renewables	=	resources that can generally be freely used without net depletion or that have the potential to renew in a reasonable period of time
RES	=	renewable energy source
RFP	=	request for proposal
ROI	=	return on investment
sensible load	=	thermal load due to temperature but not moisture effects
sg	=	specific gravity
SHW	=	space hot water
skin	=	building envelope
SMACNA	=	Sheet Metal and Air Conditioning Contractors National Association
SR	=	synthetic refrigeration
SS	=	sustainable sites
standard	=	within ASHRAE, a document that defines properties, processes, dimensions, materials, relationships, concepts, nomenclature, or test methods for rating purposes or code enforcement
sustainability	=	providing for the needs of the present without detracting from the ability to fulfill the needs of the future
TC	=	technical committee (an ASHRAE group with a common interest in a particular technical subject)
TCLP	=	toxicity characteristic leaching procedure
TES	=	thermal energy storage
Title 24	=	shortened form for California's Building Energy Efficiency Standards (Title 24, Part 6 of the California State Building Code)
ton	=	cooling capacity, equal to 12,000 Btu/h
TRACI	=	Tool for the Reduction and Assessment of Chemical and Other Environmental Impacts
USGBC	=	U.S. Green Building Council
VAV	=	variable air volume
VE	=	value engineering
VFD	=	variable-frequency drive
VOC	=	volatile organic compound
VRF	=	variable refrigeration flow
WE	=	water efficiency
windward	=	the upwind side, or side the wind blows toward

INDEX

A

F

G

U

V

W